Gibbs Random Fields

Mathematics and Its Applications (*Soviet Series*)

Volume 44

Gibbs Random Fields

Cluster Expansions

by

V. A. Malyshev and R. A. Minlos

Department of Mathematics,
Moscow State University,
Moscow, U.S.S.R.

KLUWER ACADEMIC PUBLISHERS

DORDRECHT / BOSTON / LONDON

Library of Congress Cataloging-in-Publication Data

Malyshev, V. A. (Vadim Aleksandrovich)
[Gibbsovskie sluchaĭnye poli͡a. English]
Gibbs random fields : cluster expansions / by V.A. Malyshev and R.A. Minlos.
p. cm. -- (Mathematics and its applications (Soviet Series) : v. 44)
Translation of: Gibbsovskie sluchaĭnye poli͡a.
Includes bibliographical references and index.
ISBN 0-7923-0232-X (alk. paper)
1. Random fields. 2. Cluster analysis. I. Minlos, R. A. (Robert Adol'fovich) II. Title. III. Series: Mathematics and its applications (Kluwer Academic Publishers). Soviet series ; 44.
QA274.45.M3513 1991
519.2--dc20
91-15172
CIP

ISBN 0-7923-0232-X

Published by Kluwer Academic Publishers,
P.O. Box 17, 3300 AA Dordrecht, The Netherlands.

Kluwer Academic Publishers incorporates
the publishing programmes of
D. Reidel, Martinus Nijhoff, Dr W. Junk and MTP Press.

Sold and distributed in the U.S.A. and Canada
by Kluwer Academic Publishers,
101 Philip Drive, Norwell, MA 02061, U.S.A.

In all other countries, sold and distributed
by Kluwer Academic Publishers Group,
P.O. Box 322, 3300 AH Dordrecht, The Netherlands.

Translated from the Russian by
R. Kotecky and P. Holicky

Printed on acid-free paper

Printed in the Netherlands

SERIES EDITOR'S PREFACE

'Et moi, ..., si j'avait su comment en revenir, je n'y serais point allé.'

Jules Verne

The series is divergent; therefore we may be able to do something with it.

O. Heaviside

One service mathematics has rendered the human race. It has put common sense back where it belongs, on the topmost shelf next to the dusty canister labelled 'discarded non-sense'.

Eric T. Bell

Mathematics is a tool for thought. A highly necessary tool in a world where both feedback and non-linearities abound. Similarly, all kinds of parts of mathematics serve as tools for other parts and for other sciences.

Applying a simple rewriting rule to the quote on the right above one finds such statements as: 'One service topology has rendered mathematical physics ...'; 'One service logic has rendered computer science ...'; 'One service category theory has rendered mathematics ...'. All arguably true. And all statements obtainable this way form part of the raison d'être of this series.

This series, *Mathematics and Its Applications*, started in 1977. Now that over one hundred volumes have appeared it seems opportune to reexamine its scope. At the time I wrote

> "Growing specialization and diversification have brought a host of monographs and textbooks on increasingly specialized topics. However, the 'tree' of knowledge of mathematics and related fields does not grow only by putting forth new branches. It also happens, quite often in fact, that branches which were thought to be completely disparate are suddenly seen to be related. Further, the kind and level of sophistication of mathematics applied in various sciences has changed drastically in recent years: measure theory is used (non-trivially) in regional and theoretical economics; algebraic geometry interacts with physics; the Minkowsky lemma, coding theory and the structure of water meet one another in packing and covering theory; quantum fields, crystal defects and mathematical programming profit from homotopy theory; Lie algebras are relevant to filtering; and prediction and electrical engineering can use Stein spaces. And in addition to this there are such new emerging subdisciplines as 'experimental mathematics', 'CFD', 'completely integrable systems', 'chaos, synergetics and large-scale order', which are almost impossible to fit into the existing classification schemes. They draw upon widely different sections of mathematics."

By and large, all this still applies today. It is still true that at first sight mathematics seems rather fragmented and that to find, see, and exploit the deeper underlying interrelations more effort is needed and so are books that can help mathematicians and scientists do so. Accordingly MIA will continue to try to make such books available.

If anything, the description I gave in 1977 is now an understatement. To the examples of interaction areas one should add string theory where Riemann surfaces, algebraic geometry, modular functions, knots, quantum field theory, Kac-Moody algebras, monstrous moonshine (and more) all come together. And to the examples of things which can be usefully applied let me add the topic 'finite geometry'; a combination of words which sounds like it might not even exist, let alone be applicable. And yet it is being applied: to statistics via designs, to radar/sonar detection arrays (via finite projective planes), and to bus connections of VLSI chips (via difference sets). There seems to be no part of (so-called pure) mathematics that is not in immediate danger of being applied. And, accordingly, the applied mathematician needs to be aware of much more. Besides analysis and numerics, the traditional workhorses, he may need all kinds of combinatorics, algebra, probability, and so on.

In addition, the applied scientist needs to cope increasingly with the nonlinear world and the extra mathematical sophistication that this requires. For that is where the rewards are. Linear models are honest and a bit sad and depressing: proportional efforts and results. It is in the non-linear world that infinitesimal inputs may result in macroscopic outputs (or vice versa). To appreciate what I am hinting at: if electronics were linear we would have no fun with transistors and computers; we would have no TV; in fact you would not be reading these lines.

There is also no safety in ignoring such outlandish things as nonstandard analysis, superspace and anticommuting integration, p-adic and ultrametric space. All three have applications in both electrical engineering and physics. Once, complex numbers were equally outlandish, but they frequently proved the shortest path between 'real' results. Similarly, the first two topics named have already provided a number of 'wormhole' paths. There is no telling where all this is leading - fortunately.

Thus the original scope of the series, which for various (sound) reasons now comprises five subseries: white (Japan), yellow (China), red (USSR), blue (Eastern Europe), and green (everything else), still applies. It has been enlarged a bit to include books treating of the tools from one subdiscipline which are used in others. Thus the series still aims at books dealing with:

- a central concept which plays an important role in several different mathematical and/or scientific specialization areas;
- new applications of the results and ideas from one area of scientific endeavour into another;
- influences which the results, problems and concepts of one field of enquiry have, and have had, on the development of another.

A random field is simply a collection of random variables one for each $x \in \mathbb{Z}^n$ or $\mathbb{R}^n$ (of finite chunks of these spaces). A free field is one with no interactions and until fairly recently they were the only ones which had been studied in real depth. A Gibbs random field is obtained from a free one by a 'Gibbsean perestroika' (the authors do indeed use this word in the Russian text) which defines a new measure (in various ways) by specifying the Radon-Nikodyn derivative of the new measure, μ, with respect to the old free one, μ_0, in terms of an interaction energy function. This is the main theme of the book: the construction and study of random fields defined by such a finite volume Gibbs specification (modification).

The interest in Gibbs random fields is great because a large number of spatial interaction phenomena can be statistically described by such fields. The subject started in 1925 (with the Ising model) and by now applications include ecology, statistical mechanics, image modelling and analysis. Perhaps I should say potential applications because the underlying deeper theory has only comparatively recently been developed: something to which the present authors have greatly contributed. That deeper theory is most important; for instance the presence (or absence) of phase transitions affects statistical inference and identifiability in image processing applications.

The new tool, one of the most powerful methods of contemporary mathematical physics, goes by the name of cluster expansions. This allows one to obtain expressions for local random field characteristics in terms of expansions depending only on a finite set of random field variables (the clustors).

With this book (and its planned sequel) an additional substantial chunk of mathematics is in place and ready to be applied. I am pleased to be able to offer the scientific community the chance to peruse this English language version.

The shortest path between two truths in the real domain passes through the complex domain.

J. Hadamard

La physique ne nous donne pas seulement l'occasion de résoudre des problèmes ... elle nous fait pressentir la solution.

H. Poincaré

Never lend books, for no one ever returns them; the only books I have in my library are books that other folk have lent me.

Anatole France

The function of an expert is not to be more right than other people, but to be wrong for more sophisticated reasons.

David Butler

Amsterdam, January 1991

Michiel Hazewinkel

Contents

PREFACE

This book is intended for a wide range of readers: specialists in probability theory will see new and fresh topics concerning random fields; specialists in mathematical physics (quantum field theory or statistical physics) will be interested in a thorough presentation of the cluster technique and its various devices; for the remaining mathematicians, the book offers a number of new algebraic, combinatorial, and analytic problems.

We will be occupied with explicit constructions of random fields. They are based on a single method: the so-called Gibbs modification. What is its meaning? The probability distribution of a finite system of random variables $\{\xi_1, \ldots, \xi_n\}$ (of a finite field) is most often given in terms of its density $p(x_1, \ldots, x_n)$ with respect to some (usually Lebesque) measure on R^n. Gibbs modification is a natural generalization of this method to the case of infinite fields. The simplest infinite random fields are independent and Gaussian fields and their functionals (until recently, they remained the only well-studied class of fields). As to the Gibbs method of construction, one first introduces fields with a distribution prescribed by a finite (local) density with respect to independent or Gaussian fields (a finite or local Gibbs modification) and then passes to the weak limit of such distributions (a limit Gibbs modification); this limit measure is already singular with respect to the original distribution, independent or Gaussian one. Random fields originating in this limit (the thermodynamic limit) actually constitute the main subject of study in the theory of Gibbs fields.

We will present one of the most powerful methods of investigation of Gibbs fields, namely, the method of cluster expansions, and its numerous applications. With this method, any local characteristic of a field (finite-dimensional distributions, mean values of local functions, etc.) can be represented in terms of a series where each term depends on a finite group of field variables ("a cluster") and is expressed in terms of them in an explicit way. The idea of the method goes back to the so-called virial expansions in statistical physics and, on the other hand, to the diagram expansion in quantum field theory; in its present-day form the method of cluster expansions represents a mathematicized development of these methods. Unfortunately, to apply a cluster expansion, one needs a "small parameter," i.e., the method appears in the form of a version of the perturbation theory up to now, even though by its nature it seems to be more universal and, apparently, should have broader applications. We hope that in the future, this limitation will be overcome.

Chapter 1 is an introduction to Gibbs fields; Section 0, in particular, is written in an extremely elementary manner. The reader interested directly in cluster expansions may skip this chapter. In Chapter 2 some subsidiary material is collected: the properties and estimates of semi-invariants, the diagram technique, and the most important combinatorial lemmas. The general scheme of cluster expansions is presented in Chapter 3, and numerous examples of these expansions are given in Chapters 4 and 5. These are the central chapters in the book; they contain the basic methods as well as various refinements of cluster expansions.

Chapters 6 and 7 are devoted to applications, both traditionally probability-theoretic ones and also those associated with mathematical physics.

Let us notice that our presentation includes two types of random fields: the point (labelled) ones and those on a countable set, with special emphasis on the latter. This stems, in particular, from the fact that to apply the cluster technique to continuous (functional) fields (that are not included in the book due to its limited extent), one has to reduce them beforehand to fields in a discrete set.

We have not enough space here to indicate explicitly the connection of our constructions with statistical physics and quantum field theory even though their spirit is invisibly present in the book and the terminology betrays their influence.

The limited scope of this book did not allow us to present one of the most important applications of cluster expansions, namely, an investigation of the spectrum of the transfer matrix of a Gibbs field. Also, numerous other methods used in the theory of Gibbs fields are not included in this book (correlation inequalities, reflection positivity, duality transformation, and other special methods). We are planning to write a sequel in which some of these omissions will be rectified.

NOTATIONS

(Ω, Σ, μ)—a probability space, i.e., a triple consisting of a set Ω, a σ-algebra Σ of subsets of Ω, and a probability measure μ defined on Σ.

If Ω is a topological space, Σ denotes its Borel σ-algebra $\mathfrak{B}(\Omega)$, i.e., the σ-algebra generated by open sets in Ω.

We shall usually denote by Ω a set of configurations of a random field.

μ_0—a "free" (nonperturbed) measure on Ω (usually independent or Gaussian).

For any random variable, i.e., a measurable function ξ on a probability space (Ω, Σ, μ), its mean (mathematical expectation) is denoted by

$$\langle \xi \rangle = \langle \xi \rangle_\mu = \int_\Omega \xi d\mu.$$

Whenever $\{\xi_1, \dots, \xi_n\}$ is a family of random variables,

$$\langle \xi_1, \dots, \xi_n \rangle_\mu = \left\langle \prod_{i=1}^{n} \xi_i' \right\rangle_\mu = \langle \xi_{\mathcal{N}}' \rangle_\mu$$

denotes their semi-invariant, $\mathcal{N} = \{1, \dots, n\}$ (we shall often omit the subscript μ); in addition,

$$\langle \xi_1'^{k_1}, \dots, \xi_n'^{k_n} \rangle = \langle \underbrace{\xi_1, \dots, \xi_1}_{k_1 \text{ times}}, \dots, \underbrace{\xi_n, \dots, \xi_n}_{k_n \text{ times}} \rangle.$$

G—a graph, i.e., a set V (of vertices) and a collection of unordered (possibly coinciding) pairs of elements of V (edges).

T—a tree, i.e., a connected graph without loops.

$(A_1, \dots, A_n)$—an ordered, and $\{A_1, \dots, A_n\}$—an unordered collection of sets A_i, $i = 1, \dots, n$ (similarly for collections of points).

A partition $\alpha = \{T_1, \dots, T_k\}$ of a set A is an unordered collection of nonempty mutually disjoint subsets $T_i \subset A$, $i = 1, \dots, k$, whose union is A, $\cup_{i=1}^{k} T_i = A$.

$\mathfrak{A}_{\mathcal{N}}$—the lattice of all partitions of the set $\mathcal{N} = \{1, \dots, n\}$.

$\mathfrak{A}_{\mathcal{N}}^{G} \subset \mathfrak{A}_{\mathcal{N}}$—(with a connected graph G with the set of vertices $\mathcal{N}$) the lattice of partitions of $\mathcal{N}$ whose elements form connected subgraphs of G.

$\mu_{\mathfrak{a}}$— the Möbius function of the lattice $\mathfrak{a}$.

T—a countable (or finite) set supporting a random field.

Q—a space supporting a point field.

S—a space of values of a field (a space of "spins" or "charges").

Λ, A, R—finite (bounded) subsets of T (or Q).

Ω_Λ—the space of configurations of a field in $\Lambda (\Lambda \subset T$ or $\Lambda \subset Q)$.

Σ_Λ—a σ-algebra in Ω_Λ.

U_Λ—a Hamiltonian (energy) in Λ.

$U_{\Lambda,y} = U(\cdot/y)$—a Hamiltonian (energy) in Λ under the boundary configuration y.

μ_Λ—a finite Gibbs modification (in Λ).

$\mu_{\Lambda,y}$—a Gibbs modification under the boundary configuration y.

Z_Λ—a partition function (of a Gibbs modification) in Λ.

$Z_{\Lambda,y}$—a partition function in Λ under the boundary configuration y.

$\Gamma, \gamma, \xi, \dots$—"clusters": i.e., connected collections of sets (or "labelled" sets).

k_A (or k_Γ)—quantities defining cluster representation of partition functions.

$b_R(F)$, $b_R^{(\Lambda)}(F)$, b_R, $b_R^{(\Lambda)}$—quantities appearing in cluster expansions of means (correlation functions, free energy, etc.).

$f_A^{(\Lambda)}, f_\Lambda$—correlation functions.

$:p(\xi_1, \dots, \xi_n):$—the Wick polynomial (of a Gaussian family of random variables $\{\xi_1, \dots, \xi_n\}$).

Let $\mathfrak{A}$ be a family of subsets of T. Then $d_R(\mathfrak{A}; A_1, \dots, A_n)$, where $A_i \subset T$, $i = 1, \dots, n$, and $R \subset T$, denotes the minimal cardinality of a collection $\{B_1, \dots, B_s\}$ of sets $B_i \in \mathfrak{A}$ such that the collection $\{B_1, \dots, B_s, A_1, \dots, A_n\}$ is connected and $\cup_{i=1}^{s} B_i = R$.

Further, $d(\mathfrak{A}; A_1, \dots, A_n) = \min_R d_R(\mathfrak{A}; A_1, \dots, A_n)$, i.e., it is the minimal cardinality of a collection $\{B_1, \dots, B_s\}$, $B_i \in \mathfrak{A}$, $i = 1, \dots, s$, forming together with $\{A_1, \dots, A_n\}$ a connected collection.

Finally $\hat{d}_A(\mathfrak{A})$, where $A \subset T$, denotes the minimal cardinality of connected collections $\{B_1, \dots, B_s\}$ of sets $B_i \in \mathfrak{A}$, $i = 1, \dots, s$, such that $A \subset \cup_{i=1}^{s} B_i$; $\hat{d}_A = \hat{d}_A(\mathfrak{A})$ if $\mathfrak{A}$ is the set of pairs of neighbouring sites $\{t, t'\}$ (i.e., $\rho(t, t') = 1$).

For referring to formulae and theorems, the following convention will be used:

(1) denotes the formula (1) in the same section, (2.1) denotes the formula (2) in Section 1 of the same chapter, and (3.2.1) denotes the formula (3) in Section 2 of Chapter 1; similarly for references to theorems and lemmas.

CHAPTER 1

Gibbs Fields (Basic Notions)

§0 First Acquaintance with Gibbs Fields

In this introductory section, we present the basic notions and problems of the theory of Gibbs fields and display some typical ways of argumentation by considering a simple and well-examined example of the so-called Ising model.

0.1. Ising Model

Consider the lattice Z^ν of points $t = (t^{(1)}, \dots, t^{(\nu)}) \in R^\nu$ of the ν-dimensional real space with integer coordinates. Let $\Lambda_N \equiv \Lambda$ be a "cube" in Z^ν centered at the origin, i.e., the set of points in Z^ν whose coordinates have absolute values not greater than N (with an integer $N > 0$). Each function $\sigma^\Lambda = \{\sigma_t, t \in \Lambda\}$, defined on the set Λ and taking values $\sigma_t = \pm 1$, is called a *configuration* (in the cube Λ), and the set of all such configurations is denoted by Ω_Λ. Obviously, the number of configurations in Λ equals $2^{|\Lambda|}$, where $|\Lambda|$ stands for the number of lattice sites in Λ.

We define, on Ω_Λ, a function

$$U_\Lambda \equiv U_\Lambda(\sigma^\Lambda) = -\left(h \sum_{t \in \Lambda} \sigma_t + \beta \sum_{\langle t,t' \rangle} \sigma_t \sigma_{t'} \right), \tag{1}$$

called the *energy* of the configuration σ^Λ. The summation in the second sum in (1) is taken over all unordered pairs $\langle t, t' \rangle, t, t' \in \Lambda$, such that $\rho(t,t') = 1$ (pairs of "nearest neighbours"), with

$$\rho(t,t') = \sum_{i=1}^{\nu} \left| t^{(i)} - t'^{(i)} \right|,$$

$$t = \left(t^{(1)}, \dots, t^{(\nu)}\right), \qquad t' = \left(t'^{(1)}, \dots, t'^{(\nu)}\right). \tag{1'}$$

A physical system with the configuration space Ω_Λ of configurations in Λ and a configuration energy of the form (1) is usually called the *Ising model.* The real numbers h and β in (1) are fixed (parameters of the model). We refer to the case $\beta > 0$, which will be investigated here, as the *ferromagnetic* Ising model.

We introduce a probability distribution on the space Ω_Λ defining the probability of a configuration σ^Λ by

$$P_\Lambda(\sigma^\Lambda) = Z_\Lambda^{-1} \exp\left\{-U_\Lambda(\sigma^\Lambda)\right\}. \tag{2}$$

The normalization factor Z_Λ is defined by the condition

$$\sum_{\sigma^\Lambda \in \Omega_\Lambda} P_\Lambda(\sigma^\Lambda) = 1,$$

and thus,

$$Z_\Lambda = \sum_{\sigma^\Lambda \in \Omega_\Lambda} \exp\left\{-U_\Lambda(\sigma^\Lambda)\right\}. \tag{3}$$

The quantity Z_Λ is called a *partition function* and plays a key role in what follows. The probability distribution (2) is called a *Gibbs* probability distribution in Λ corresponding to the Ising model. In general, a model is defined by the choice of a set of values of configurations and of a particular form of energy function (see p. 19).

Having introduced probability distributions on the configuration space, the values σ_t of these configurations may be considered as random variables and the formula (2) as the joint probability distribution of these random variables. Later on, the mean (value) of an arbitrary function f on the space Ω_Λ under the distribution (2) will be denoted by $\langle f\rangle_\Lambda$. The means $\langle\sigma_T\rangle_\Lambda$ of random variables

$$\sigma_T = \prod_{t\in T} \sigma_t, \qquad \sigma_\emptyset = 1, \tag{4}$$

with $T \subset \Lambda$ being an arbitrary subset of Λ, are called *correlation functions* (or moments) of the distribution (2).

For any $T \subset \Lambda$, we use $P_\Lambda^{(T)}$ to denote the joint distribution of the system of random variables $\{\sigma_t, t \in T\}$, i.e., the collection of probabilities

$$P_\Lambda^{(T)}(\bar\sigma_{t_1},\dots,\bar\sigma_{t_n}) = \Pr(\sigma_{t_1} = \bar\sigma_{t_1},\dots,\sigma_{t_n} = \bar\sigma_{t_n}), \tag{5}$$

with $T = \{t_1,\dots,t_n\}$ and $\{\overline{\sigma}_{t_1},\dots,\overline{\sigma}_{t_n}\}$ being an arbitrary collection of values $\overline{\sigma}_{t_i} = \pm 1$, $i = 1,2,\dots,n$. The probabilities (5) may be very simply expressed by means of correlation functions $\langle\sigma_T\rangle_\Lambda$. Indeed,

$$\begin{aligned} P_\Lambda^{(T)}(\bar\sigma_{t_1},\dots,\bar\sigma_{t_n}) &= \frac{1}{2^n}(-1)^k \left\langle \prod_{i=1}^{n} (\sigma_{t_i} + \bar\sigma_{t_i}) \right\rangle_\Lambda \\ &= \frac{(-1)^k}{2^n} \sum_{T'\subseteq T} C_{T'} \langle\sigma_{T'}\rangle_\Lambda, \end{aligned} \tag{6}$$

with k being the number of values $\bar{\sigma}_{t_i}$ that equal -1, and

$$C_{T'} = \prod_{t \in T \setminus T'} \bar{\sigma}_t.$$

0.2. Thermodynamic Limit

Now we fix T and let Λ expand to Z^ν, $\Lambda \nearrow Z^\nu$, i.e., put $N \to \infty$. If we succeed in proving the existence of the limits

$$\lim_{\Lambda \nearrow Z^\nu} \langle \sigma_T \rangle_\Lambda, \tag{7}$$

we may conclude that correlation functions (and finite-dimensional distributions) almost do not depend on Λ for Λ sufficiently large in comparison with T. Such a passage to the limit is called the *thermodynamic* limit (the limit of a large number of degrees of freedom σ_t). The limits (7) are denoted by $\langle \sigma_t \rangle$ and are called *limit correlation functions.* It follows from (6) that finite-dimensional distributions also have limits, which form a compatible family of finite-dimensional distributions. By the Kolmogorov theorem (see [51]), this family defines a system of random variables $\{\sigma_t,\ t \in Z^\nu\}$, called a (limit) *Gibbs random field* (for the Ising model), as well as their distribution P (a measure) on the space $\Omega = \{-1, 1\}^{Z^\nu}$ of infinite configurations in the lattice Z^ν.

The methods developed in this book enable us to establish the existence of the limits (7) for a sufficiently general class of models. However, in the case of the ferromagnetic Ising model, we shall use a special device of the so-called Griffith correlation inequalities to control the passage to the limit (7). The existence of the limit distribution P follows from the following theorem.

Theorem 1. *The thermodynamic limit (7) of correlation functions $\langle \sigma_T \rangle_\Lambda$ exists for $\beta \geqslant 0$ and every finite T.*

Remark 1. In the case of $\beta = 0$, it is not difficult to calculate $\langle \sigma_t \rangle_\Lambda$:

$$\langle \sigma_T \rangle_\Lambda = \left(\frac{e^h - e^{-h}}{e^h + e^{-h}} \right)^{|T|}. \tag{8}$$

Consequently, $\langle \sigma_T \rangle_\Lambda$ does not depend on Λ (for $T \subset \Lambda$). Thus the thermodynamic limit $\langle \sigma_T \rangle_\Lambda$ exists in this case and equals (8). The random variables σ_t are mutually independent, both with respect to the distributions in finite Λ and with respect to the limit distribution.

Proof of Theorem 1. Notice that it is sufficient to consider the case $h \geqslant 0$. This is obvious from the following symmetry property of the Ising model (in the notations introduced below, β and h as subscripts indicate the dependence of Gibbs distributions on these parameters):

$$P_{\Lambda,\beta,h}(\sigma^\Lambda) = P_{\Lambda,\beta,-h}(-\sigma^\Lambda), \tag{9}$$

with $-\sigma^\Lambda$ denoting the configuration whose values have an opposite sign to those of the configuration σ^Λ.

It follows from (9) that

$$\langle\sigma_T\rangle_{\Lambda,\beta,h} = \begin{cases} \langle\sigma_T\rangle_{\Lambda,\beta,-h}, & |T| \text{ even}, \\ -\langle\sigma_T\rangle_{\Lambda,\beta,-h}, & |T| \text{ odd}. \end{cases} \tag{10}$$

In particular,

$$\langle\sigma_T\rangle_{\Lambda,\beta,0} = 0 \tag{11}$$

for odd $|T|$. The proof of the relations (9) and (10) is simple, and is left to the reader.

We need two inequalities to prove the theorem. It is suitable to consider a more general situation. Let Λ be an arbitrary subset of Z^ν, Ω_Λ a set of all configurations $\sigma^\Lambda = \{\sigma_t,\ t \in \Lambda\}$, $\sigma_t = \pm 1$, in Λ, and the energy $U_\Lambda(\sigma^\Lambda)$ of the configuration σ^Λ be of the form

$$U_\Lambda(\sigma^\Lambda) = -\left(\sum_{t\in\Lambda} h_t\sigma_t + \sum_{t,t'\in\Lambda} \beta_{t,t'}\sigma_t\sigma_{t'}\right), \tag{12}$$

with $h_t \geqslant 0$ and $\beta_{tt'} \geqslant 0$ (in general, the sum in (12) is taken over all pairs of points $t, t' \in \Lambda$). The distribution P_Λ on Ω_Λ is given as before by the formula (2), and $\langle\ \rangle_\Lambda$ again denotes the mean under this distribution.

Lemma 2. *The inequalities*

$$\langle\sigma_T\rangle_\Lambda \geqslant 0 \tag{13}$$

(the first Griffith inequality) and

$$\langle\sigma_T\sigma_{T'}\rangle_\Lambda - \langle\sigma_T\rangle_\Lambda\langle\sigma_{T'}\rangle_\Lambda \geqslant 0 \tag{14}$$

(the second Griffith inequality) are valid.

Proof. To prove (13), it is sufficient to verify

$$\sum_{\sigma^\Lambda\in\Omega_\Lambda} \sigma_T \exp\{-U_\Lambda(\sigma^\Lambda)\} \geqslant 0. \tag{15}$$

By expanding the exponential function $\exp\{-U_\Lambda(\sigma_\Lambda)\}$ in the series $\sum_{n=0}^\infty (-U_\Lambda)^n/n!$, by removing the parentheses in each term of this series, and by taking into account that $\sigma_t^2 = 1$, we express the left-hand side of the inequality (15) in the form of the sum

$$\sum_{B\subseteq\Lambda} C_B \sum_{\sigma^\Lambda\in\Omega_\Lambda} \sigma_B \tag{16}$$

with $C_B \geqslant 0$. Since

$$\sum_{\sigma_t=\pm 1} \sigma_t = 0 \tag{17}$$

for any $t \in \Lambda$, the sum (16) is equal to $C_\emptyset$, which proves (13).

To prove (14) we investigate two independent samples of the distribution P_Λ, i.e., a distribution on the space $\Omega_\Lambda \times \Omega_\Lambda$ of pairs $\{\sigma^\Lambda, \tilde{\sigma}^\Lambda\}$ of configurations of the form

$$\begin{aligned} &\hat{P}_\Lambda(\sigma^\Lambda, \tilde{\sigma}^\Lambda) \\ &\quad = (Z_\Lambda^{-1})^2 \exp\left\{\sum_{t\in\Lambda} h_t(\sigma_t + \tilde{\sigma}_t) + \sum_{t,t'\in\Lambda} \beta_{t,t'}(\sigma_t\sigma_{t'} + (\tilde{\sigma}_t\tilde{\sigma}_{t'})\right\}. \end{aligned} \tag{18}$$

We introduce new variables

$$\xi_t = \sigma_t + \tilde{\sigma}_t, \qquad \eta_t = \sigma_t - \tilde{\sigma}_t, \qquad t \in \Lambda,$$

taking values $(\xi_t, \eta_t) = (2,0), (-2,0), (0,2), (0,-2)$. By means of these variables, the probability (18) may be rewritten in the form

$$Z_\Lambda^{-2} \exp\left\{\sum_{t\in\Lambda} h_t\xi_t + \frac{1}{2}\sum_{t,t'\in\Lambda} \beta_{t,t'}(\xi_t\xi_{t'} + \eta_t\eta_{t'})\right\}.$$

Taking into account that

$$\xi_t\eta_t = 0 \qquad \text{and} \qquad \sum_{\xi_t=-2,0,2} \xi_t^k = \sum_{\eta_t=-2,0,2} \eta_t^k \geqslant 0$$

for each $t \in \Lambda$ and each integer $k \geqslant 0$, and by repeating the proof of the inequality (13), we get

$$\langle \xi_T \eta_{T'} \rangle_{\Lambda,\Lambda} \geqslant 0 \tag{19}$$

for all T and T', with ξ_T and η_T being defined as in (4) and the mean $\langle\ \rangle_{\Lambda,\Lambda}$ evaluated under the distribution (18).

Notice that

$$\langle \sigma_T\sigma_{T'} \rangle_\Lambda - \langle \sigma_T \rangle_\Lambda \langle \sigma_{T'} \rangle_\Lambda = \frac{1}{2}\langle (\sigma_T - \tilde{\sigma}_T)(\sigma_{T'} - \tilde{\sigma}_{T'}) \rangle_{\Lambda,\Lambda}. \tag{20}$$

We shall show that the difference $\sigma_T - \tilde{\sigma}_T$ and the sum $\sigma_T + \tilde{\sigma}_T$ can be represented in the form

$$\sigma_T \pm \tilde{\sigma}_T = \sum_{A,B\subseteq T} C^{\pm}_{A,B} \xi_A \eta_B \tag{21}$$

with $C^{\pm}_{A,B} \geqslant 0$. The relations (19), (20), and (21) imply the inequality (15). The expression (21) can be proved by induction on $|T|$ if we notice that

$$\sigma_{T\cup\{t\}} + \tilde{\sigma}_{T\cup\{t\}} = \frac{1}{2}\left[(\sigma_T + \tilde{\sigma}_T)\xi_t + (\sigma_T - \tilde{\sigma}_T)\eta_t\right],$$
$$\sigma_{T\cup\{t\}} - \tilde{\sigma}_{T\cup\{t\}} = \frac{1}{2}\left[(\sigma_T + \tilde{\sigma}_T)\eta_t + (\sigma_T - \tilde{\sigma}_T)\xi_t\right]$$

for $t \notin T \subset \Lambda$.

The lemma is proved.

Now we return to the proof of the theorem.

The derivatives $\frac{\partial}{\partial h_t}\langle\sigma_T\rangle_\Lambda$ and $\frac{\partial}{\partial \beta_{t,t'}}\langle\sigma_T\rangle_\Lambda$ are equal to

$$\frac{\partial}{\partial h_t}\langle\sigma_T\rangle_\Lambda = \langle\sigma_T\sigma_t\rangle_\Lambda - \langle\sigma_T\rangle_\Lambda\langle\sigma_t\rangle_\Lambda \geqslant 0,$$
$$\frac{\partial}{\partial \beta_{t,t'}}\langle\sigma_T\rangle_\Lambda = \langle\sigma_T\sigma_t\sigma_{t'}\rangle_\Lambda - \langle\sigma_T\rangle_\Lambda\langle\sigma_t\sigma_{t'}\rangle_\Lambda \geqslant 0, \tag{22}$$

and thus the correlation functions increase when increasing the parameters h_t and $\beta_{t,t'}$. It follows that, in the case of the Ising model,

$$\langle\sigma_T\rangle_{\Lambda_1} \leqslant \langle\sigma_T\rangle_{\Lambda_2} \tag{23}$$

for $T \subset \Lambda_1 \subset \Lambda_2$. Indeed, the mean $\langle\sigma_T\rangle_{\Lambda_1}$ coincides with the mean under the distribution of the form (12) in Λ_2, with parameters

$$h_t = \begin{cases} h, & t \in \Lambda_1, \\ 0, & t \in \Lambda_2 \backslash \Lambda_1 \end{cases}$$

and

$$\beta_{t,t'} = \begin{cases} \beta & \text{if } t, t' \text{ are nearest neighbours in } \Lambda_1, \\ 0 & \text{otherwise.} \end{cases}$$

Using the monotonicity of $\langle\sigma_T\rangle$ with respect to the parameters h_t and $\beta_{t,t'}$, we get (23). Since $|\langle\sigma_T\rangle| \leqslant 1$, the statement of the theorem follows from (23).

0.3. Markov Property

Let $A \subset Z^\nu$ be a set; its *boundary* ∂A is defined to be the set of all lattice sites of distance 1 from A:

$$\partial\Lambda = \{t \in Z^\nu : \rho(t, A) = 1\}. \tag{24}$$

Let $\Lambda \subset Z^\nu$ be a cube, and let $A, B \subseteq \Lambda$ be such that $A \cap B = \emptyset$ and $\partial A \subset B$. We use

$$P_\Lambda^{(A)}(\bar{\sigma}^A/\tilde{\sigma}^B) = \Pr\{\sigma_t = \bar{\sigma}_t, t \in A/\sigma_{t'} = \tilde{\sigma}_{t'}, t' \in B\}$$

to denote the conditional probability that σ^Λ equals $\bar{\sigma}^A = \{\bar{\sigma}_t,\ t \in A\}$ on the set A under the condition that its values on the set B equal $\tilde{\sigma}^B = \{\tilde{\sigma}_{t'},\ t' \in B\}$.

Lemma 3. *The following equalities hold true:*

$$\begin{aligned} P_\Lambda^{(A)}(\bar{\sigma}^A/\tilde{\sigma}^B) &= P_\Lambda^{(A)}(\bar{\sigma}^A/\tilde{\sigma}^{\partial A}) \\ &= Z_A^{-1}(\tilde{\sigma}^{\partial A}) \exp\left\{-(U_A(\bar{\sigma}^A) + U_{A,\partial A}(\bar{\sigma}^A, \tilde{\sigma}^{\partial A}))\right\}. \end{aligned} \tag{25}$$

Here $U_A(\bar{\sigma}^A)$ is the energy of the configuration $\bar{\sigma}^A$ defined as in (1), $U_{A,\partial A}(\bar{\sigma}^A, \tilde{\sigma}^{\partial A})$ is the energy of the interaction between the configurations $\bar{\sigma}^A$ and $\tilde{\sigma}^{\partial A}$:

$$U_{A,\partial A}(\bar{\sigma}^A, \tilde{\sigma}^{\partial A}) = -\beta \sum_{\substack{t \in A, t' \in \partial A \\ \rho(t,t')=1}} \bar{\sigma}_t \tilde{\sigma}_{t'}, \tag{26}$$

and $Z_A(\tilde{\sigma}^{\partial A})$ is the conditional partition function

$$Z_A(\tilde{\sigma}^{\partial A}) = \sum_{\bar{\sigma}^A} \exp\left\{-(U_A(\bar{\sigma}^A) + U_{A,\partial A}(\bar{\sigma}^A, \tilde{\sigma}^{\partial A}))\right\}. \tag{27}$$

The first equality in (25) is called the *Markov* property of the distribution P_Λ, and the other equality expresses its Gibbs property: the conditional distribution $P_\Lambda^{(A)}$ is similar in form to the distribution (2), except that the energy $U_{A,\partial A}$ of the interaction with the "boundary" configuration $\tilde{\sigma}^{\partial A}$ was added to the energy U_A. Usually, the distribution given by the formula on the right-hand side of (25) is called the *Gibbs distribution in A with the boundary configuration $\tilde{\sigma}^{\partial A}$*.

Proof of Lemma 3. It is carried out by a direct computation: The formula (2) implies that

$$\begin{aligned} P_\Lambda^{(A)}(\bar{\sigma}^A/\tilde{\sigma}^B) &= \frac{P_\Lambda^{(A\cup B)}(\bar{\sigma}^A, \tilde{\sigma}^B)}{P_\Lambda^{(B)}(\tilde{\sigma}^B)} \\ &= \frac{\sum_{\sigma^\Lambda\setminus\{A\cup B\}} \exp\left\{-U_\Lambda(\bar{\sigma}^A, \tilde{\sigma}^B, \sigma^{\Lambda\setminus(A\cup B)})\right\}}{\sum_{\sigma^\Lambda\setminus(A\cup B), \bar{\sigma}^A} \exp\left\{-U_\Lambda(\bar{\sigma}^A, \tilde{\sigma}^B, \sigma^{\Lambda\setminus(A\cup B)})\right\}}, \end{aligned} \tag{28}$$

where $\sigma^\Lambda = (\bar\sigma^A, \tilde\sigma^B, \sigma^{\Lambda\setminus(A\cup B)})$ and $\sigma^{\Lambda\setminus(A\cup B)}$ is a configuration in the set $\Lambda\setminus(A\cup B)$. Further,

$$U_\Lambda(\sigma^\Lambda) = U_A(\bar\sigma^A) + U_{A,B}(\bar\sigma^A, \tilde\sigma^B) + U_B(\tilde\sigma^B) + U_{\Lambda\setminus(A\cup B)}(\sigma^{\Lambda\setminus(A\cup B)}) \\ + U_{B,\Lambda\setminus(A\cup B)}(\tilde\sigma^B, \sigma^{\Lambda\setminus(A\cup B)}),$$

with the energies $U_{A,B}$ and $U_{B,\Lambda\setminus(A\cup B)}$ given analogously to (26). Accordingly, the nominator on the right-hand side of the equalities (28) equals

$$\exp\left\{-(U_A(\bar\sigma^A) + U_{A,B}(\bar\sigma^A, \tilde\sigma^B) + U_B(\tilde\sigma^B))\right\} Z_{\Lambda\setminus(A\cup B)}(\tilde\sigma^B),$$

and the denominator is

$$\exp\left\{-U_B(\tilde\sigma^B)\right\} Z_{\Lambda\setminus(A\cup B)}(\tilde\sigma^B) Z_A(\tilde\sigma^B),$$

with $Z_{\Lambda\setminus(A\cup B)}(\tilde\sigma^B)$ and $Z_A(\tilde\sigma^B)$ defined analogously to (27). Inserting these expressions into (28) and noticing that $U_{A,B}(\bar\sigma^A, \tilde\sigma^B) = U_{A,\partial A}(\bar\sigma_A, \tilde\sigma_{\partial A})$ and $Z_A(\tilde\sigma^B) = Z_A(\tilde\sigma^{\partial A})$, we arrive at (25) after obvious cancellations. The lemma is proved.

It is obvious from (25) that the conditional distribution $P_\Lambda^{(A)}(\cdot/\tilde\sigma^B)$ does not depend on Λ. This observation lays the foundation for the following definition of the Gibbs random field in Z^ν.

Definition 1. A probability distribution P on the space Ω is said to determine a *Gibbs random field* $\{\sigma_t,\ t \in Z^\nu\}$ (for the Ising model) if the conditional distribution $P^{(A)}(\bar\sigma^A/\tilde\sigma^B)$, generated by the distribution P, coincides with the Gibbs distribution in A, with the boundary configuration $\tilde\sigma^{\partial A}$ (see the second equality in (25)) for arbitrary finite subsets A, $B \subset Z^\nu$ such that $A \cap B = \emptyset$ and $\partial A \subset B$.

Notice that, according to the first equality in (28) and the definition (7), the limit Gibbs distribution constructed above defines a Gibbs random field in Z^ν also in the sense of Definition 1. Are there still other Gibbs fields in Z^ν for the Ising model? It turns out that this depends on the dimension of the lattice Z^ν and on the parameters (h, β). The values of parameters (h, β) for which there exist more than one Gibbs field in Z^ν define points of the first-order phase transition in the plane (h, β).

Theorem 4. *For a ferromagnetic Ising model:*
1) for $\nu = 1$, there is a unique Gibbs field;
2) for $\nu \geqslant 2$ and $h \neq 0$, or $h = 0$ and β sufficiently small, $0 \leqslant \beta \leqslant \beta_0(\nu)$, there is a unique Gibbs field;
3) for $\nu \geqslant 2$, the points $(0, \beta)$ with β sufficiently large, $\beta > \beta_1(\nu)$, are points of the first-order phase transition.

We shall prove only the statements 1) and 3) of this theorem.

Before proving the theorem, we investigate the possible ways of construction of Gibbs fields in Z^ν for the Ising model. Let $\Lambda \subset Z^\nu$ be a cube, $\tilde{\sigma}^{\partial\Lambda}$ be a configuration in the boundary $\partial\Lambda$ of the cube Λ, and let $P_{\Lambda,\tilde{\sigma}^{\partial\Lambda}}(\sigma^\Lambda)$ denote the Gibbs distribution in Λ (on the space Ω_Λ) with the boundary configuration $\tilde{\sigma}^{\partial\Lambda}$ (see (25)). Let $q^{\partial\Lambda}$ be an arbitrary probability distribution on the set $\Omega_{\partial\Lambda}$ of boundary configurations $\tilde{\sigma}^{\partial\Lambda}$. We use the notation $P_{\Lambda,q^{\partial\Lambda}}$ for the distribution

$$P_{\Lambda,q^{\partial\Lambda}}(\sigma^\Lambda) = \langle P_{\Lambda,\tilde{\sigma}^{\partial\Lambda}}(\sigma^\Lambda)\rangle_{q^{\partial\Lambda}}, \tag{29}$$

on the space Ω_Λ, arising by averaging of distributions $P_{\Lambda,\tilde{\sigma}_{\partial\Lambda}}$ over all boundary configurations $\tilde{\sigma}^{\partial\Lambda}$. The distribution (29) is called the Gibbs distribution in Λ with *a random boundary configuration.* Finally, besides $P_{\Lambda,\tilde{\sigma}^{\partial\Lambda}}$ and $P_{\Lambda,q^{\partial\Lambda}}$, the Gibbs distribution P_Λ^{per} with the so-called *periodic* boundary conditions is often considered. It is defined analogously to the distribution P_Λ (see (2)) except for replacing the "cube" Λ by the "torus" (by identifying the opposite sides of Λ) and the energy U_Λ in (2) by the energy U_Λ^{per} of the interaction of the nearest neighbours on this torus. The Gibbs distribution (2) is often called the Gibbs distribution in Λ under the "empty boundary conditions."

Following the proof of Lemma 3, we can easily convince ourselves that the distributions

$$P_{\Lambda,\tilde{\sigma}^{\partial\Lambda}}, \quad P_{\Lambda,q^{\partial\Lambda}}, \quad P_\Lambda^{\text{per}} \tag{30}$$

have the Gibbs property (25).

As in the case of Gibbs distributions with the empty boundary conditions, we conclude that the limit $P = \lim_{\Lambda_n \nearrow Z^\nu} P_{\Lambda_n}$ of the sequence P_{Λ_n} of distributions of the form (30), with Λ_n being an increasing sequence of cubes, $\Lambda_1 \subset \Lambda_2 \subset \ldots \subset \Lambda_n \subset \ldots \subset \cup\Lambda_n = Z^\nu$, defines a Gibbs field in Z^ν. The converse is also true.

Lemma 5. *Every probability distribution P on the space Ω that is a Gibbs random field in Z^ν is the thermodynamic limit of a sequence $P_{\Lambda_n,q_n^{\partial\Lambda_n}}$ for some choice of $q_n^{\partial\Lambda_n}$.*

Proof. For every cube $\Lambda \subset Z^\nu$, we choose $q^{\partial\Lambda}$ to be the probability distribution on $\Omega_{\partial\Lambda}$ induced by the distribution P. It is obvious that $P_{\Lambda,q^{\partial\Lambda}}$ coincides in this case with the distribution induced by P on Ω_Λ. Therefore, $P_{\Lambda,q^{\partial\Lambda}} \to P$ (in the sense (7)) as $\Lambda \nearrow Z^\nu$.

Now we return to the proof of the theorem.

Proof of the statement 1) of Theorem 4. *Transfer matrix.* To simplify the formulae, we put $h = 0$. The 2×2 matrix $J = \|j_{\sigma\sigma'}\|$ with matrix elements $j_{\sigma,\sigma'} = e^{\beta\sigma\sigma'}, \sigma\sigma' = \pm 1$,

$$J = \begin{Vmatrix} e^\beta & e^{-\beta} \\ e^{-\beta} & e^\beta \end{Vmatrix} \tag{31}$$

is called the *transfer matrix* of the Ising model.

Let $\Lambda = [-N, N] \subset Z^1$ and P_Λ be the Gibbs distribution in Λ (under the empty boundary conditions).

Lemma 6. *The following equalities hold:*

$$Z_\Lambda = (J^{2N} e, e), \tag{32}$$

$$\begin{aligned} P_\Lambda^{\{t_1,\dots,t_n\}}&(\bar\sigma_{t_1},\dots,\bar\sigma_{t_n}) \\ &= \frac{\left(e^{(\bar\sigma_{t_1})}, J^{N_1} e\right)\left(e^{(\bar\sigma_{t_2})}, J^{t_2-t_1} e^{(\bar\sigma_{t_1})}\right)\dots\left(e, J^{N_2} e^{\bar\sigma_{t_4}}\right)}{(e, J^{2N} e)}, \end{aligned} \tag{33}$$

with $e = (1,1), e^{(1)} = (1,0), e^{(-1)} = (0,1), N_1 = t_1 + N, N_2 = N - t_n$, $-N \leqslant t_1 < t_2 < \dots < t_n \leqslant N$. *The proof is obvious.*

Now let $g^{(1)}$ and $g^{(2)}$ be two normalized eigenvectors of the transfer matrix J, with eigenvalues λ_1 and $\lambda_2, \lambda_1 > |\lambda_2| \geqslant 0$. Using the decompositions

$$e = C_1 g^{(1)} + C_2 g^{(2)}, \qquad e^{(\pm 1)} = B_1^{(\pm 1)} g^{(1)} + B_2^{(\pm 1)} g^{(2)},$$

we get

$$\begin{aligned} (J^{2N} e, e) &\sim C_1^2 \lambda_1^{2N}, \\ \left(e^{(\bar\sigma_{t_1})}, J^{N_1} e\right) &\sim B_1^{(\bar\sigma_{t_1})} C_1 \lambda_1^{N_1}, \\ \left(e, J^{N_2} e^{(\bar\sigma_{t_n})}\right) &\sim B_1^{(\bar\sigma_{t_n})} C_1 \lambda_1^{N_2}, \end{aligned}$$

for large N and fixed $\{t_1,\dots,t_n\}$, and thus

$$\lim_{N\to\infty} P_\Lambda^{\{t_1,\dots,t_n\}}(\bar\sigma_{t_1},\dots,\bar\sigma_{t_n}) = B_1^{(\bar\sigma_{t_1})} B_1^{(\bar\sigma_{t_n})} \prod_{k=2}^{n} \frac{\left(e^{(\bar\sigma_{t_k})}, J^{t_k - t_{k-1}} e^{(\bar\sigma_{t_{k-1}})}\right)}{\lambda_1^{t_k - t_{k-1}}}.$$

Similarly, it can be shown that the probabilities $P_{\Lambda_n, q_n^{\partial\Lambda_n}}^{\{t_1,\dots,t_n\}}$ have the same limit for any sequence of Gibbs distributions $P_{\Lambda_n, q_n^{\partial\Lambda_n}}, \Lambda_n \nearrow Z^1$. The proof of the first statement is finished.

Remark. From our considerations, one may easily derive that the limit Gibbs field $\{\sigma_t, t \in Z^1\}$ is a stationary Markov chain with the matrix of transition probabilities

$$P_{\sigma_1,\sigma_2} = \frac{J_{\sigma_1\sigma_2} g_{\sigma_2}^{(1)}}{\lambda_1 g_{\sigma_1}^{(1)}}, \qquad \sigma_1, \sigma_2 = \pm 1,$$

and the stationary distribution $\pi_\sigma = (g_\sigma^{(1)})^2, \sigma = \pm 1$, where $g_1^{(1)}$, $g_{-1}^{(1)}$ are the components of the eigenvector $g^{(1)}$.

Proof of the statement 3) of Theorem 4. We denote the Gibbs distribution in Λ with the boundary configuration $\tilde{\sigma}_t \equiv +1, t \in \partial\Lambda$ ((+)-boundary conditions) by $P_{\Lambda,(+)}$.

Lemma 7. *The inequality*

$$\Pr_{\Lambda,(+)}(\sigma_0 = -1) < 1/3 \tag{34}$$

holds uniformly with respect to all cubes $\Lambda \subset Z^\nu$, $0 \in \Lambda$, for all sufficiently large $\beta, \beta > \beta_1(\nu)$.

First, we shall derive our statement from this lemma. Consider the Gibbs distribution $P_{\Lambda,(-)}$ with the boundary configuration $\tilde{\sigma}_t \equiv -1$, $t \in \partial\Lambda$ (the (−)-boundary conditions). By virtue of the symmetry valid for $h = 0$, we have

$$P_{\Lambda,(+)}(\sigma^\Lambda) = P_{\Lambda,(-)}(-\sigma^\Lambda),$$

and hence,

$$\Pr_{\Lambda,(-)}(\sigma_0 = 1) < 1/3$$

for every Λ, and consequently,

$$\Pr_{\Lambda,(-)}(\sigma_0 = -1) > 2/3. \tag{35}$$

The inequalities (34) and (35) ensure that there are at least two different Gibbs distributions in Z^ν.

Proof of Lemma 7. To make the proof more transparent, we restrict ourselves to the case $\nu = 2$. Let $\widetilde{Z}^2$ be the dual lattice obtained from the lattice Z^2 by shifting it by the vector $(1/2, 1/2)$. For any configuration σ^Λ, we use $\gamma = \gamma(\sigma^\Lambda)$ to denote the collection of those bonds of $\widetilde{Z}^2$ that separate two neighbouring sites $t, t' \in \Lambda \cup \partial\Lambda$ with $\sigma_t \neq \sigma_{t'}$, ($\sigma_t = 1$ for $t \in \partial\Lambda$). It is easy to see that the number of bonds from $\gamma(\sigma^\Lambda)$ attached to a lattice site from $\widetilde{Z}^2$ is always even. Thus, the connected components of γ are closed polygons (possibly self-intersecting). We shall denote them by $\Gamma_1, \ldots, \Gamma_n$ and call them *contours.* We shall show that there is a configuration σ^Λ with $\gamma = \gamma(\sigma^\Lambda)$ for each collection $\gamma = \{\Gamma_1, \ldots, \Gamma_n\}$ of mutually disjoint contours. Indeed, it is enough to put $\sigma_t = 1$ for $t \in \Lambda$ that are outside all contours. Further, we put $\sigma_t = -1$ for the sites that are inside one contour Γ only, $\sigma_t = 1$ for the sites that are encircled by two contours, and so on. Thus, there is a one-to-one correspondence between the configurations σ^Λ and the collections of contours γ. In addition,

$$U_{\Lambda,(+)}(\sigma^\Lambda) = U_\Lambda(\sigma^\Lambda) + U_{\Lambda,\partial\Lambda}(\sigma^\Lambda, \tilde{\sigma}^{\partial\Lambda} \equiv 1) = 2\beta|\gamma| - \beta|\tilde{\Lambda}|,$$

$$Z_{\Lambda,(+)} = Z_\Lambda(\tilde{\sigma}^{\partial\Lambda} \equiv 1) = \exp\left\{\beta|\tilde{\Lambda}|\right\} \sum_{\nu} e^{-2\beta|\nu|},$$

where $|\gamma|$ is the number of bonds in γ (the length of γ) and $|\tilde{\Lambda}|$ is the number of bonds from $\widetilde{Z}^2$ adjacent to at least one site from Λ.

Lemma 8. *The probability $P_{\Lambda,(+)}(\Gamma)$ of the event that Γ is contained in the collection γ can be estimated by*

$$P_{\Lambda,(+)}(\Gamma) \leqslant e^{-2\beta|\Gamma|}.$$

Proof. The probability $P_{\Lambda,(+)}(\Gamma)$ equals

$$\begin{aligned} P_{\Lambda,(+)}(\Gamma) &= \sum_{\gamma:\Gamma\in\gamma} P_{\Lambda,(+)}(\gamma) \\ &= \frac{\sum_{\gamma:\Gamma\in\gamma} e^{-2\beta|\gamma|}}{\sum_{\gamma} e^{-2\beta|\gamma|}} = \frac{e^{-2\beta|\Gamma|}\sum'_{\gamma} e^{-2\beta|\gamma|}}{\sum_{\gamma} e^{-2\beta|\gamma|}} < e^{-2\beta|\Gamma|}, \end{aligned}$$

where $\sum'_{\gamma}$ denotes the sum over all γ that do not contain Γ or intersect it. The lemma is proved.

Further, it is easy to convince ourselves that the number of contours Γ of the length n encircling a given site $t_0 \in Z^2$ is not greater than $n^2 3^n$. Since the event $\sigma_0 = -1$ under the (+)-boundary conditions implies the existence of at least one contour Γ encircling the point 0, we have

$$\mathrm{Pr}_{\Lambda,(+)}(\sigma_0 = -1) \leqslant \sum_{\Gamma:\Gamma \text{ encircles } 0} P_{\Lambda,(+)}(\Gamma) \leqslant \sum_{n\geqslant 4} n^2 3^n e^{-2\beta n} < 1/3$$

for β large enough. Lemma 7 and, at the same time, the statement 3) of Theorem 4, are proved.

0.4. Other Models

To enrich the reader's idea of what the theory of Gibbs fields is all about, we shall add a few more examples of physical lattice models often met in the literature.

I. Rotator model

Let $\Lambda \subset Z^\nu$ be a finite subset, the configurations $\sigma^\Lambda = \{\sigma_t,\ t \in \Lambda\}$ take values in the n-dimensional sphere, $\sigma_t \in R^n, |\sigma_t| = 1$, and the energy of a configuration be given by

$$U_\Lambda(\sigma^\Lambda) = \sum_{t,t'\in\Lambda} J_{t-t'} \cdot (\sigma_t, \sigma_{t'}) + \sum_{t\in\Lambda}(h, \sigma_t),$$

with $(.\,,.)$ being the scalar product in R^n, and J_t being some finite function of $t \in Z^\nu$, $t \neq 0$, $h \in R^n$. In the case of this model, the Gibbs distribution P is given by

$$p_\Lambda(\sigma^\Lambda) = Z_\Lambda^{-1} \exp\left\{-\beta U_\Lambda(\sigma^\Lambda)\right\}, \tag{36}$$

where $p_\Lambda(\sigma^\Lambda)$ is the density of the distribution P_Λ with respect to the measure $\lambda_0^\Lambda = \underbrace{\lambda_0 \times \ldots \times \lambda_0}_{|\Lambda| \text{times}}$ of the space Ω_Λ of configurations, with λ_0 a uniform distribution on the sphere $S \subset R^n$ and Z_Λ the normalization factor.

II. Model with a global symmetry (G-model)

Let G be a group, and let the configuration $g^\Lambda = \{g_t,\ t \in \Lambda\}$, $\Lambda \subset Z^\nu$, take values from G. The energy of a configuration g^Λ is of the form

$$U_\Lambda(g^\Lambda) = \sum_{\substack{\rho(t,t')=1 \\ t,t' \in \Lambda}} \psi(g_t g_{t'}^{-1}),$$

where ψ is some even function on the group $G : \psi(g) = \psi(g^{-1})$. The Gibbs distribution is defined according to the preceding by a formula similar to (2) in the case of the discrete group G, or to (36) in the case of the continuous group G with λ_0 standing for the normalized Haar measure on G.

III. Gauge G-model (a model with a local G-symmetry)

In this case, the configurations $g^\Lambda = \{g_\tau,\ \tau \in \Lambda\}$ are defined on a finite set of oriented bonds $\tau = (t_1, t_2)$, t_1, $t_2 \in Z^\nu$, $\rho(t_1, t_2) = 1$, and take values from G as before. Moreover, $g_\tau = (g_{-\tau})^{-1}$, where $-\tau = (t_2, t_1)$ denotes the bond that differs from τ by its orientation. The energy of the configuration g^Λ is given in the form

$$U_\Lambda^{\text{gauge}}(g^\Lambda) = \sum_p \psi(g_p), \tag{37}$$

with the summation taken over all two-dimensional (nonoriented) plaquettes p of the lattice Z^ν whose boundary bonds $\partial p = (\tau_1, \tau_2, \tau_3, \tau_4)$ belong to Λ, and

$$g_p = g_{\tau_1} g_{\tau_2} g_{\tau_3} g_{\tau_4}.$$

In addition, the numbering of the bonds τ_i and their orientation are chosen so that they form a path around p in some direction. The function ψ in (37) is a function on the group G so that

$$\psi(g) = \psi(g^{-1}), \qquad \psi(g g_0 g^{-1}) = \psi(g_0). \tag{38}$$

Under these conditions the value $\psi(g_p)$ does not depend on the orientation and the starting point of the path around p. Most often, the function ψ is chosen equal to

$$\psi(g) = \operatorname{Re} \chi_\alpha(g),$$

with χ_α being the character of some irreducible unitary representation of the group G. The Gibbs distribution for this model is introduced analogously to the preceding cases.

IV. $P(\varphi)_\nu$-model

Configurations $x^\Lambda = \{x_t,\ t \in \Lambda\}$, $\Lambda \subset Z^\nu$, take arbitrary real values, the energy $U_\Lambda(x^\Lambda)$ is given by

$$U_\Lambda(x^\Lambda) = \sum_{\substack{(t,t')\in\Lambda \\ \rho(t,t')=1}} (x_t - x_{t'})^2 + \sum_{t\in\Lambda}(P(x_t) + m x_t^2),$$

where $m \geqslant 0$ and $P(\cdot)$ is some polynomial of an even degree with a positive leading coefficient. The density of the Gibbs distribution p_Λ on the configuration space $\Omega_\Lambda = R^{|\Lambda|}$ with respect to the Lebesgue measure $(dx)^{|\Lambda|}$ on $R^{|\Lambda|}$ is defined by a formula similar to (36). Moreover, it may be easily shown that the normalization factor Z_Λ is finite.

We confine ourselves to this list of models. What kind of questions arise in the theory of Gibbs fields in connection with these (and other) models?

1. The first question is: Does there exist a limit Gibbs field (i.e., a limit probability distribution for infinite configurations)?

2. Second question: Is it unique? This question is very interesting because the existence of several limit distributions (phases) is connected with the so-called phase transitions.

3. If the interaction depends on one or more parameters, the question is: Are the various characteristics of the limit field (correlation functions, semi-invariants, and so on) analytic with respect to these parameters (or at least continuous or differentiable)? In addition, the singularities of correlation functions with respect to these parameters (discontinuities, discontinuities of their derivatives, and so on) also indicate phase transitions.

4. Furthermore, questions concerning the ergodic properties of the limit Gibbs fields arise; mixing, decay of correlations, estimates of semi-invariants, limit distributions of sums of local variables, etc.

These are the basic topics of the theory of Gibbs fields touched upon in this book.

Let us now proceed with a systematic presentation of this theory.

§1 Gibbs Modifications

1.1. Random Fields

In this book the following classes of random fields will be studied:

(1) *Random fields in a countable set T with values in a metric (complete and separable) space S.* The probability space (Ω, Σ, μ) is represented in this case

by the set $S^T = \Omega$ of functions (also called *configurations*) $x = \{x_t,\ t \in T\}$ defined on T, with values in S (S is often called the set of "spins"). The space S^T is endowed with the (metrizable) Tikhonov (cf. [16]) topology.

We shall investigate probability distributions μ defined on the Borel σ-algebra $\mathfrak{B}(S) = \Sigma$ of the space S^T. The collection of random variables x_t, $t \in T$, arising in this way, i.e., the values of the random configuration x at points $t \in T$, forms a *random field.*

The simplest example of such a field is the field of independent and identically distributed variables. In this case, the measure μ on $\mathfrak{B}(S^T)$ is defined to be the product of countably many identical copies of some probability measure λ_0 on the space S.

(2) *Random point fields in a separable metric space Q with values in a space S.* The role of the probability space is played by the set Ω of all locally finite subsets $x \subset Q$. The subset x (at most countable) is called *locally finite* if any bounded set $\Lambda \subset Q$ contains only a finite number of points from x. Below, we shall introduce a metrizable topology in Ω. Every probability measure defined on the Borel (with respect to that topology) σ-algebra $\mathfrak{B}(\Omega)$ is called a *random point field* in Q (sometimes we shall say pure point field, emphasizing its distinctness from the labelled point field to be introduced later).

Now suppose that a metrizable space S, also called the space of "charges" (or "labels"), is given. We use Ω^S to denote the space of pairs $\{x, s_x\}$, with $x \in \Omega$ and s_x being a function on x taking values from S. Such pairs will be called *configurations.* In the space Ω^S, as well as in Ω, a metrizable topology can be introduced. Every probability measure on $\mathfrak{B}(\Omega^S)$ determines a *labelled* random field in Q with values in the space S of charges.

The theory of Gibbs fields includes the investigation of:

(3) *Ordinary or generalized fields in R^ν.* The probability space is, in this case, a topological vector (locally convex) space Ω of functions or distributions (in the sense of Schwartz) defined on R^ν. As before, a random field is given by a definition of a probability measure on the Borel σ-algebra $\mathfrak{B}(\Omega)$.

In what follows, we emphasize that, by mentioning any random field, we implicitly understand a measure on the space of its configurations.

1.2 Method of Gibbs Modifications[1]

Gibbs modification (of a random field) is an important device for the construction of new measures from an originally given measure μ_0 (or from a family of measures). We shall first describe a general scheme of dealing with

[1] Translators remark: The authors stress, by introducing a new term "gibbsovskaya perestroĭka," the fact that the new measure arises by modifying the original one. To reflect this stress, and since the notion is crucial in the whole book, we avoided the term "specification" sometimes used in the literature, and follow the authors in introducing a new term. In the present context the translation "Gibbs modification" is preferable to, also possible, "Gibbs reconstruction."

Gibbs modifications.

Finite Gibbs modifications. Let (Ω, Σ, μ_0) be a measurable space with a finite or σ-finite measure μ_0 (called usually a "*free*" measure), and let $U(x)$, $x \in \Omega$, be a real function on Ω (taking possibly the value $+\infty$), often called the *interaction energy* (or *Hamiltonian*).

The measure μ with the density

$$\frac{d\mu}{d\mu_0}(x) = Z^{-1} \exp\{-U(x)\} \tag{1}$$

with respect to the measure μ_0 will be called the Gibbs modification of the measure μ_0 by means of the interaction U.

It is moreover assumed that the normalization factor Z (called the *partition function*) fulfills the *stability condition*

$$Z = \int_\Omega \exp\{-U(x)\} d\mu_0(x) \neq 0, \infty. \tag{2}$$

Using finite Gibbs modifications, the measures absolutely continuous with respect to μ_0 arise. A more interesting class of measures, already singular with respect to the original measure μ_0, arises when passing to the weak limit of finite Gibbs modifications.

1.3. Weak Convergence of Measures

Let Ω be a topological space, $\mathfrak{B} = \mathfrak{B}(\Omega)$ its Borel σ-algebra, and $\Sigma \subset \mathfrak{B}$ some of its sub-σ-algebras.

Definition 1. Let a directed family $\mathcal{F} = \{\Lambda\}$ of indices be given. We call a measure μ, defined on the σ-algebra $\Sigma \subset \mathfrak{B}$, the *weak limit* of the sequence of measures μ_Λ, $\Lambda \in \mathcal{F}$, defined on Σ if

$$\int_\Omega f(x) d\mu_\Lambda \to \int_\Omega f(x) d\mu \tag{3}$$

for any bounded continuous Σ-measurable function f given on Ω.

A more general situation may be examined. Let a complete family $\{\Sigma_\Lambda,\ \Lambda \in \mathcal{F}\}$, $\Sigma_{\Lambda_1} \subset \Sigma_{\Lambda_2}$, $\Lambda_1 < \Lambda_2$, of sub-σ-algebras of the σ-algebra $\mathfrak{B}$ (i.e., such that $\mathfrak{B}$ coincides with the smallest σ-algebra containing the algebra $\mathfrak{A} = \cup_{\Lambda \in \mathcal{F}} \Sigma_\Lambda$) be given; the σ-algebras Σ_Λ will be called *local* σ-algebras and any function f, defined on Ω and measurable with respect to some of the *local* algebras, will be called a *local*[2] function (function f, measurable with respect to a σ-algebra Σ_A, $A \in \mathcal{F}$, will often be denoted by f_A).

[2] The notion cylinder function is commonly used for local functions (translators remark).

A finitely additive measure μ defined on the algebra $\mathfrak{A}$ so that its restriction $\mu|_{\Sigma_\Lambda}$ to any σ-algebra Σ_Λ is a σ-additive measure on Σ_Λ is called a *cylinder measure*. If Ω is a complete separable metric space, and the cylinder measure μ is a probability ($\mu(A) \geqslant 0$, $A \in \mathfrak{A}$, $\mu(\Omega) = 1$), μ can be extended to a σ-additive (probability) measure defined on the σ-algebra $\mathfrak{B}$ (the Kolmogorov theorem; cf. [51]).

The following definition generalizes Definition 1.

Definition 2. Let a finite or σ-finite measure be given on each σ-algebra Σ_Λ. A cylinder measure μ on $\mathfrak{A}$ will be called the *weak local limit* of the measures μ_Λ if

$$\lim_\Lambda \int_\Omega f(x)d\mu_\Lambda = \int_\Omega f(x)d\mu \tag{4}$$

for any bounded continuous local function f defined on Ω.

In other words, a cylinder measure (or its extension to a measure on the σ-algebra $\mathfrak{B}$) is the weak local limit of measures $\{\mu_\Lambda,\ \Lambda \in \mathcal{F}\}$ if, for each $\Lambda_0 \in \mathcal{F}$, the restrictions $\mu_\Lambda|_{\Sigma_{\Lambda_0}} = \mu_\Lambda^{\Lambda_0}$, $\Lambda_0 < \Lambda$, $\Lambda \in \mathcal{F}$, of the measures μ_Λ to the σ-algebra Σ_{Λ_0} weakly converge to $\mu|_{\Sigma_{\Lambda_0}} = \mu_{\Lambda_0}$.

In the case $\Omega = S^T$ (with T a countable set and S a metric space; see Section 1), the index Λ runs over finite subsets of T, and $\Sigma_\Lambda = \varphi_\Lambda^{-1}(\mathfrak{B}(S^\Lambda))$, where $\varphi_\Lambda : S^T \to S^\Lambda$ is the restriction mapping (cf. p. 19), the convergence (4) is called the *weak convergence of finite-dimensional distributions* if, moreover, μ_Λ are probability measures.

The relationship between Definitions 1 and 2 is established by:

Proposition 1. *Let a family $\{\Sigma_\Lambda\ \Lambda \in \mathcal{F}\}$ of σ-algebras be such that the set $C_0(\Omega)$ of bounded continuous local functions is dense everywhere in the space $C(\Omega)$ of all bounded continuous functions defined on Ω (in the uniform metric in $C(\Omega)$). Then the necessary and sufficient condition for a measure μ on $\mathfrak{B}(\Omega)$ to be the local limit of probability measures $\{\mu_\Lambda\}$ (defined each on the σ-algebra Σ_Λ) is that their arbitrary extensions $\tilde{\mu}_\Lambda$ to probability measures on the σ-algebra $\mathfrak{B}(\Omega)$ weakly converge to μ.*

The proof is obvious.

1.4. Limit Gibbs Modifications

Let a complete directed family $\{\Sigma_\Lambda,\ \Lambda \in \mathcal{F}\}$ of sub-σ-algebras of the σ-algebra $\mathfrak{B}(\Omega)$ be given, and let a free measure μ_Λ^0 and a Hamiltonian U_Λ be defined for each Λ so that the stability condition (2) is satisfied. A cylinder measure μ on the algebra $\mathfrak{A} = \cup\Sigma_\Lambda$ (or its σ-additive extension to the σ-algebra $\mathfrak{B}(\Omega)$) is called a *limit Gibbs* measure (or a limit Gibbs modification) if it is the weak local limit of the Gibbs modifications μ_Λ of the measures μ_Λ^0 (by means of the energies U_Λ).

We note that the theory of Gibbs measures becomes meaningful only with a special choice of σ-algebras Σ_Λ, measures μ_Λ^0, and Hamiltonians U_Λ. Now we are going to describe the respective ways of such a choice of Σ_Λ, μ_Λ^0, and U_Λ in connection with the three types of random fields listed above.

(1) Gibbs modifications of fields in a countable set T. We introduce a set of configurations $S^\Lambda = \{x^\Lambda = (x_t,\ t \in \Lambda)\}$ defined in finite $\Lambda \subset T$, and endow it with a Tikhonov topology and Borel σ-algebra $\mathcal{B}(S^\Lambda)$. The restriction mapping $\varphi_\Lambda : x \mapsto x^\Lambda = x|_\Lambda$ defines a σ-algebra $\Sigma_\Lambda = \varphi_\Lambda^{-1}(\mathfrak{B}(S^\Lambda)) \subset \mathfrak{B}(S^T)$ that will be often identified with $\mathfrak{B}(S^\Lambda)$. Obviously, $\{\Sigma_\Lambda,\ \Lambda \subset T\}$ is complete in $\mathfrak{B}(S^T)$.

Remark 1. The set $C_0(S^T) \subset C(S^T)$ of bounded continuous local functions on S^T is, according to Stone's theorem (cf.[16]), dense everywhere in $C(S^T)$, and hence Proposition 1 applies in the case considered.

Hamiltonians U_Λ are usually defined with the help of a *potential* $\{\Phi_A;\ A \subset T,\ |A| < \infty\}$, i.e., a family of functions Φ_A on Ω that are measurable with respect to σ-algebras Σ_A (i.e., Φ_A can be viewed as a function defined on the space S^A).

For any finite A, we put

$$U_\Lambda = \sum_{A \subseteq \Lambda} \Phi_A. \tag{5}$$

Often, instead of the explicit formula for the potential, the formal Hamiltonian (formal sum)

$$U = \sum_A \Phi_A \tag{6}$$

is used.

Remark 2. In many cases, the free measures μ_Λ^0 are restrictions of some probability measure μ_0 defined on S^T to the respective σ-algebras $\Sigma_\Lambda \subset \mathfrak{B}$. In such cases, instead of a Gibbs modification μ_Λ defined by the formula (1), the measure $\hat{\mu}_\Lambda$ given on the σ-algebra $\mathfrak{B}(S^T)$ by

$$\frac{d\hat{\mu}_\Lambda}{d\mu_0}(x) = Z_\Lambda^{-1} \exp\{-U_\Lambda(x)\} \tag{7}$$

is investigated. The measure $\hat{\mu}_\Lambda$ is a "natural" extension of the measure μ_Λ to the whole σ-algebra $\mathfrak{B}(S^T)$. This measure is also called a finite (i.e., given in a finite volume) Gibbs modification of the measure μ_0. By virtue of Remark 1, a limit Gibbs measure μ on the space S^T (i.e., the weak limit of the measures μ_Λ) is the weak limit of the measures $\hat{\mu}_\Lambda$, $\Lambda \nearrow T$.

(2) Gibbs modifications of point fields. Let $\Lambda \subset Q$ be a domain in Q (cf. (2) in Section 1), $\Omega^S(\Lambda, n) \subset (\Lambda \times S)^n/\Pi_n$ be the set of sequences of pairs

$$\{(q_1, s_1), \dots, (q_n, s_n)\}, \quad q_i \in Q, \quad q_i \neq q_j, \quad i \neq j, \quad s_i \in S, \tag{7'}$$

factorized with respect to the group Π_n of permutations of n elements (two sequences (7′) are considered to be equivalent if one arises from the other by means of some permutation). In this way, $\Omega^S(\Lambda, n)$ is endowed with a metrizable topology. We use the notation $\Omega^S(\Lambda) = \bigcup_{n=0}^{\infty} \Omega^S(\Lambda, n)$, $\Omega^S(\Lambda, 0) = \emptyset$, and we introduce on $\Omega^S(\Lambda)$ the topology of the direct sum of topological spaces. By means of the restriction mapping

$$\varphi_\Lambda : (x, s_x) \mapsto (x \cap \Lambda, s_x|_{x\cap\Lambda}) \in \Omega^S(\Lambda), \tag{7''}$$

where Λ is a bounded domain in Q, the topology on Ω^S is defined as the weakest topology making all mappings φ_Λ continuous. We define, for any bounded domain $\Lambda \subset Q$, the sub-σ-algebra of the Borel σ-algebra $\mathfrak{B}(\Omega^S)$ by

$$\Sigma_\Lambda = \varphi_\Lambda^{-1}\left[\mathfrak{B}(\Omega^S(\Lambda))\right].$$

The family of local σ-algebras Σ_Λ generates the whole Borel σ-algebra $\mathfrak{B}(\Omega^S)$, and the set $C_0(\Omega^S)$ of bounded continuous local functions is dense everywhere in $C(\Omega^S)$ (due to Stone's Theorem; cf. Remark 1).

Poisson field. In the role of the free measure μ_0 on Ω^S, the distribution of the so-called *labelled Poisson* field in Q is chosen. It is defined in the following way. Let a positive σ-finite (or finite) measure $d\lambda_0$ such that $\lambda_0(\Lambda) < \infty$ for each bounded domain, Λ be given on the space Q, and let a probability measure ds be given on the space S. The measure $(d\lambda_0 \times ds)^n$, defined on the space $(Q \times S)^n$, induces, on the space $\Omega^S(Q, n) \equiv \Omega_n^S$, the factor-measure

$$d\nu_n = (d\lambda_0 \times ds)^n/n!, \qquad n > 0, \qquad \nu_0(\emptyset) = 1. \tag{8}$$

We shall consider a measure ν on the space $\Omega_{\text{fin}}^S = \bigcup_{n\geqslant 0}\Omega_n^S$ of finite configurations in Q, coinciding on each set Ω_n^S with the measure ν_n, $n = 0, 1, \ldots$.

Now let $\Lambda \subset Q$ be a bounded domain and μ_Λ^0 be a probability measure on $\Omega^S(\Lambda)$ equal to

$$\mu_\Lambda^0 = e^{-\lambda_0(\Lambda)}\nu \tag{9}$$

(since $\Omega^S(\Lambda) \subset \Omega_{\text{fin}}^S$, the measure ν is defined on the space $\Omega^S(\Lambda)$, too). Notice that $\mu_\Lambda^0(\Omega^S(\Lambda, n))$, i.e., the probability of the occurrence of exactly n points of the labelled field in Λ (with arbitrary values of charges), equals $\lambda_0^n(\Lambda)e^{-\lambda_0(\Lambda)}/n!$. Each measure μ_Λ^0 can be considered as defined on the σ-algebra Σ_Λ. Furthermore, one may easily verify that there is a unique measure μ^0 on the space Ω^S such that its restrictions to sub-σ-algebras Σ_Λ coincide with the measures μ_Λ^0. The labelled point field in Q generated by this measure is called the *Poisson field with independent charges.*

Any function $\Phi[(x, s_x)]$ defined on the set Ω_{fin}^S of finite configurations (x, s_x) is called *a potential*. For each bounded domain $\Lambda \subset Q$, we put

$$U_\Lambda[(x, s_x)] = \sum_{y \subseteq x\cap\Lambda} \Phi[(y, s_y)],$$

with $s_y = s_x|_y$ being the restriction of the function s_x to $y \subset x$.

The Gibbs modification μ_Λ of the Poisson field μ^0 is defined by the density

$$\frac{d\mu_\Lambda}{d\mu^0} = Z_\Lambda^{-1} \exp\{-U_\Lambda\}, \qquad Z_\Lambda = \int_{\Omega^S} \exp\{-U_\Lambda\} d\mu^0. \tag{10}$$

Primarily, the case of the proper point field in the space R^ν is investigated. At the same time, the Poisson measure μ^0 is defined with the help of the Lebesgue measure $d\lambda_0 = d^\nu x$ on R^ν, and the energies U_Λ are defined by means of a two-point (or two-particle) translation-invariant potential Φ, i.e.,

$$\Phi(x) = \begin{cases} \hat{\mu}, & |x| = 1, \\ \beta\varphi(q_1 - q_2), & x = (q_1, q_2), \\ 0, & |x| > 2; \end{cases} \tag{11}$$

where $\hat{\mu} \in R^1$ (so-called chemical potential), φ is an even function defined on the space R^ν, and $\beta > 0$.

We formulate here one stability condition for Z_Λ in the case of a two-particle potential of the form (11).

Theorem 2. *Let φ be a real even upper semi-continuous function on R^ν. Then the following conditions are equivalent:*

(a) the inequality

$$\sum_{i=1}^{n} \sum_{j=1}^{n} \varphi(q_i - q_j) \geqslant 0 \tag{12}$$

is fulfilled for any n and $q_i \in R^\nu$, $i = 1, \dots, n$;

(b) there is a $B \geqslant 0$ such that

$$U_\Lambda(x) \geqslant -B|x| \tag{13}$$

for any $x \in \Omega_{\text{fin}}$ and $\Lambda \subset R^\nu$.

(c) the partition functions Z_Λ are finite for all bounded domains Λ.

For proof see [60]. An example when the condition (a) implies the stability of Z_Λ is the case of positive definite function $\varphi(q)$, $q \in R^\nu$.

(3) *Gibbs modifications of measures on function spaces.* Let Ω be some locally convex space of functions $x(t) = \{x_1(t), \dots, x_n(t)\}$, $t \in R^\nu$, defined on the space R^ν, with values in R^n.

Suppose that the topology on Ω is such that the functionals of the form $F_{t_0}(x) = x_k(t_0)$, $t_0 \in R^\nu$, $k = 1, 2, \dots, n$, are continuous with respect to it (i.e., the convergence of a sequence of functions in Ω implies their pointwise convergence). For each bounded open or closed set $\Lambda \subset R^\nu$, we define the σ-algebra Σ_Λ to be the smallest sub-σ-algebra of the Borel σ-algebra $\mathfrak{B}(\Omega)$ making all the functionals $\{F_{t_0}, t_0 \in \Lambda\}$ measurable. Suppose that the family of σ-algebras Σ_Λ is generating for the σ-algebra $\mathfrak{B}(\Omega)$.

We suppose that a probability measure μ_0 (free measure) is defined on the Borel σ-algebra $\mathfrak{B}(\Omega)$, and a functional $U_\Lambda(x)$ is given to each bounded open or closed set $\Lambda \subset R^\nu$, so that:

1) $U_\Lambda = 0$ if $|\Lambda| = 0$, where $|\Lambda|$ is the Lebesgue measure of Λ;
2) $U_{\Lambda_1 \cup \Lambda_2} = U_{\Lambda_1} + U_{\Lambda_2}$ if $|\Lambda_1 \cap \Lambda_2| = 0$;
3) U_Λ is Σ_Λ-measurable.

A family $\{U_\Lambda\}$ of functionals fulfilling conditions 1), 2), and 3) is called a *local additive* functional.

Suppose that the stability condition

$$0 < \int_\Omega \exp\{-U_\Lambda(x)\} d\mu_0 < \infty \tag{14}$$

is fulfilled for each bounded domain $\Lambda \subset R^\nu$ and define a Gibbs modification μ_Λ of the measure μ_0 with the help of the formula (7). In what follows, a limit Gibbs modification of the measure μ_0 is defined as before.

Here is a typical example of a local additive functional (in the case when the space Ω contains only smooth locally bounded functions $x(t)$):

$$U_\Lambda(x) = \int_\Lambda \Phi\left[x_i(t), \frac{\partial x_i}{\partial t^{(j)}}\right] d^\nu t, \qquad t = (t^{(1)}, \dots, t^{(\nu)}),$$

with Φ a real function of $n(\nu+1)$ variables that is bounded from below.

Remark 1. In some cases, local additive functionals may also be defined on the space of distributions (called the *Schwartz* space $\mathcal{D}'(R^\nu)$) such that they fulfill the stability condition (14) (μ_0 is a probability measure on $\mathcal{D}'(R)$), and the Gibbs modifications μ_ν and the limit Gibbs modification μ may be defined with the help of them.

Remark 2. We have examined Gibbs modifications of measures on function spaces with the help of additive local functionals in such detail only because this case covers most of the known examples of such modifications. Of course, it is also possible to investigate nonlocal functionals U_Λ, e.g., functionals of the form

$$U_\Lambda(x) = \iint\limits_{\Lambda\Lambda} \Phi\left[x(t), x(t')\right] d^\nu t d^\nu t',$$

with Φ being a bounded real function of $2n$ variables.

Discretization. The *discretization* method is used when applying the theory of cluster expansions to functional fields. By discretization a reduction to a field in a countable set is to be understood. Namely, the space R^ν is broken up into congruent cubes, and the field in Z^ν (the centers of those cubes), with values in the space S of spins coinciding with some space of functions defined on a cube, is investigated. At the same time, the energy is rewritten in terms of the newly introduced field in Z^ν with values in S, and the free measure of the original field determines a measure on S^{Z^ν}. Even though we do not investigate fields in this book, many statements may be immediately, and in an obvious way, transferred to the case of functional fields. Notice that no method of cluster expansion that would exclude discretization exists at present (except for the expansion in a series of semi-invariants; see Section 6.IV).

1.5. Weak Compactness of Measures. The Concept of Cluster Expansion

Let $\mathcal{A}$ be some collection of measures defined on the whole Borel σ-algebra $\mathfrak{B}(\Omega)$ of a topological space Ω or on its sub-σ-algebra $\Sigma \subset \mathfrak{B}(\Omega)$. By weak compactness of the set $\mathcal{A}$, its sequential compactness is always understood, i.e., there is a weakly converging sequence $\mu_n \to \mu$, $n \to \infty$, $\mu_n \in \mathcal{B}$, in any infinite subset $\mathcal{B} \subset \mathcal{A}$.

Lemma 3. *In the case of a complete separable metric space Ω and $\Sigma = \mathfrak{B}(\Omega)$, each of the conditions stated below is sufficient for weak compactness of the set $\mathcal{A}$.*

1) Each measure $\mu \in \mathcal{A}$ is a probability measure, and there is a compact function $h > 0$ defined on Ω such that

$$\int_\Omega h(x)d\mu < C$$

for any measure $\mu \in \mathcal{A}$ where C does not depend on μ. A function h on Ω is called compact if the set $\{x \in \Omega,\ h(x) < a\}$ is compact for any $a > 0$.

2) There are a nonnegative measure μ_0 on $\mathfrak{B}(\Omega)$ and a μ_0-integrable function $\varphi(x) \geqslant 0$ such that any measure $\mu \in \mathcal{A}$ is absolutely continuous with respect to μ_0 and

$$\left|\frac{d\mu}{d\mu_0}(x)\right| < \varphi(x), \qquad x \in \Omega.$$

The statement 1) is a simple corollary of the known weak compactness criterion of Prokhorov (cf. [6]). A derivation of the statement 2) can be found in [16].

Definition 3. Let $\{\mu_\Lambda,\ \Lambda \in \mathcal{F}\}$ be a family of measures, each of which is defined on the σ-algebra Σ_Λ from a complete family $\{\Sigma_\Lambda,\ \Lambda \in \mathcal{F}\}$ of sub-σ-algebras of the σ-algebra $\mathfrak{B}(\Omega)$ (here, as above, $\mathcal{F}$ is a directed family of indices). A family $\{\mu_\Lambda,\ \Lambda \in \mathcal{F}\}$ is called *weakly locally compact* if the set $\{\mu_\Lambda^{\Lambda_0},\ \Lambda_0 < \Lambda\}$ of restrictions of measures $\{\mu_\Lambda\}$ to the σ-algebra Σ_{Λ_0} is weakly compact for any $\Lambda_0 \in \mathcal{F}$.

Lemma 4. *Let a family $\{\mu_\Lambda,\ \Lambda \in \mathcal{F}\}$ of measures be locally compact. Then, in any increasing sequence $\Lambda_1 < \Lambda_2 < \dots < \Lambda_n < \dots$ of indices ensuring that the sequence of σ-algebras Σ_{Λ_n}, $n = 1, 2, \dots$, is complete, there* *is a subsequence possessing the same property and a cylinder measure* μ *on* $\mathfrak{A} = \cup\Sigma_\Lambda$ *such that*

$$\mu = \lim_{k\to\infty} \mu_{i_k} \qquad (\mu_n = \mu_{\Lambda_n}). \tag{15}$$

The proof is obvious.

Now we shall introduce the concept of cluster expansion of measures in a way applicable to the case of fields in a countable set T (with values in a space S).

Let $G \subset C_0(S^T)$ be some set of bounded continuous local functions whose linear hull is dense everywhere in the space $C(S^T)$ of all bounded continuous functions. Let the mean $\langle F\rangle_\mu$ of an arbitrary function $F \in G$ under a measure (or cylinder measure) μ be expanded in the form

$$\langle F\rangle_\mu = \sum_{R\subset T, |R|<\infty} b_R(F), \tag{16}$$

with $b_R(F)$ being some quantities depending on F and finite subsets $R \subset T$. Every such expansion is usually called a *cluster expansion* of the measure μ. Of course, the extent of "constructiveness" of definitions of $b_R(F)$ determines the significance of such an expansion. In the following, we shall construct the expansions (16) explicitly each time, and the statement that this or that measure admits a cluster expansion is to be understood only in the sense of (16).

Definition 4. Let $\{\mu_\Lambda,\ \Lambda \subset T\}$ be a family of measures defined each on the σ-algebra $\Sigma_\Lambda = \mathfrak{B}(S^\Lambda)(\Lambda \subset T, |\Lambda| < \infty)$. The family $\{\mu_\Lambda\}$ is said to admit a *cluster expansion* if

1) it is weakly locally compact;

2) there is a set $G \subset C_0(S^T)$ of bounded continuous functions whose linear hull is dense everywhere in the space $C(S^T)$ such that the mean $\langle F\rangle_{\mu_\Lambda} \equiv \langle F\rangle_\Lambda$ of any function $F \in G$ admits an expansion

$$\langle F\rangle_\Lambda = \sum_{R\subseteq\Lambda} b_R^{(\Lambda)}(F) \tag{17}$$

with the quantities $b_R^{(\Lambda)}(F)$ fulfilling the following conditions:

a) there is a majorant

$$\left|b_R^{(\Lambda)}(F)\right| < C_R(F), \qquad \sum_{R\subset T} C_R(F) < \infty; \tag{18}$$

b) there are limits

$$\lim_{\Lambda\nearrow T} b_R^{(\Lambda)}(F) = b_R(F). \tag{19}$$

Lemma 5. *Let a family $\{\mu_\Lambda\}$ of measures admit a cluster expansion. Then the weak local limit*

$$\mu = \lim_{\Lambda \nearrow T} \mu_\Lambda \tag{20}$$

exists and μ admits a cluster expansion.

In the case of probability measures $\{\mu_\Lambda\}$, the cylinder measure μ is also a probability, hence it can be extended to a probability measure on the σ-algebra $\mathfrak{B}(\Omega)$.

Proof. The condition 1) and Lemma 4 imply the existence of at least one weak limit point of the set $\{\mu_\Lambda\}$. Its uniqueness, and (20), follow from the condition 2). The cluster expansion (16) of the limit measure is obvious.

The cluster expansion of measures in the case of point fields is defined in Section 6.3.

§2 Gibbs Modifications under Boundary Conditions and Definition of Gibbs Fields by Means of Conditional Distributions

The definition of limit Gibbs modifications presented above does not include all interesting cases of fields of that kind, and we offer a more general definition of the limit Gibbs field. We confine ourselves only to the case of fields in a countable set T (on which a metric ρ is given) with values in a (metric) space S. Further, we suppose that a finite or σ-finite measure is defined on S, and we choose, for any finite set $\Lambda \subset T$, the measure $\mu_\Lambda^0 = \lambda_0^\Lambda$, i.e., the product of $|\Lambda|$ copies of the measure λ_0, as the free measure μ_Λ^0 on the space S^Λ.

Finally, we suppose that we are given a potential $\{\Phi_A;\ A \subset T,\ |A| < \infty\}$ of a finite range, i.e., $\Phi_A \equiv 0$ if $\operatorname{diam} A \equiv \max_{t_1,t_2 \in A} \rho(t_1, t_2) > d$ for some constant $d > 0$, and that the Hamiltonian $U_\Lambda = \Sigma_{A \subseteq \Lambda} \Phi_A$ determined by it fulfills the stability condition

$$0 < \int_{S^\Lambda} \exp\{-U_\Lambda(x)\}\, d\lambda_0^\Lambda < \infty$$

for any finite $\Lambda \subset T$. Let μ_Λ be a Gibbs modification of the measure λ_0^Λ, and for any $\Lambda_0 \subset \Lambda$, we use $\mu_\Lambda^{\Lambda_0}(\cdot/\bar{x}^{\Lambda\setminus\Lambda_0})$ to denote the conditional probability distribution on the set of configurations $x^{\Lambda_0} \in S^{\Lambda_0}$ under the condition that a configuration $\bar{x}^{\Lambda\setminus\Lambda_0} \in S^{\Lambda\setminus\Lambda_0}$, in the set $\Lambda\setminus\Lambda_0$, is fixed. An elementary computation (cf. Section 0) shows that the density of the measure $\mu_\Lambda^{\Lambda_0}(\cdot/\bar{x}^{\Lambda\setminus\Lambda_0})$ with respect to the measure $\lambda_0^{\Lambda_0}$ equals

$$\frac{d\mu_\Lambda^{\Lambda_0}\left(x^{\Lambda_0}/\bar{x}^{\Lambda\setminus\Lambda_0}\right)}{d\lambda_0^{\Lambda_0}} = Z_{\Lambda_0}^{-1}\left(\bar{x}^{\Lambda\setminus\Lambda_0}\right) \exp\left\{-U_{\Lambda_0}\left(x^{\Lambda_0}/\bar{x}^{\Lambda\setminus\Lambda_0}\right)\right\} \tag{1}$$

with

$$Z_{\Lambda_0}\left(\bar{x}^{\Lambda\setminus\Lambda_0}\right) = \int_{S^{\Lambda_0}} \exp\left\{-U_{\Lambda_0}\left(x^{\Lambda_0}/\bar{x}^{\Lambda\setminus\Lambda_0}\right)\right\} d\lambda_0^{\Lambda_0},$$

$$U_{\Lambda_0}\left(x^{\Lambda_0}|\bar{x}^{\Lambda\setminus\Lambda_0}\right) = U_{\Lambda_0}\left(x^{\Lambda_0}\right) + \sum_{\substack{A:A\cap\Lambda_0\neq\emptyset \\ A\cap(\Lambda\setminus\Lambda_0)\neq\emptyset}} \Phi_A\left(x^{\Lambda_0}\cup\bar{x}^{\Lambda\setminus\Lambda_0}\right). \qquad (1')$$

Here, $x^{\Lambda_0}\cup\bar{x}^{\Lambda\setminus\Lambda_0}$ denotes the configuration in Λ whose restrictions to Λ_0 and $\Lambda\setminus\Lambda_0$ are equal to x^{Λ_0} and $\bar{x}^{\Lambda\setminus\Lambda_0}$, respectively. The second summand (1′) is called the energy of the interaction with an external (boundary) configuration ("the boundary term").

Notice that, for a fixed Λ_0 and a sufficiently large $\Lambda \supset \Lambda_0$ (so that $\rho(\Lambda_0, T\setminus\Lambda) > d$), the energy $U_{\Lambda_0}(x^{\Lambda_0}|\bar{x}^{\Lambda\setminus\Lambda_0})$ does not depend on the whole configuration $\bar{x}^{\Lambda\setminus\Lambda_0}$, but only on its restriction $\bar{x}^{\partial_d\Lambda_0}$ to the d-neighbourhood of Λ_0, i.e., $\partial_d\Lambda_0 = \{t \in T\setminus\Lambda_0,\ \rho(t,\Lambda_0) \leqslant d\}$.

We denote this energy by

$$U_{\Lambda_0}\left(x^{\Lambda_0}/\bar{x}^{\partial_d\Lambda_0}\right), \qquad (1'')$$

and let $\mu^{\Lambda_0}_{\bar{x}^{\partial_d\Lambda_0}}$ denote the Gibbs modification of the measure $\lambda_0^{\Lambda_0}$ by means of the Hamiltonian (1″). The measure $\mu^{\Lambda_0}_{\bar{x}^{\partial_d\Lambda_0}}$ is called the *Gibbs distribution on* Λ_0 *with the boundary configuration* $\bar{x}^{\partial_d\Lambda_0}$ in the neighbourhood $\partial_d\Lambda_0$.

The formula (1) suggests the following:

Definition. A probability measure μ on the space S^T is called a *Gibbs distribution in* T (for a given potential $\{\Phi_\Lambda\}$) if, for any finite $\Lambda \subset T$ and any configuration $\bar{x} \in S^{T\setminus\Lambda}$, the conditional distribution $\mu(\cdot/x^{T\setminus\Lambda} = \bar{x})$ on the set S^Λ coincides, under the condition that the external configuration $x^{T\setminus\Lambda}$ is fixed and equal to $\bar{x}$, with the measure $\mu_{\bar{x}^{\partial_d\Lambda}}$ given by

$$\mu\left(\cdot/x^{T\setminus\Lambda} = \bar{x}\right) = \mu^{\Lambda}_{\bar{x}^{\partial_d\Lambda}}, \qquad (2)$$

with $\bar{x}^{\partial_d\Lambda}$ being a restriction of $\bar{x}$ to $\partial_d\Lambda$.

This definition is due to Dobrushin, Lanford, and Ruelle (DLR), and equation (2) representing it is sometimes called the DLR equation.

From the definition (2), in particular, the so-called *d-Markov property* of a Gibbs measure μ follows. Namely, the conditional measure $\mu(\cdot/x^{T\setminus\Lambda} = \bar{x})$ depends only on the values of the configuration $\bar{x}$ in the set $\partial_d\Lambda$ for arbitrary $\Lambda \subset T$ and $\bar{x} \in S^{T\setminus\Lambda}$.

Let $\Lambda \subset T$ be a finite set and let some probability distribution $q = q^{\partial_d\Lambda}$ on the set $S^{\partial_d\Lambda}$ of boundary configurations $\bar{x} = \bar{x}^{\partial_d\Lambda}$ be given. The measure

$$\mu_q^\Lambda = \int_{S^{\partial_d\Lambda}} \mu_{\bar{x}}^\Lambda dq(\bar{x}) \qquad (3)$$

on S^Λ is called a Gibbs distribution with a *q-random boundary configuration* in Λ.

Proposition 1. *For a measure μ on the space S^T to be Gibbsian, it is necessary that, for any increasing sequence $\Lambda_n \nearrow T$, $n \to \infty$, of finite sets Λ_n, there is a sequence of distributions $q_n = q^{\partial_d \Lambda_n}$ defined each on the set $S^{\partial_d \Lambda_n}$ of boundary configurations, so that the weak local limit of measures $\mu_{q_n}^{\Lambda_n}$ coincides with μ, i.e.,*

$$\lim_{n\to\infty} \mu_{q_n}^{\Lambda_n} = \mu, \tag{4}$$

and it is sufficient that the condition (4) is fulfilled for some increasing sequence $\Lambda_n \nearrow T$.

Corollary. *Let a family $\{\mu_{\bar{x}}^{\Lambda}\}$ of Gibbs modifications be such that there is a unique limit*

$$\mu = \lim_{\Lambda \nearrow T} \mu_{\bar{x}}^{\Lambda}$$

for any sequence $\Lambda \nearrow T$ and any choice of boundary configurations $\bar{x} \in S^{\partial_d \Lambda}$. Then μ is the unique Gibbs measure on S^T.

Proof of Proposition 1.
Necessity. We choose a distribution $q = \mu|_{S^{\partial_d \Lambda}}$ on the set $S^{\partial_d \Lambda}$ induced by the measure μ for any $\Lambda \subset T$. Then it is obvious that $\mu_q^{\Lambda} = \mu|_{S_\Lambda}$, and (4) is fulfilled.
Sufficiency. For any $\Lambda_0 \subset \Lambda$ such that $\rho(\Lambda_0, T - \Lambda) > d$ and any distribution q of boundary configurations $\bar{x} \in S^{\partial_d \Lambda}$, the conditional distribution $\mu_q^{\Lambda}(\cdot/\bar{x}^{\Lambda \setminus \Lambda_0})$ on S^{Λ_0} generated by the Gibbs measure μ_q^{Λ} on Λ with random boundary configuration coincides (as is easily seen from (1) and (3)) with the measure $\mu_{\bar{x}^{\partial_d \Lambda}}^{\Lambda_0}$, where $\bar{x}^{\partial_d \Lambda_0}$ is the restriction of the configuration $\bar{x}^{\Lambda \setminus \Lambda_0}$ to $\partial_d \Lambda_0$, i.e.,

$$\mu_q^{\Lambda}\left(\cdot/\bar{x}^{\Lambda \setminus \Lambda_0}\right) = \mu_{\bar{x}^{\partial_d \Lambda_0}}^{\Lambda_0}. \tag{5}$$

Now let $\Lambda_n \nearrow T$ and q_n be sequences such that (4) is fulfilled. Since the equality (5) is satisfied for any fixed $\Lambda_0 \subset T$ and all sufficiently large Λ_n, it is also valid for the limit measure μ.

Remark 1. A Hamiltonian U_Λ of the form (5.1), and the Gibbs modification μ_Λ of the measure λ_0^{Λ} generated by it, are sometimes called, for uniformity, the energy (and, accordingly, the Gibbs measure) with the "empty" boundary conditions. Moreover, their weak local limit, the limit Gibbs modification, is the Gibbs distribution in the sense of the definition (2) as easily follows from the proved proposition.

Remark 2. Definition (2), presented here for the case of a finite-range potential and the free measure $\mu_\Lambda^0 = \lambda_0^{\Lambda}$, can be applied as well (with some refinements) to the case of an arbitrary "rapidly decreasing" potential $\{\Phi_A\}$, and also to the case of any d-Markov free measure μ^0 (or even a measure μ^0 with a rapid "decrease of memory"). In addition, a definition of the Gibbs field in an infinite space, analogous to the definition (2), is possible both for the case of point fields and for the case of fields in R^ν (with ordinary or generalized configurations). However, they are rarely studied at present.

CHAPTER 2

SEMI-INVARIANTS AND COMBINATORICS

§1 Semi-Invariants and Their Elementary Properties

Let a family $\{\xi_1, \ldots, \xi_n\}$ of random variables be given (repeated occurrence of identical ones among them is not excluded). In the following, we shall suppose that all moments of these variables are finite:

$$\langle |\xi_j^k| \rangle < \infty \tag{1}$$

for each j and $k = 1, 2, \ldots$.

For any nonempty subset $T \subset \mathcal{N} = \mathcal{N}_n = \{1, \ldots, n\}$, we use the notation

$$\xi_T = \prod_{i \in T} \xi_i, \quad \xi'_T = \{\xi_i : i \in T\},$$

i.e., ξ'_T is a subsystem of the system $\{\xi_1, \ldots, \xi_n\}$. If $T = \emptyset$, we put $\xi_\emptyset = 1$ and introduce the symbol $\xi'_\emptyset$. For any random variable ξ, we denote the system of k samples of the random variable ξ by ξ'^k, and $\xi'^0 = \xi'_\emptyset$. An analogous meaning is given to the notation

$$\xi^\kappa_{\mathcal{N}} = \prod_{i=1}^n \xi_i^{k_i}, \quad \xi'^\kappa_{\mathcal{N}} = \left\{\xi_1'^{k_1}, \ldots, \xi_n'^{k_n}\right\},$$

with $\kappa = (k_1, \ldots, k_n)$, $k_i \geqslant 0$, a multiindex.

We shall consider the function

$$f(\lambda_1, \ldots, \lambda_n) = \langle \exp(\lambda_1 \xi_1 + \cdots + \lambda_n \xi_n) \rangle,$$

defined for all purely imaginary $\lambda_1, \ldots, \lambda_n$. It coincides with the characteristic function of the family $\{\xi_1, \ldots, \xi_n\}$ with real variables $t_j = \lambda_j / i$.

Lemma 1. *The function $f(\lambda_1, \ldots, \lambda_n)$ is infinitely differentiable for all purely imaginary λ_j. Mixed moments $\langle \xi_{\mathcal{N}}^{\kappa} \rangle$ exist for an arbitrary multiindex κ. Moreover,*

$$\langle \xi_{\mathcal{N}}^{\kappa} \rangle = \frac{\partial^{k_1+\cdots+k_n}}{\partial \lambda_1^{k_1} \ldots \partial \lambda_n^{k_n}} f(\lambda_1, \ldots, \lambda_n) \bigg|_{\lambda_1=\lambda_2=\cdots=\lambda_n=0} \tag{2}$$

(we put $\partial^0 / \partial \lambda_0^0 f = f$).

The proof can be found in [58].

It follows from Lemma 1 that $\ln f$ is an infinitely differentiable function in a neighbourhood of the point $(0, \ldots, 0)$.

Definition 1. The number

$$\langle \xi_1, \ldots, \xi_n \rangle \stackrel{\text{def}}{=} \frac{\partial^n}{\partial \lambda_1 \ldots \partial \lambda_n} \ln f(\lambda_1, \ldots \lambda_n) \bigg|_{\lambda_1=\lambda_2=\cdots=\lambda_n=0} \tag{3}$$

is called the *semi-invariant* of the family $\{\xi_1, \ldots, \xi_n\}$ of random variables.

The abbreviated notation for it is $\langle \xi'_{\mathcal{N}} \rangle$.

Accordingly, the semi-invariant of the family $\xi_{\mathcal{N}}'^{\kappa}$ is denoted by

$$\langle \xi_{\mathcal{N}}'^{\kappa} \rangle = \langle \xi_1'^{k_1}, \ldots, \xi_n'^{k_n} \rangle.$$

In addition, let us agree that

$$\langle \xi'_{\emptyset} \rangle = 0, \quad \langle \xi'_{\emptyset}, \xi_1'^{k_1}, \ldots \rangle = \langle \xi_1'^{k_1}, \ldots \rangle.$$

Properties of semi-invariants.

A.

$$\langle \xi_1'^{k_1}, \ldots, \xi_n'^{k_n} \rangle = \frac{\partial^{k_1+\cdots+k_n}}{\partial \lambda_1^{k_1} \ldots \partial \lambda_n^{k_n}} \ln f(\lambda_1, \ldots, \lambda_n) \bigg|_{\lambda_1=\lambda_2=\cdots=\lambda_n=0}. \tag{4}$$

We have

$$\left\langle \exp\left(\sum_{i=1}^{n} \sum_{j=1}^{k_i} \lambda_{ij} \xi_i \right) \right\rangle = f(\lambda_1, \ldots, \lambda_n),$$

with $\lambda_i = \sum_{j=1}^{k_i} \lambda_{ij}$ for $k_i > 0$ and $\lambda_i = 0$ for $k_i = 0$.

Using the definition (2) we get (4).

B. *Symmetry and multilinearity.*

$$\langle \xi_1, \ldots, \xi_n \rangle = \langle \xi_{i_1}, \ldots, \xi_{i_n} \rangle$$

for any permutation

$$\begin{pmatrix} 1 & 2 & \ldots & n \\ i_1 & i_2 & \ldots & i_n \end{pmatrix}$$

and

$$\langle a'\xi_1' + a''\xi_1'', \xi_2, \dots, \xi_n\rangle = a'\langle \xi_1', \xi_2, \dots, \xi_n\rangle + a''\langle \xi_1'', \xi_2, \dots, \xi_n\rangle.$$

The symmetry is obvious and the multilinearity follows from the equality

$$\langle \exp\{\lambda_1(a'\xi_1' + a''\xi_1'') + \lambda_2\xi_2 + \dots + \lambda_n\xi_n\}\rangle = f(a'\lambda_1, a''\lambda_1, \lambda_2, \dots, \lambda_n)$$

by a direct application of (3). Here, $f(\lambda_1', \lambda_1'', \lambda_2, \dots, \lambda_n)$ is the characteristic function of the family $\{\xi_1', \xi_1'', \xi_2, \dots, \xi_n\}$.

C. If $\mathcal{N}$ can be split up into two nonempty nonintersecting subsets, $\mathcal{N} = A \cup B$, so that the family ξ_A' does not depend on ξ_B', then

$$\langle \xi_1, \dots, \xi_n\rangle = 0. \tag{5}$$

Proof. For $\lambda_1, \dots, \lambda_n$ sufficiently small, we have

$$\ln\langle \exp(\lambda_1\xi_1 + \dots + \lambda_n\xi_n)\rangle = \ln\left\langle \exp\left(\sum_{i\in A}\lambda_i\xi_i\right)\right\rangle + \ln\left\langle \exp\left(\sum_{i\in B}\lambda_i\xi_i\right)\right\rangle.$$

By applying the definition (3), we get (5).

In particular, even if only one variable ξ_j = const., the semi-invariant $\langle \xi_1, \dots, \xi_n\rangle$ equals zero.

Property C suggests that the semi-invariant characterizes the degree of dependence of random variables. Notice also that, for $n = 2$, the semi-invariant coincides with the covariance

$$\langle \xi_1, \xi_2\rangle = \langle \xi_1 \cdot \xi_2\rangle - \langle \xi_1\rangle\langle \xi_2\rangle,$$

as easily follows from the definition.

D. *Expression of moments by means of semi-invariants.*

$$\langle \xi_{\mathcal{N}}\rangle = \sum_{\alpha=\{T_1,\dots,T_k\}} \langle \xi_{T_1}'\rangle \cdots \langle \xi_{T_k}'\rangle, \tag{6}$$

with the sum taken over all partitions of the set $\mathcal{N}$, i.e., over all unordered collections $\alpha = \{T_1, \dots, T_k\}$ of mutually disjoint nonempty subsets (blocks) $T_i \subset \mathcal{N}$ such that $\mathcal{N}$ is their union.

Proof. Consider formal Taylor series[1] for functions f and $\ln f$ in the point $\lambda_1 = \dots = \lambda_n = 0$:

$$\begin{aligned} f(\lambda_1, \dots, \lambda_n) &= \sum_{\kappa} \frac{\lambda_{\mathcal{N}}^{\kappa}}{\kappa!}\langle \xi_{\mathcal{N}}^{\kappa}\rangle, \\ \ln f(\lambda_1, \dots, \lambda_n) &= \sum_{\kappa} \frac{\lambda_{\mathcal{N}}^{\kappa}}{\kappa!}\langle \xi_{\mathcal{N}}'^{\kappa}\rangle, \end{aligned} \tag{7}$$

[1] For the formal series, see [63].

with the sum taken over all multiindices κ; here,

$$\kappa! = k_1! \dots k_n!, \quad \lambda_{\mathcal{N}}^{\kappa} = \lambda_1^{k_1} \dots \lambda_n^{k_n}.$$

Now consider $\xi = \lambda_1 \xi_1 + \dots + \lambda_n \xi_n$. We notice that

$$\langle e^{\xi} \rangle = \exp(\ln \langle e^{\xi} \rangle) = \exp\left(\sum_{s=1}^{\infty} \frac{1}{s!} \langle \xi'^s \rangle \right) = \sum_{k=0}^{\infty} \frac{1}{k!} \left(\sum_{s=1}^{\infty} \frac{1}{s!} \langle \xi'^s \rangle \right)^k. \tag{8}$$

We fix k and compute the coefficient of $\lambda_1 \dots \lambda_n$ in the formal series of the k-th summand of the right-hand side of (8). It is easy to see that this is the sum of expressions of the form

$$\langle \xi'_{T_1} \rangle \dots \langle \xi'_{T_k} \rangle$$

over all partitions $\{T_1, \dots, T_k\}$.

E. *Expression of semi-invariants by means of moments.*

$$\langle \xi'_{\mathcal{N}} \rangle = \sum_{\alpha} (-1)^{k-1} (k-1)! \langle \xi_{T_1} \rangle \dots \langle \xi_{T_k} \rangle, \tag{9}$$

with the summation taken over partitions $\alpha = (T_1, \dots, T_k)$ of the set $\mathcal{N}$.

The proof can be carried out analogously to the proof of property D. It will be proved below as a particular case of the Möbius inversion formula.

F. *Characterization of Gaussian families.* A system is Gaussian if and only if

$$\langle \xi_{i_1}, \dots, \xi_{i_s} \rangle = 0, \quad s > 2,$$

for any (possibly coinciding) indices $i_1, \dots, i_s \in \mathcal{N}$.

This property follows from the fact that the logarithm of the characteristic function of Gaussian families (and only of such) is a quadratic polynomial.

The set of all partitions of $\mathcal{N}$ is denoted by $\mathfrak{A}_{\mathcal{N}} = \mathfrak{A}$. We define in $\mathfrak{A}$ an ordering by putting $\alpha < \beta$ if each element of the partition α is contained in some element of the partition β (α "refines" β). We denote the partition to points (the minimal partition) by $\underline{0}$, and the partition consisting of the only element $\mathcal{N}$ (the maximal partition) by $\underline{1}$. The smallest partition $\delta \in \mathfrak{A}$ such that $\alpha < \delta$ and $\beta < \delta$ is denoted by $\alpha \vee \beta$ (for a more explicit account on the lattice of partitions, see Section 6).

A partition β is called *connected with respect to the partition* α if $\alpha \vee \beta = \underline{1}$. This obviously means that $\mathcal{N}$ cannot be split up into two subsets, each being a union of elements of β as well as of α.

G. *Generalized expansion over connected groups.* For any partition $\alpha = \{T_1, \dots, T_k\} \in \mathfrak{A}_{\mathcal{N}}$, it is

$$\langle \xi_{T_1}, \dots, \xi_{T_k} \rangle = \sum_{\beta : \alpha \vee \beta = 1} \langle \xi'_{Q_1} \rangle \dots \langle \xi'_{Q_m} \rangle \tag{10}$$

with the sum taken over partitions $\beta = \{Q_1, \dots, Q_m\}$ connected with respect to α.

Proof. Consider first the particular case when $|T_1| = \dots = |T_{k-1}| = 1$, $|T_k| = 2$. In other words, we shall prove that $(n = k+1)$

$$\langle \xi_1, \dots, \xi_{n-2}, \xi_{n-1} \cdot \xi_n \rangle = \langle \xi_1, \dots, \xi_n \rangle + \sum_{T \subseteq \{1,\dots,n-2\} = \mathcal{N}_{n-2}} \langle \xi_{n-1}, \xi'_T \rangle \langle \xi_n, \xi'_{\mathcal{N}_{n-2} \setminus T} \rangle. \tag{11}$$

We shall proceed by induction on n. Denoting $\eta = \xi_{n-1} \cdot \xi_n$ and using the property D, we have

$$\langle \xi_1, \dots \xi_n \rangle = \langle \xi_1, \dots, \xi_n \rangle + \sum_{\substack{\{T_1,\dots,T_k\} \\ k \geqslant 2}} \langle \xi'_{T_1} \rangle \dots \langle \xi'_{T_k} \rangle,$$

$$\langle \xi_1 \dots \xi_{n-2} \eta \rangle = \langle \xi_1, \dots, \xi_{n-2}, \eta \rangle + \sum_{\substack{\{S_1,\dots,S_k\} \\ k \geqslant 2}} \left[\langle \xi'_{S_1} \rangle \dots \langle \xi'_{S_k}, \eta \rangle + \langle \xi'_{S_1} \rangle \dots \langle \xi'_{S_k} \rangle \langle \eta \rangle \right] \tag{12}$$

with $\{T_i\}$ a partition of $\mathcal{N}_n$ and $\{S_i\}$ a partition of $\mathcal{N}_{n-2}$. Hence,

$$\langle \xi_1, \dots, \xi_{n-2}, \eta \rangle - \langle \xi_1, \dots, \xi_n \rangle = \sum_{k \geqslant 2} \left[\langle \xi'_{T_1} \rangle \dots \langle \xi'_{T_k} \rangle \right] - \sum_{k \geqslant 2} \left[\langle \xi'_{S_1} \rangle \dots \langle \xi'_{S_k}, \eta \rangle + \langle \xi'_{S_k} \rangle \langle \eta \rangle \right].$$

Since $|S_k| < n-2$, we infer from the induction hypothesis that

$$\langle \xi'_{S_k}, \eta \rangle = \langle \xi'_{S_k}, \xi_{n-1}, \xi_n \rangle + \sum_{S' \cup S'' = S^k} \langle \xi'_{S'}, \xi_{n-1} \rangle \langle \xi'_{S''}, \xi_n \rangle.$$

Moreover,

$$\langle \eta \rangle = \langle \xi_{n-1}, \xi_n \rangle + \langle \xi_{n-1} \rangle \langle \xi_n \rangle.$$

Inserting the last two formulas into (12), we can easily convince ourselves that only the terms

$$\sum_{T \subseteq \mathcal{N}_{n-2}} \langle \xi_{n-1}, \xi'_T \rangle \langle \xi_n, \xi'_{\mathcal{N}_{n-2} \setminus T} \rangle$$

will remain on the right-hand side of (12).

We pass to the general case. Let the statement (10) be proved for all partitions of the sets $\mathcal{N}_k$ with a number of elements less than n and for all partitions $\alpha' \in \mathfrak{A}_{\mathcal{N}}$ less than the partition α. We shall prove its validity for the partition α. Let $n \in T_k$ and

$$\eta_1 = \xi_{T_1}, \dots, \eta_{k-1} = \xi_{T_{k-1}}, \quad \eta_k = \xi_{T_k \setminus \{n\}}, \quad \eta_{k+1} = \xi_n.$$

Using (11) we get

$$\langle \xi_{T_1}, \dots, \xi_{T_{k-1}}, \xi_{T_k \setminus \{n\}} \cdot \xi_n \rangle = \langle \xi_{T_1}, \dots, \xi_{T_k \setminus \{n\}}, \xi_n \rangle + \sum_{S \subseteq \mathcal{N}_{k-1}} \langle \eta'_S, \xi_n \rangle \langle \eta'_{\mathcal{N}_{k-1} \setminus S}, \xi_{T_k \setminus \{n\}} \rangle. \tag{12a}$$

By the induction hypothesis, each semi-invariant on the right-hand side can be expanded over connected groups. Inserting these expansions into (12a), we may convince ourselves that one gets a sum of terms of the form $\langle \xi'_{Q_1} \rangle \dots \langle \xi'_{Q_m} \rangle$ for each partition β connected with respect to α.

H. *Cauchy inequalities.* Suppose that the function $\langle \exp(\lambda_1 \xi_1 + \dots + \lambda_n \xi_n) \rangle$ may be analytically continued to a closed polydisc $|\lambda_1| \leqslant r_1, \dots, |\lambda_n| \leqslant r_n$ and differs from zero on it. Denote $\Gamma = \{\lambda_1, \dots, \lambda_n;\ |\lambda_i| = r_i,\ i = 1, \dots, n\}$. Then

$$\left| \langle \xi_1'^{k_1}, \dots, \xi_n'^{k_n} \rangle \right| \leqslant k_1! \dots k_n! \frac{C}{r_1^{k_1} \dots r_n^{k-n}} \tag{13}$$

with $C = \sup_\Gamma |\ln \langle \exp(\lambda_1 \xi_1 + \dots + \lambda_n \xi_n) \rangle|$.

This property is a reformulation of the Cauchy inequality (cf. [66]) and can be derived from the integral Cauchy representation of a function which is analytic in a polydisc.

The conditions of Proposition H are clearly fulfilled in the case when all ξ_i are bounded.

Even for one random variable, the estimate (13) cannot, in general, be improved: it is not difficult to find examples so that $|\langle \xi'^n \rangle| \geqslant C^n \cdot n!$ for any constant C and infinitely many values of n.

Remark. In many problems of probability theory and mathematical sta- tistics, an important role is played by asymptotic estimates of semi-invariants $\langle \xi_1'^{k_1}, \dots, \xi_n'^{k_n} \rangle$ for fixed n and $k_1, \dots, k_n \to \infty$. These estimates are usually obtained by standard methods of the asymptotic analysis and theory of functions of several complex variables. Notice that the specific character of estimates obtained in this book is related to the case $n \to \infty$.

I. *Expression of semi-invariants in the form of moments.* This useful device consists of the following. Let a family $\{\xi_1, \dots, \xi_n\}$ of random variables be given. We consider n independent collections of random variables: $\{\xi_1^{(1)}, \dots, \xi_n^{(1)}\}, \dots, \{\xi_1^{(n)}, \dots, \xi_n^{(n)}\}$, each of them having the same distribution as the system $\{\xi_1, \dots, \xi_n\}$. Let $\omega = \exp\{2\pi i/n\}$ and $\tilde{\xi}_k = \sum_{j=1}^n \omega^j \xi_k^{(j)}$. Then

$$\langle \xi_1, \dots, \xi_n \rangle = \frac{1}{n} \langle \tilde{\xi}_1 \dots \tilde{\xi}_n \rangle. \tag{14}$$

Proof. We notice that the cyclic mapping

$$\xi_k^{(j)} \mapsto \xi_k^{(j+1)}, \quad j = 1, \dots, n-1, \quad \xi_k^{(n)} \mapsto \xi_k^{(1)}$$

preserves the probability distributions, and hence the families $(\tilde{\xi}_1, \ldots, \tilde{\xi}_n)$, $(\omega\tilde{\xi}_1, \ldots, \omega\tilde{\xi}_n)$ have the same distribution. It follows from this, in particular, that

$$\langle \tilde{\xi}_T \rangle = \omega^{|T|} \langle \tilde{\xi}_T \rangle$$

for $T \subset \mathcal{N}$, and thus

$$\langle \tilde{\xi}_T \rangle = 0, \quad \langle \tilde{\xi}'_T \rangle = 0$$

for $T \neq \mathcal{N}$. From this and property D, one may conclude that

$$\langle \tilde{\xi}'_{\mathcal{N}} \rangle = \langle \tilde{\xi}_{\mathcal{N}} \rangle. \tag{15}$$

Properties B and C imply that $\langle \tilde{\xi}'_{\mathcal{N}} \rangle = n \langle \xi'_{\mathcal{N}} \rangle$, and this, together with (15), gives (14).

J. *Formal semi-invariants.* Let for each $T \subset \mathcal{N}$ a number $f(T)$ be given. These numbers will be called (mixed) moments of a virtual field (on $\mathcal{N}$).

Define the semi-invariants $g(T)$, $T \subset \mathcal{N}$ of a virtual field by the inductive formula

$$g(T) = f(T) \quad \text{for} \quad |T| = 1$$

and

$$g(T) = f(T) - \sum_{\{B_1, \ldots, B_k\}} g(B_1) \ldots g(B_k) \quad \text{for} \quad |T| > 1, \tag{16}$$

with the sum taken over all partitions $\{B_1, \ldots, B_k\}$ of the set T, with $k \geqslant 2$. Notice that formula (16) coincides exactly with formula (6). For the numbers $g(T)$ the formula

$$g(\mathcal{N}) = \sum_{\alpha} (-1)^{k-1} (k-1)! f(T_1) \ldots f(T_k), \tag{16'}$$

analogous to formula (9) (see the property E), is valid with the sum taken over all partitions $\alpha = \{T_1, \ldots, T_k\}$ of the set $\mathcal{N}$. Formula (16′), as well as formula (9), can be obtained with the help of the comparison of the expansions of the generating functions

$$S_g(\lambda_1, \ldots, \lambda_n) = \sum_{T \subseteq \mathcal{N}} g(T) \lambda^T$$

and

$$S_f(\lambda_1, \ldots, \lambda_n) = \sum_{T \subseteq \mathcal{N}} f(T) \lambda^T$$

with

$$\lambda^T = \prod_{i \in T} \lambda_i.$$

We shall need an analogy of the suitably reformulated property C. Let $\mathcal{N}$ be the set of vertices of some graph G. We denote the subgraph of the graph G spanned by a set A of vertices by G_A for $A \subset \mathcal{N}$. For $B \subset A$ we call the graph G_B a component of G_A if, in G_A, there are no edges connecting B with $A \backslash B$.

A system (virtual field) $\{f(T), T \subset \mathcal{N}\}$ is called *independent with respect to the graph* G if

$$f(A) = f(A_1)f(A_2) \tag{17}$$

for any $A \subset \mathcal{N}$ and all partitions $\{A_1, A_2\}$ of the set A such that $G_{A_i}, i = 1, 2$, are components of G_A.

Lemma 2. *If the graph* G *is not connected and the system* $f(T)$ *is independent with respect to* G*, then*

$$g(\mathcal{N}) = 0. \tag{18}$$

The proof is carried out by induction on n. Let $\xi = \{A_1, A_2\}$ be a partition of $\mathcal{N}$ such that G_{A_1} and G_{A_2} are components of the graph $G = G_{\mathcal{N}}$ (ξ exists since the graph is disconnected). From (16) we have

$$g(\mathcal{N}) = f(\mathcal{N}) - \Sigma'_{\dots} - \Sigma''_{\dots}, \tag{19}$$

where we sum over all $\alpha < \xi$ in Σ' and over $\alpha \neq \underline{1}$ in Σ''. All the summands in Σ'' equal 0 according to the induction hypothesis (since in each such α there is an element intersecting A_1 and A_2). At the same time, according to (17),

$$\Sigma' = f(A_1)f(A_2),$$

which implies the lemma.

K. *Semi-invariants and modifications of measures.* Let (Ω, Σ, μ_0) be a probability space, $\xi_1, \dots, \xi_k$ be random variables with finite moments, and U be a bounded random variable. Consider a new measure (the Gibbs modification of μ_0)

$$\mu(A) = \frac{\langle \chi_A e^{zU} \rangle_{\mu_0}}{\langle e^{zU} \rangle_{\mu_0}}, \quad A \in \Sigma, \tag{20}$$

with z a real parameter, and χ_A the indicator of the set A. Then the semi-invariant $\langle \xi_1, \dots, \xi_k \rangle_\mu$ evaluated with respect to the measure μ is an analytic function of z in a small neighbourhood of zero in the z-plane, and

$$\langle \xi_1, \dots, \xi_k \rangle_\mu = \sum_{n=0}^{\infty} \frac{z^n}{n!} \langle \xi_1, \dots, \xi_k, U'^n \rangle_{\mu_0}. \tag{21}$$

Proof.

$$\begin{aligned}\langle \xi_1, \dots, \xi_k \rangle_\mu &= \frac{\partial_k}{\partial \lambda_1 \dots \partial \lambda_k} \ln \langle \exp\{\lambda_1 \xi_1 + \dots + \lambda_k \xi_k\} \rangle_\mu \Big|_{\lambda_1 = \dots = \lambda_k = 0} \\ &= \frac{\partial^k}{\partial \lambda_1 \dots \partial \lambda_k} \ln \langle \exp\{\lambda_1 \xi_1 + \dots + \lambda_k \xi_k + zU\} \rangle_{\mu_0} \Big|_{\lambda_1 = \dots = \lambda_k = 0}.\end{aligned}$$

Expanding the last logarithm in z, we obtain (21).

§2 Hermite–Itô–Wick Polynomials. Diagrams. Integration by Parts

1. Wick polynomials

Let ξ be a Gaussian random variable defined on some probability space (Ω, Σ, μ). The polynomials in ξ (for brevity called Wick polynomials[2] and denoted by the notation : : introduced by him) arise by orthogonalizing the system of monomials $1, \xi, \xi^2, \ldots$ in $L_2(\Omega, \Sigma, \mu)$.

Definition 1 (Wick exponential function). We put

$$:\exp(a\xi): \stackrel{\text{def}}{=} \frac{\exp(a\xi)}{\langle \exp(a\xi) \rangle} = \exp\left\{ a(\xi - \langle \xi \rangle) - \frac{1}{2} a^2 \langle \xi^2 \rangle \right\}, \tag{1}$$

with a being an arbitrary complex number. Expanding the right-hand side of (1) in a (converging) power series in a, we denote the coefficient of $a^n/n!$ by $:\xi^n:$, i.e.,

$$:\exp(a\xi): \stackrel{\text{def}}{=} \sum_{n=0}^{\infty} \frac{a^n}{n!} :\xi^n:. \tag{2}$$

It is easy to see that $:\xi^n:$ is a polynomial of the n-th degree in ξ with the leading coefficient 1, i.e.,

$$:\xi^n: = \xi^n + Q(\xi), \tag{3}$$

with Q being a polynomial of the degree not higher than $n-1$. The polynomial $:\xi^n:$ is called the *Wick power*. For example,

$$:\xi^0: = :1: = 1, \quad :\xi: = \xi - \langle \xi \rangle.$$

If $\langle \xi \rangle = 0$, then

$$\begin{aligned} &:\xi^2: = \xi^2 - \langle \xi^2 \rangle, \quad :\xi^3: = \xi^3 - 6\xi \langle \xi^2 \rangle, \\ &:\xi^4: = \xi^4 - 6\xi^2 \langle \xi^2 \rangle + 3(\langle \xi^2 \rangle)^2. \end{aligned} \tag{4}$$

For any polynomial $P(\xi) = a_n \xi^n + \cdots + a_0$ in the random variable ξ, we define the polynomial $:P(\xi):$ by the linear relation

$$:P(\xi): = a_n :\xi^n: + \cdots + a_0.$$

We emphasize that the operation $P \mapsto :P:$ depends on the distribution of ξ (more exactly, on $\langle \xi \rangle$ and $\langle \xi^2 \rangle$).

In the following, we shall consider the case of $\langle \xi \rangle = 0$.

---

[2] These polynomials, as we shall see below, coincide with the known Hermite polynomials (cf. [62]); in the form of stochastic integrals, they appeared in an article by Itô (cf. [23]).

Lemma 1.

1) The polynomials $:\xi^n:$ *and* $:\xi^m:$ *are orthogonal in the space* $L_2(\Omega, \Sigma, \mu)$ *for* $n \neq m$ *and*

$$\langle :\xi^n: :\xi^n: \rangle = n! \langle \xi^2 \rangle^n. \tag{5}$$

2) The polynomial $:\xi^n:$ *is orthogonal to all powers* ξ^m *with* $m < n$.

3) The system $\{:\xi^n:; n = 0, 1, \dots\}$ *of polynomials is complete in the space* $L_2(\Omega, \Sigma(\xi), \mu|_{\Sigma(\xi)})$, *with* $\Sigma(\xi) \subset \Sigma$ *being the smallest* σ*-algebra making* ξ *measurable. In other words, the polynomials*

$$h_n(\xi) = \frac{1}{\sqrt{n!} \langle \xi^2 \rangle^{n/2}} :\xi^n: \tag{6}$$

form an orthonormal basis in $L_2(\Omega, \Sigma(\xi), \mu|_{\Sigma(\xi)})$.

Proof. We write

$$\sum_{n,m=0}^{\infty} \frac{\alpha^n \beta^m}{n! m!} \langle :\xi^n: :\xi^m: \rangle = \langle :\exp(\alpha\xi): :\exp(\beta\xi): \rangle = \frac{\langle \exp[(\alpha+\beta)\xi] \rangle}{\langle \exp(\alpha\xi) \rangle \langle \exp(\beta\xi) \rangle}$$

$$= \exp(\alpha\beta\langle \xi^2 \rangle) = \sum_n \frac{\alpha^n \beta^n}{n!} \langle \xi^2 \rangle^n. \tag{7}$$

Comparing the coefficients of the monomials $\alpha^n \beta^m$ in the first and the last sums of this equality, we get statement 1). It follows from it and from (3) that the polynomials $1, :\xi:, \dots, :\xi^n:$ form an orthogonal basis in the space $L^{(n)} \subset L_2(\Omega, \Sigma, \mu)$ of polynomials in ξ of a degree not higher than n. From this and 1), we get the statement 2). Therefore, the system $:\xi^n:$ is an orthogonal system of polynomials with the leading coefficient being equal to one and, consequently,

$$:\xi^n: = \langle \xi^2 \rangle^{n/2} H_n(\xi / \langle \xi^2 \rangle^{1/2}),$$

with $\{H_n(x)\}$ being the Hermite polynomials (with leading coefficients one; cf. [62]).

Statement 3) is valid since the system $H_n(x)$ is complete in the space $L_2(R^1, (2\pi)^{-1/2} e^{-x^2/2} dx)$ (cf. [62]).

Corollary 1. *In the decomposition (3), the polynomial* $Q \in L^{(n-1)}$ *coincides up to the sign with the projection of the vector* ξ^n *into the subspace* $L^{(n-1)}$. *In other words,* $:\xi^n:$ *is the perpendicular, in* $L_2(\Omega, \Sigma, \mu)$, *dropped from the vector* ξ^n *to this subspace.*

Definition 1 easily yields

$$\frac{d}{d\xi} :\xi^n: = n :\xi^{n-1}:. \tag{8}$$

Definition 2. In the case of an arbitrary Gaussian family $\{\xi_1, \dots, \xi_n\}$ with $\langle \xi_1 \rangle = \langle \xi_2 \rangle = \dots = \langle \xi_n \rangle = 0$, the *Wick polynomial* $:P:$, for a homogeneous polynomial of the degree k in $\xi_1, \ldots, \xi_n$, is defined to be the perpendicular in the space $L_2(\Omega, \Sigma, \mu)$ dropped from the vector P to the closed subspace $L^{(k-1)}$ generated by all polynomials in $\xi_1, \ldots, \xi_n$ of a degree lower than k.

For a monomial $\xi_1^{k_1} \ldots \xi_n^{k_n} \equiv \xi_{\mathcal{N}}^{\kappa}$, $\kappa = (k_1, \ldots, k_n)$, the polynomial $:\xi_{\mathcal{N}}^{\kappa}:$ is called the *Wick* monomial.

Notice that, in the case $n = 1$, Definition 2 of the operation : : is identical to the preceding Definition 1 if $\langle \xi \rangle = 0$. Therefore, the formula

$$:\exp\left(\sum_i a_i \xi_i\right): = \exp\left\{\sum_i \xi_i a_i - \frac{1}{2}\sum_{i,j} a_i a_j \langle \xi_i \xi_j \rangle\right\} = \sum_{\kappa} \frac{a_{\mathcal{N}}^{\kappa}}{\kappa!} :\xi_{\mathcal{N}}^{\kappa}: \qquad (9)$$

is valid, with $\kappa! = k_1! \ldots k_n!$, $a_{\mathcal{N}}^{\kappa} = a_1^{k_1} \ldots a_n^{k_n}$. An explicit decomposition of $:\xi_{\mathcal{N}}^{\kappa}:$ into powers of $\xi_1, \ldots, \xi_n$ may be derived. An analogous decomposition of $:P:$, for any polynomial $P = P(\xi_1, \ldots, \xi_n)$, follows from the linearity.

Lemma 2. *1) If $\xi_1, \ldots, \xi_n$ are mutually orthogonal (and thus independent), then*

$$:\xi_1^{k_1} \ldots \xi_n^{k_n}: = :\xi_1^{k_1}: \ldots :\xi_n^{k_n}:, \qquad (10)$$

and all Wick monomials are pairwise orthogonal in $L_2(\Omega, \Sigma, \mu)$.

2) In the case of an arbitrary Gaussian family $\{\xi_1, \ldots, \xi_n\}$, the system $\{:\xi_{\mathcal{N}}^{\kappa}:,\ \kappa$ arbitrary multiindex$\}$ of polynomials is complete in the space $L_2(\Omega, \Sigma', \mu|_{\Sigma'})$ with Σ' being the smallest σ-algebra making $\xi_1, \ldots, \xi_n$ measurable.

Proof. The equality (10) follows from the equality

$$:\exp\left(\sum_{i=1}^{n} a_i \xi_i\right): = \prod_{i=1}^{n} :\exp(a_i \xi_i):, \qquad (11)$$

implied by (1). As for the orthogonality, it follows from Lemma 1.

For the purpose of the proof of 2) we choose an orthonormal basis $\eta_1, \ldots, \eta_n$ in $L^{(1)}$. In addition, each polynomial $:P(\xi_1, \ldots, \xi_n):$ can be decomposed into (Wick) monomials $:\eta_1^{k_1} \eta_2^{k_2} \ldots \eta_n^{k_n}:$ and, conversely, the monomials $:\eta_{\mathcal{N}}^{\kappa}:$ are represented in the form of linear combinations of the monomials $:\xi_{\mathcal{N}}^{\kappa}:$. As for the completeness of the monomials $:\eta_{\mathcal{N}}^{\kappa}:$ in $L_2(\Omega, \Sigma', \mu|_{\Sigma'})$, it follows from statement 1) and Lemma 1.

As before (cf. (8)), the formula

$$\frac{\partial}{\partial \xi_i} :\xi_1^{k_1} \ldots \xi_n^{k_n}: = k_i :\xi_1^{k_1} \ldots \xi_i^{k_i - 1} \ldots \xi_n^{k_n}:, \quad i = 1, \ldots, n, \qquad (12)$$

is valid.

Remark. The preceding constructions can be generalized also to the case of a countable collection $\{\xi_n; n = 1, 2 \ldots\}$ of Gaussian random variables. In particular, if all ξ_n have zero means, $\langle\xi_n\rangle = 0$, and $\langle\xi_n\xi_{n'}\rangle = \delta_{nn'} = \left\{\begin{smallmatrix}1, n=n'\\0, n\neq n'\end{smallmatrix}\right.$, then the normalized Wick monomials

$$h_{\{k_1,\dots,k_n,\dots\}} = \frac{1}{\prod_i (k_i!)^{1/2}} :\xi_1^{k_1}\dots\xi_n^{k_n}\dots: = \prod_i \frac{:\xi_i^{k_i}:}{(k_i!)^{1/2}}, \tag{12a}$$

with $(k_1,\dots,k_n,\dots)$ being any sequence of integers $k_1 \geqslant 0$ such that $k_i = 0$ for $i > i_0$, form an orthonormal basis in the space $L_2(\Omega, \Sigma', \mu|_{\Sigma'})$.

2. Diagrams

The diagram technique is useful in the computations with functionals of Gaussian families. It is geometrically intuitive and gets rid of complicated formulas with a large number of indices. We shall explain this technique in the case of Gaussian families $\{\xi_1,\dots,\xi_n\}$ of random variables with $\langle\xi_i\rangle = 0$, $i = 1,\dots,n$.

Let a partition $\alpha = \{T_1,\dots,T_k\} \in \mathfrak{A}_{\mathcal{N}}$ be given. We shall show how to compute the following moments and semi-invariants:

$$\begin{aligned} &1. \quad \langle\xi_{T_1}\xi_{T_2}\dots\xi_{T_k}\rangle = \langle\xi_{\mathcal{N}}\rangle;\\ &2. \quad \langle\xi_{T_1},\xi_{T_2},\dots,\xi_{T_k}\rangle;\\ &3. \quad \langle:\xi_{T_1}:\dots:\xi_{T_k}:\rangle;\\ &4. \quad \langle:\xi_{T_1}:,\dots,:\xi_{T_k}:\rangle. \end{aligned} \tag{13}$$

Notice that, for an odd n, all the variables in (13) equal zero. Indeed, we may pass to a system of random variables $\xi_i' = -\xi_i$ and notice that the family $\{\xi_1',\dots,\xi_n'\}$ has the same joint distribution as the original one. Hence, all the variables (13) for the new system are the same as for the old one and, on the other hand, they differ by the factors $(-1)^n$.

Thus, we shall only examine the case of an even n.

Denote the set of all partitions $\beta = \{(i_1,i_1'),\dots,(i_k,i_k')\}$ of the set $\mathcal{N}$ into pairs by $\mathfrak{A}_{\mathcal{N}}^2 \subset \mathfrak{A}_{\mathcal{N}}$.

Each pair $G = (\alpha,\beta)$, with $\alpha = \{T_1,\dots,T_s\} \in \mathfrak{A}_{\mathcal{N}}$, and $\beta = \{(i_1,i_1'),\dots,(i_k,i_k')\} \in \mathfrak{A}_{\mathcal{N}}^2$, is called a (vacuum) *diagram* of the Gaussian family $\{\xi_1,\dots,\xi_n\}$. With each diagram, we shall associate a graph with the set of vertices V being the elements of the partition α, and the set of edges E consisting of the elements of the partition β. We then say that the edge $(i,i') \in E$ connects the vertices T_j, $T_{j'} \in \alpha$ if $i \in T_j$, $i' \in T_{j'}$.

We shall call a diagram *connected, without loops*, etc., if the corresponding graph is connected, has no loops, etc.

Notice that a diagram is connected if and only if the partition β is connected with respect to the partition α (cf. p. 33).

With each diagram G, we associate a quantity $I(G)$ called the contribution of G and defined by

$$I(G) = \prod_{(i,i')\in E} \langle\xi_i\xi_{i'}\rangle, \tag{14}$$

with the multiplication taken over all elements (i, i') of the partition β.

Lemma 3. *The values of the quantities (13) equal*

$$\sum I(G),$$

with the sums taken for the fixed $\alpha = (T_1, \dots, T_k)$ over the classes of diagrams of the following type, respectively:

1) all diagrams;

2) all connected diagrams;

3) all diagrams without loops;

4) all connected diagrams without loops.

Proof. 1) is a reformulation of property D of semi-invariants;

2) is a reformulation of property G of semi-invariants. It will be suitable to postpone the proofs of statements 3) and 4) and to introduce first diagrams of a more general form. Namely, we shall call a diagram any pair $G(\alpha, \beta)$, with $\alpha \in \mathfrak{A}_{\mathcal{N}}$ being an arbitrary partition and $\beta \in \mathfrak{A}_{\mathcal{N}}$ being a partition, each element of which consists of at most two elements. With a diagram G, we associate a graph with "free legs." The set V of its vertices is defined as before, the set E of its edges is the set of elements (i, i') of the partition β consisting of pairs of elements, and the set L of its free legs is the set of one-point elements $\{l\}$ of the partition β. Moreover, the leg $\{l\} \in L$ is said to be attached to the vertex $T_j \in V$ if $l \in T_j$. The random variable

$$I(G) = \left[\prod_{(i,i') \in E} \langle \xi_i \xi_{i'} \rangle\right] \prod_{T_j \in V} : \prod_{l \in T_j : \{l\} \in L} \xi_l : \tag{15}$$

is called the contribution of the diagram.

Lemma 4. *The following formulas hold true:*

$$\xi_{\mathcal{N}} = \sum I(G), \tag{16}$$

with the sum taken over all diagrams $G = (\alpha, \beta)$ with $\alpha = \underline{1}$, for which the parity of the number $|L|$ of legs coincides with the parity of n, and

$$:\xi_{\mathcal{N}}: = \sum (-1)^{(n-|L|)/2} I(G), \tag{17}$$

with the sum taken over all diagrams $G(\alpha, \beta)$ with $\alpha = \underline{0}$ and with the parity of the number of legs coinciding with the parity of n.

Proof. From the comparison of the coefficients of $a_1 a_2 \dots a_n$ in both sides of the equality

$$\exp\left\{\sum_i a_i \xi_i - \frac{1}{2} \sum_{i,j} a_i a_j \langle \xi_i \xi_j \rangle\right\} = \sum_\kappa \frac{a_{\mathcal{N}}^\kappa}{\kappa!} :\xi_{\mathcal{N}}^\kappa: \tag{18}$$

(see formula (9)), we get the equality

$$:\xi_{\mathcal{N}}: = \sum_{\substack{T \subseteq \mathcal{N}, \\ |\mathcal{N}\setminus T| - \text{even}}} A_T \xi_T \tag{18a}$$

with

$$A_T = (-1)^{|\mathcal{N}\setminus T|/2} \sum_{\beta \in \mathfrak{A}^2_{\mathcal{N}\setminus T}} \prod_{(i,i') \in \beta} \langle \xi_i \xi_{i'} \rangle, \tag{18b}$$

where the sum is taken over all partitions of the set $\mathcal{N}\setminus T$ into pairs. The equalities (18a) and (18b) yield (17). Rewriting (18) in the form

$$:\exp\left(\sum_i a_i \xi_i\right):\exp\left(\frac{1}{2}\sum_{i,j} a_i a_j \langle \xi_i \xi_j \rangle\right) = \exp\left(\sum_i a_i \xi_i\right) = \sum_{\kappa} \frac{a^{\kappa}_{\mathcal{N}}}{\kappa!} \xi^{\kappa}_{\mathcal{N}}$$

and repeating the preceding considerations, we arrive at (16).

Remark. From (18a), (18b), and statement 1) of Lemma 3, we get

$$:\xi_{\mathcal{N}}: = \sum_{\substack{T \subseteq \mathcal{N}, \\ |\mathcal{N}\setminus T| - \text{even}}} (-1)^{|\mathcal{N}\setminus T|/2} \langle \xi_{\mathcal{N}\setminus T} \rangle \xi_T.$$

Similarly, we get

$$\xi_{\mathcal{N}} = \sum_{\substack{T \subseteq \mathcal{N}, \\ |\mathcal{N}\setminus T| - \text{even}}} \langle \xi_{\mathcal{N}\setminus T} \rangle :\xi_T:.$$

Lemma 5. *For any partition $\alpha = \{T_1, \ldots, T_k\}$, it is*

$$:\xi_{T_1}: \ldots :\xi_{T_k}: = \sum_{G = (\underline{0}, \beta)} (-1)^{(n - |L|)/2} I(G) \tag{19}$$

and

$$\xi_{T_1} \ldots \xi_{T_k} = \sum_{G = (\alpha, \beta)} I(G), \tag{20}$$

with the sum taken over all diagrams of the form $G = (\underline{0}, \beta)$ in the former case, and of the form $G = (\alpha, \beta)$ in the latter, and such that, in both cases,

1) *$\beta \leqslant \alpha$;*
2) *for all $i = 1, \ldots, k$, the number $|T_i| - |L_i|$ is even, where $L_i \subset L$ is the set of legs $\{l\} \in \beta$ attached to T_i, i.e., $l \in T_i$.*

The proof of both, formulas (19) and (20), is obtained by multiplying out formulas (17) for $:\xi_{T_i}:$ and (16) for ξ_{T_i}.

Lemma 6. *Let a Gaussian family $\{\xi_1, \dots, \xi_n, \eta_1, \dots, \eta_m\}$ of random variables and a differentiable function F of n real variables be given so that the random variables $F(\xi_1, \dots, \xi_n)$ and $F_i'(\xi_1, \dots, \xi_n)$ belong to $L_2(\Omega, \Sigma, \mu)$ for all i. (F_i' stands for the partial derivative F with respect to its i-th argument.) Then*

$$\langle :\eta_1 \dots \eta_m : F(\xi_1, \dots, \xi_n)\rangle = \sum_{i=1}^{n} \langle \eta_1 \xi_i \rangle \langle :\eta_2 \dots \eta_m : F_i'(\xi_1, \dots, \xi_n)\rangle. \tag{21}$$

Proof. In the particular case when $F(\xi_1, \dots, \xi_n) = :\xi_1 \dots \xi_n:$, formula (21), as follows from (12), is of the form

$$\langle :\eta_1 \dots \eta_m : \, :\xi_1 \dots \xi_n : \rangle = \sum_i \langle \eta_1 \xi_i \rangle \langle :\eta_2 \dots \eta_m : \, :\xi_1 \dots \check{\xi}_i \dots \xi_n : \rangle \tag{21a}$$

(the sign ˇ above the variable ξ_i means it has to be omitted). Formula (21a) is obtained as follows: we use the equality (19) for $k = 2$, take the mean value $\langle \; \rangle$ of it, and, with the help of statement 1) of Lemma 3, we represent this mean in the form of the sum of the correlations $\langle \xi_i \xi_{i'} \rangle, \langle \eta_j \eta_{j'} \rangle$, and $\langle \xi_i \eta_j \rangle$. Separating the coefficient of $\langle \eta_1 \xi_i \rangle$, we again convince ourselves, with the help of the equality (19) and Lemma 3, that it is of the same form as in (21a). In the general case, any function F can be expanded in a series in the Wick monomials, and, applying the formula (21a) to each summand, we get (21).

Proof of statements 3) and 4) of Lemma 3. We shall prove both statements 3) and 4) by induction on the number n of random variables $\{\xi_1, \dots, \xi_n\}$. Let statement 3) be fulfilled for all $n < m$. Then, applying formula (21) to the mean $\langle :\xi_{T_1} : \dots : \xi_{T_k} : \rangle$, with $\alpha = \{T_1, \dots, T_k\}$ a partition of the set $\mathcal{N} = (1, \dots, n)$, and using the induction hypothesis, we verify statement 3) for $n = m$. We now pass to statement 4) and suppose again that it is true for $n < m$. Writing $\eta_i = :\xi_{T_i}:$ and applying formula (6.1), we get

$$\langle \eta_1, \dots \eta_k \rangle = \langle \eta_1 \dots \eta_k \rangle - \sum_{\substack{\delta = \{Q_1, \dots, Q_p\} \\ p > 1}} \langle \eta'_{Q_1} \rangle \dots \langle n'_{Q_p} \rangle. \tag{21b}$$

By virtue of statement 3), the first summand equals the sum of the diagrams without loops. By the induction hypotheses applied to each factor $\langle \eta'_{Q_i} \rangle$, the sum $\sum_\delta$ equals the sum of all disconnected diagrams without loops, i.e., only connected diagrams without loops remain in (21b). Lemma 3 is proved.

3. Formulas of integration by parts

One of such formulas is formula (21). We will establish two other such formulas.

Lemma 7. *Let a Gaussian family* $\{\xi_T,\dots,\xi_n,\eta_1,\dots,\eta_m\}$ *be given, and let the functions* $F_0 = {:}\eta_1\dots\eta_m{:}, F_1(\xi_1,\dots,\xi_n),\dots,F_s(\xi_1,\dots,\xi_n)$ *obey the following property: all the mixed moments of them and their first derivatives exist. Then*

$$\langle F_0,F_1,\dots,F_s\rangle = \sum_{i=1}^{n}\sum_{j=1}^{s}\langle\eta_1\xi_i\rangle\left\langle\frac{\partial F_0}{\partial\eta_1},F_1,\dots,\frac{\partial F_j}{\partial\xi_i},\dots,F_s\right\rangle. \tag{22}$$

Proof. We expand the semi-invariant from the left-hand side of (22) into the moments with the help of (9.1) and use the formula of the integration by parts for the moment containing F_0. Using formula (9.1) again, we transform the obtained expression to the right-hand side of (22).

Remark 1. The reason why the term "integration by parts" is used for (21) and (22) is explained, for example, by the fact that formula (21) for $m = 1$ can be obtained by the usual integration by parts of the integral

$$\int\limits_{R^{n+1}} yF(x_1,\dots,x_n)p(y,x_1,\dots,x_n)\,dy\,dx_1\dots dx_n$$

with p being a Gaussian density.

Lemma 8. *Let a family of Gaussian measures* μ_s *on* R^n, $0\leqslant s\leqslant 1$, *be given so that their means vanish and the covariances* c_s *depend smoothly on* s. *Write* $\langle\cdot\rangle_s = \langle\cdot\rangle_{\mu_s}$.

For any twice-differentiable function $F(x_1,\dots,x_n)$, *for which the means* $\langle F\rangle_s, \langle\frac{\partial F}{\partial x_i}\rangle_s, \langle\frac{\partial^2 F}{\partial x_i\partial x_j}\rangle_s$ *and dispersions of the function* F *and its partial derivatives up to the second order are finite for all* s, *the equality*

$$\langle F\rangle_1 - \langle F\rangle_0 = \frac{1}{2}\sum_{(i,j)}\int\limits_0^1\left[\frac{d}{ds}\langle x_ix_j\rangle_s\right]\left\langle\frac{\partial^2 F}{\partial x_i\partial x_j}\right\rangle_s ds \tag{23}$$

holds true.

Proof. As in the proof of formula (21), it is sufficient to consider the case $F = x_1\dots x_n$. Apply the formula

$$\langle F\rangle_1 - \langle F\rangle_0 = \int\limits_0^1\left[\frac{d}{ds}\langle F\rangle_s\right]ds \tag{24}$$

and notice that, as follows from the statement of Lemma 3,

$$\frac{d}{ds}\langle x_1\dots x_n\rangle_s = \sum_{i<j}\left[\frac{d}{ds}\langle x_ix_j\rangle_s\right]\langle x_1\dots\check{x}_i\dots\check{x}_j\dots x_n\rangle_s. \tag{25}$$

From (24) and (25), we get (23).

§3 Estimates on Moments and Semi-Invariants of Functionals of Gaussian Families

Let T and A be countable (or finite) sets and let a Gaussian family $\{\xi_{t\alpha}, t \in T, \alpha \in A\}$ be given such that for all t and α,

$$\langle \xi_{t\alpha} \rangle = 0,$$
$$\sum_{t':t' \neq t} \sum_{\alpha' \in A} |\langle \xi_{t\alpha} \xi_{t'\alpha'} \rangle| < d < \infty. \tag{1}$$

Let γ denote an arbitrary integer-valued nonnegative finite function defined on the set $A : \gamma = \{m(\alpha), \alpha \in A\}$, $m(\alpha) \geqslant 0$ be integers, and the quantity $|\gamma| = \sum_{\alpha \in A} m(\alpha)$. We write

$$\gamma! = \prod_{\alpha \in A} m(\alpha)! \quad \text{and} \quad \gamma!! = \prod_{\alpha \in A} m(\alpha)!!,$$

with $m!! = m(m-2) \ldots$.

We use $:\xi_t^\gamma:$ to denote the Wick polynomial

$$:\xi_t^\gamma: = :\prod_{\alpha \in A} \xi_{t\alpha}^{m(\alpha)}:. \tag{1a}$$

We use the notation $\sum_{(t_2,\ldots,t_n)} (\sum_{\{t_2,\ldots,t_n\}})$ for the sum over all ordered (or unordered, respectively) collections of mutually distinct $t_2, \ldots, t_n \in T$.

Lemma 1. *Let the conditions (1) be fulfilled. Then, for any finite ordered collection of functions* $(\gamma_1, \ldots, \gamma_n)$, $\gamma_i = \{m_i(\alpha), \alpha \in A\}$ *and* $t = t_1 \in T$, *the estimate*

$$\sum_{(t_2,\ldots,t_n)} |\langle :\xi_{t_1}^{\gamma_1}:, :\xi_{t_2}^{\gamma_2}:, \ldots, :\xi_{t_n}^{\gamma_n}: \rangle| \leqslant (n-1)! d^{N/2} \prod_{i=1}^{n} \gamma_i!! \tag{2}$$

is valid, with $N = \sum_{i=1}^{n} |\gamma_i|$ *(N is an even number).*

Proof. In accordance with Lemma 3.2,

$$\sum_{(t_2,\ldots,t_n)} |\langle :\xi_{t_1}^{\gamma_1}:, \ldots, :\xi_{t_n}^{\gamma_n}: \rangle| \leqslant \sum |I(G)|, \tag{3}$$

where the summation on the right-hand side is taken over some collection of connected vacuum diagrams without loops. We are going to describe these diagrams. They are given by:

1) a sequence of vertices $V = (t_1, \ldots, t_n)$, with t_1 being fixed and all points t_i being mutually distinct; a set of legs $\{i, \alpha, k\}$, $\alpha \in \operatorname{supp}\gamma_i$, $k = 1, \ldots, m_i(\alpha)$, corresponds to each t_i;

2) a partition of the set of legs into pairs (edges)

$$\{(i_l, \alpha_l, k_l), (i'_l, \alpha'_l, k'_l)\}, \quad l = 1, \ldots, N/2.$$

We order all the edges of a given diagram G according to the following rule. First step: starting from the point t_1 along the leg $(1, \alpha_1, 1) = (i_1, \alpha_1, k_1)$, with α_1 being the smallest $\alpha \in \operatorname{supp}\gamma_1$, we arrive, along the corresponding edge $\{(i_1, \alpha_1, k_1), (i'_1, \alpha'_1, k'_1)\}$, to the point t'_{i_1}. We say that the legs (i_1, α_1, k_1) and (i'_1, α'_1, k'_1) contained in this edge are "used." We will now describe the general induction step. Let the s edges already be ordered,

$$\{(i_1, \alpha_1, k_1), (i'_1, \alpha'_1, k'_1)\}, \ldots, \{(i_s, \alpha_s, k_s), (i'_s, \alpha'_s, k'_s)\}.$$

When constructing the $(s+1)$-st edge, two cases arise.

First case. Let nonused legs exist among the legs (i'_s, α'_s, k), $k = 1, \ldots, m_{i'_s}, (a'_s)$. We choose such a leg with the smallest $k = k_{s+1}$ and put $(i_{s+1}, \alpha_{s+1}, k_{s+1}) = (i'_s, \alpha'_s, k_{s+1})$.

Second case. If there is no such leg, we take the first one (in the lexicographical order) from the nonused legs in the visited points (they exist because the diagram is connected). We denote it by $(i_{s+1}, \alpha_{s+1}, k_{s+1})$.

As a result of such an ordering of the edges of G, we get the corresponding sequence of pairs

$$\{(t_1, \alpha_1), (t_{i'_1}, \alpha_{i'_1})\}, \ldots, \{(t_{i_{N/2}}, \alpha_{i_{N/2}}), (t_{i'_{N/2}}, \alpha_{i'_{N/2}})\}. \tag{4}$$

Moreover, the contribution $I(G)$ corresponding to the diagram G equals

$$I(G) = \langle \xi_{t_1\alpha_1} \cdot \xi_{t_{i'_1}\alpha_{i'_1}} \rangle \ldots \langle \xi_{t_{i_{N/2}}\alpha_{i_{N/2}}} \xi_{t_{i'_{N/2}}\alpha_{i'_{N/2}}} \rangle.$$

We notice that, for fixed $(t_1, \ldots, t_n)$ and $(\gamma_1, \ldots, \gamma_n)$, there are at most

$$\prod_{i,\alpha} m_i(\alpha)(m_i(\alpha) - 2) \ldots \tag{5}$$

diagrams leading to the same sequence (4). Indeed, we may enter any pair (t_i, α) for the first time in the course of our walk along the edges through one of $m_i(\alpha)$ legs. Since we leave this pair each time along the exactly defined leg, we can enter into it for the second time by $(m_i(\alpha) - 2)$ ways and so on.

We introduce the notations

$$(x_m, \alpha_m) = (t_{i_m}, \alpha_{i_m}), \quad (x'_m, \alpha'_m) = (t_{i'_m}, \alpha_{i'_m})$$

and investigate the collection $\mathfrak{A}$ of sequences of pairs

$$\{(x_1, \alpha_1), (x'_1, \alpha'_1)\}, \ldots, \{(x_{N/2}, \alpha_{N/2}), (x'_{N/2}, \alpha'_{N/2})\} \tag{6}$$

being obtained from all diagrams G over which the sum in (3) is taken.

The sequences from $\mathfrak{A}$ obey the following properties:

1) $x_m = x'_l$ for some $l < m$, $m = 1, \ldots, N/2$, $x_1 = t_1$;

2) the pair (x_m, α_m) is uniquely defined by all preceding pairs

$$\{(x_1, \alpha_1), (x'_1, \alpha'_1)\}, \ldots, \{(x_{m-1}, \alpha_{m-1}), (x'_{m-1}, \alpha'_{m-1})\};$$

3) all the sites x_m, x'_m exhaust some n-point set $S \subset T$;

4) there are at most $(n-1)!$ sequences $(t_2, \ldots, t_n)$ such that the sequence (4) coincides with them for the given sequence (6).

To prove 4), we associate with each sequence (4) the permutation π which is obtained by enumerating the points $t_2, \ldots, t_n$ in the order of their first appearance in the sequence (4). Knowing the permutation π, we can enumerate elements of the set $S = (t_1, \ldots, t_n)$ in such a way that the sequence (6) coincides with (4).

Lemma 2. *Let a collection $\mathfrak{A}$ of sequences of pairs of the form (6) fulfilling the conditions 1) and 2) be given. Then*

$$\sum_{\mathfrak{A}} \prod_{m=1}^{N/2} |\langle \xi_{x_m \alpha_m} \xi_{x'_m \alpha'_m} \rangle| = d^{N/2}. \tag{7}$$

The proof can be easily obtained by the induction on even N's.

Estimate (2) follows from the formulas (5), (7), and property 4).

Lemma 1 is proved.

Estimates of semi-invariants

Fundamental Lemma 1 enables different applications. We shall use it for estimates of semi-invariants of arbitrary functionals of Gaussian families.

Let the Gaussian family $\{\xi_{t_\alpha};\, t \in T,\, \alpha \in A\}$ fulfil, in addition to condition (1), the condition that for arbitrary $t \in T$, we have

$$\langle \xi_{t\alpha}, \xi_{t\alpha'} \rangle = \delta_{\alpha\alpha'} = \begin{cases} 1, & \alpha = \alpha', \\ 0, & \alpha \neq \alpha'. \end{cases} \tag{8}$$

It follows from condition (8), Lemma 2.2, and the remark on page 40 that the normalized monomials

$$h_{\gamma t} = \frac{:\xi_t^\gamma:}{(\gamma!)^{1/2}}, \tag{9}$$

with γ running over all the integer-valued finite functions on A, form an orthonormal basis in $L_2(\Omega, \Sigma_t, \mu_t)$, and $\Sigma_t = \Sigma(\{\xi_{t\alpha}\})$ is the smallest σ-algebra making all the variables $\xi_{t\alpha}, \alpha \in A$, measurable, and $\mu_t = \mu|_{\Sigma_t}$. We shall use F_t to denote a random variable which is measurable with respect to the σ-algebra Σ_t, and let $F_t \in L_2(\Omega, \Sigma_t, \mu_t)$, i.e., F_t have a finite mean $\langle F_t \rangle$ and finite dispersion $D_t = \langle (F_t - \langle F_t \rangle)^2 \rangle$. In this case, F_t can be expanded in a sum of the monomials $h_{\gamma t}$ which converges in the quadratic mean:

$$F_t = \sum_{\gamma} a_{\gamma t} h_{\gamma t}. \tag{10}$$

According to the Parseval equality,

$$\sum_{\gamma \neq \widehat{0}} |a_{\gamma t}|^2 = D_t, \tag{11}$$

where $\widehat{0}$ denotes the null function: $\widehat{0} = \{m(\alpha) \equiv 0\}$.

Theorem 3. *Let $\{\xi_{t\alpha}; t \in T, \alpha \in A\}$ be a Gaussian family of random variables fulfilling the conditions (1) and (8), with $d < 1$ in (1). Further, let $\{T_t, t \in T\}$ be some other family of random variables such that, for any $t \in T$, $F_t \in L_p(\Omega, \Sigma_t, \mu_t)$ for any $p \leqslant 1$ and its Fourier coefficients $\{a_{\gamma t}\}$, in the expansion (10), obey the estimate*

$$|a_{\gamma t}| < a_\gamma \tag{12}$$

for any function γ, and also

$$\sum_{\gamma \neq \widehat{0}} |a_\gamma|^2 = D < \infty. \tag{13}$$

Then, for any finite subset $Q = \{t_1, \ldots, t_n\} \subset T$, the moments $\langle \prod_{i=1}^n F_{t_i} \rangle$ are finite, and consequently, there are semi-invariants $\langle F_{t_1}, \ldots, F_{t_n} \rangle \equiv \langle F'_Q \rangle$. Moreover,

$$\langle F'_Q \rangle = \sum_{(\gamma_1, \ldots, \gamma_n)} \prod_{i=1}^{n} a_{\gamma_i t_i} \langle h_{\gamma_1 t_1}, \ldots, h_{\gamma_n t_n} \rangle, \tag{13'}$$

and the series (13′) is absolutely convergent.

For any $t_1 \in T$ and any integer $n \geqslant 1$,

$$\sum_{\substack{Q \subset T: \\ t_1 \in Q, |Q| = n}} |\langle F'_Q \rangle| \leqslant \begin{cases} \sum_{\gamma \neq \widehat{0}} |\alpha_{\gamma t_1}| \widetilde{C}(\gamma) \left(\sum_{\gamma \neq \widehat{0}} a_\gamma \widetilde{C}(\gamma) \right)^{n-1} & \text{for any } A, \quad (14) \\ D_{t_1}^{1/2} D^{(n-1)/2} C^{|A| n}, & |A| < \infty, \quad (15) \end{cases}$$

with

$$\widetilde{C}(\gamma) = \prod_{\alpha \in A} (m(\alpha) + 1)^{1/2} d^{m(\alpha)/2},$$

and

$$C = \left(\sum_{m=1}^{\infty} m d^{m-1} \right)^{1/2} .$$

Proof. By virtue of the Hölder inequality,

$$\left| \left\langle \prod_{t \in Q} |F_t| \right\rangle \right| \leqslant \prod_{i=1}^{n} \|F_{t_i}\|_{L_n} , \tag{15'}$$

and consequently, the moment $\langle \prod_{i=1}^{n} F_{t_i} \rangle$, and also the semi-invariant $\langle F'_Q \rangle$, exist. Further, for any integer N, we write the decomposition $F_{t_i} = \sum_{\gamma_1 : |\gamma_1| \leqslant N} a_{\gamma_1 t_1} h_{\gamma_1 t_1} + F_{t_1}^{(N)}$. Then

$$\langle F'_Q \rangle = \sum_{\gamma_1 : |\gamma_1| \leqslant N} a_{\gamma_1 t_1} \langle h_{\gamma_1 t_1}, F'_{Q \setminus \{t_1\}} \rangle + \langle F_{t_1}^{(N)}, F'_{Q \setminus \{t_1\}} \rangle . \tag{15''}$$

Using the Hölder inequality once more, we get

$$\left| \langle F_{t_1}^{(N)}, F'_{Q \setminus \{t_1\}} \rangle \right| < \left\| F_{t_1}^{(N)} \right\|_{L_2} \cdot C_Q ,$$

where the constant C_Q does not depend on N.

Passing, in (15''), to the limit with $N \to \infty$, we get

$$\langle F'_Q \rangle = \sum_{\gamma_1} a_{\gamma_1 t_1} \langle h_{\gamma_1 t_1}, F'_{Q \setminus \{t_1\}} \rangle ,$$

with the series converging in the sense that its partial sums

$$\sum_{\gamma_1 : |\gamma_1| \leqslant N} \to \langle F'_Q \rangle \quad \text{as} \quad N \to \infty . \tag{15'''}$$

Proceeding further in an analogous way, we obtain

$$\langle F'_Q \rangle = \sum_{\gamma_1} \cdots \sum_{\gamma_n} a_{\gamma_1 t_1} \ldots a_{\gamma_n t_n} \langle h_{\gamma_1 t_1}, \ldots, h_{\gamma_n t_n} \rangle , \tag{16a}$$

where we are summing iteratively: at first over the index γ_n (in the sense analogous to (15''')), then over the index γ_{n-1}, and so on. It follows easily from estimate (2) for $d < 1$ that the series (13') converges absolutely, and consequently, the iterated sum (16a) equals the sum of (13').

Using (13′) and applying Lemma 1, we get

$$\sum_{\substack{Q\subset T:\\ t_1\in Q, |Q|=n}} |\langle F'_Q\rangle| = \sum_{\{t_2,\ldots,t_n\}} |\langle F_{t_1}, F_{t_2}, \ldots, F_{t_n}\rangle|$$

$$\leqslant \frac{1}{(n-1)!} \sum_{(t_2,\ldots,t_n)} \sum_{\substack{(\gamma_1,\ldots,\gamma_n)\\ \gamma_i\neq \hat{0}, i=1,\ldots,n}} \frac{|a_{\gamma_1 t_1}|\ldots|a_{\gamma_n t_n}|}{\prod_{i=1}^{n}(\gamma_i!)^{1/2}}$$

$$\times \left|\langle :\xi_{t_1}^{\gamma_1}:, \ldots, :\xi_{t_n}^{\gamma_n}:\rangle\right|$$

$$\leqslant \sum_{(\gamma_1,\ldots,\gamma_n)} |a_{\gamma_1 t_1}| \prod_{i=2}^{n} a_{\gamma_i} \prod_{i=1}^{n} d^{|\gamma_i|/2} \frac{\gamma_i!!}{(\gamma_i!)^{1/2}}$$

$$< \left(\sum_{\gamma_1} |a_{\gamma_1 t_1}| \prod_{\alpha\in A} (m(\alpha)+1)^{1/2} d^{m(\alpha)/2}\right)$$

$$\times \left(\sum_{\gamma} a_\gamma \prod_{\alpha\in A} (m(\alpha)+1)^{1/2} d^{m(\alpha)/2}\right)^{(n-1)/2}, \tag{16}$$

which coincides with (14). Here, we have used the simply verifiable inequality $m!!/(m!)^{1/2} \leqslant (m+1)^{1/2}$.

Estimate (15) is obtained from (16) with the help of the Schwartz inequality:

$$\sum_{\gamma_1\neq\hat{0}} |a_{\gamma_1 t_1}| \prod (m(\alpha)+1)^{1/2} d^{m(\alpha)/2}$$

$$< \left(\sum_{\gamma_1\neq\hat{0}} |a_{\gamma_1 t_1}|^2\right)^{1/2} \left(\sum_{\gamma}\prod_{\alpha} (m(\alpha)+1) d^{m(\alpha)}\right)^{1/2}$$

$$= D_{t_1}^{1/2} \left(\sum_{m=1}^{\infty} m d^{m-1}\right)^{|A|/2}.$$

The sum from the second parenthesis in (16) can be estimated analogously.

Estimates of moments

Theorem 4. *1) Let a family* $\{\xi_{t\alpha};\, t\in T, \alpha\in A\}$ *of Gaussian random variables satisfy the conditions (1). Then, for any set of points* $\{t, \ldots, t_n\}\subset T$ *and any collection of functions* $(\gamma_1, \ldots, \gamma_n)$*, the estimate*

$$\left|\langle :\xi_{t_1}^{\gamma_1}: \ldots :\xi_{t_n}^{\gamma_n}:\rangle\right| \leqslant d^{N/2} \prod_{i=1}^{n} \gamma_i!! \tag{17}$$

holds true (the notations are those of Lemma 1).

2) Let the conditions of Theorem 3 be fulfilled. Then, for any finite $Q \subset T$,

$$\left|\left\langle \prod_{t\in Q} F_t \right\rangle\right| \leqslant \begin{cases} \prod_{t\in Q} \sum_{\gamma} |a_{\gamma t}| \widetilde{C}(\gamma), & A \text{ arbitrary}, \quad (18) \\ \left(\prod_{t\in Q} D_t^{1/2}\right) C^{|A||Q|}, & |A| < \infty, \quad (19) \end{cases}$$

with the constants $\widetilde{C}(\gamma)$ and C from the estimates (14) and (15).

Proof. Statement 2) can be proved by repeating the proof of Lemma 1. Since a moment of Wick monomials enters (17) only for a fixed collection of points $t_1, \ldots, t_n$, and not in a sum over such collections as in estimate (2), it is no longer essential that the diagrams in (3) can be also disconnected. We bring another independent proof of the estimate which is slightly weaker than (17). We use the equality

$$\left\langle \prod_{t\in Q} : \exp\left\{ \sum_{\alpha\in A} \xi_{t\alpha} z_{t\alpha} \right\} : \right\rangle = \exp\left\{ \frac{1}{2} \sum_{\substack{t\neq t' \\ \alpha,\alpha'}} z_{t\alpha} z_{t'\alpha'} \langle \xi_{t\alpha} \xi_{t'\alpha'} \rangle \right\} = \Phi(\{z_{t\alpha}\}) \tag{20}$$

with $\{z_{t\alpha}, t \in Q, \alpha \in A\}$ being an arbitrary finite collection of complex parameters.

We imply from the easily verifiable estimate

$$|\Phi(\{z_{t\alpha}\})| \leqslant \exp\left\{ \frac{1}{2} d \sum_{\substack{t\in Q \\ \alpha\in A}} |z_{t\alpha}|^2 \right\} \tag{21}$$

and from the Cauchy formula

$$\left\langle \prod_{i=1}^{n} :\xi_{t_i}^{\gamma_i}: \right\rangle = \frac{\prod \gamma_j!}{\prod_j (2\pi i)^{|\operatorname{supp}\gamma_j|}} \int_{|z_{t_i\alpha}|=r_{i\alpha}} \cdots \int \Phi(\{z_{t\alpha}\}) \prod_{i=1}^{n} \prod_{\alpha\in\operatorname{supp}\gamma_i} z_{t_i\alpha}^{-(m_i(\alpha)+1)} \times \prod dz_{t_i\alpha},$$

with the integration taking place on the set $|z_{t_i\alpha}| = r_{i\alpha}$ in the space of variables $\{z_{t_i\alpha};\ \alpha \in \operatorname{supp}\gamma_i,\ i = 1, \ldots, n\}$ and $r_{i\alpha} = \sqrt{m_i(\alpha)/d}$, that the inequality

$$\left|\left\langle \prod_{i=1}^{n} :\xi_{t_i}^{\gamma_i}: \right\rangle\right| \leqslant d^{N/2} \prod_{i=1}^{n} \left[(\gamma_i!)^{1/2} \prod_{\alpha\in\operatorname{supp}\gamma_i} (2\pi(m_i(\alpha)+1)^{1/4}) \right]$$

is true. Statement 2) can be proved in complete analogy with the proof of Theorem 3.

§4 Connectedness and Summation over Trees

In this section we introduce the concepts of connectedness and d-connectedness, often used in the following text, and we establish some simple facts related to these concepts; in particular, estimates for certain sums over collections of connected graphs.

Connectedness. Let T be a countable set and $\Gamma = \{A_1, \dots, A_n\}$ be a finite or infinite (unorderd) collection of its subsets. The collection Γ is called *connected* if the following graph G_Γ is connected: the set of its vertices consists of the elements of $(1, 2, \dots, n)$, and the set of its edges contains those pairs (i, j) for which the sets A_i and A_j intersect: $A_i \cap A_j \neq \emptyset$.

Let $\mathfrak{A}$ be a finite or countable family of subsets of T satisfying the following conditions:

1) The cardinality of all subsets $A \in \mathfrak{A}$ is bounded:

$$|A| \leqslant M, \quad A \in \mathfrak{A}. \tag{1}$$

2) Each point $t \in T$ is contained in at most K sets from $\mathfrak{A}$ (M and K are some constants).

For any point $t \in T$, the maximal number of connected collections $\Gamma = \{A_1, \dots, A_n\}$ consisting of n (possibly coinciding) sets $A_i \in \mathfrak{A}$, $i = 1, \dots, n$, such that $t \in \widetilde{\Gamma} \equiv \cup_{i=1}^{n} A_i$ is denoted by $S_{n,t}$. Let $S_n = \max_{t \in T} S_{n,t}$.

Lemma 1. *There are the constants $C = C(M, K)$ and $B = B(M, K)$ such that for any n,*

$$S_n \leqslant BC^n. \tag{2}$$

Proof. We use $\widehat{S}_n$ to denote the maximal number of connected collections $\Gamma = \{A_1, \dots, A_n\}$ consisting of n mutually different sets $A_i \in \mathfrak{A}(t \in \widetilde{\Gamma})$. The numbers $\widehat{S}_n$ satisfy the following recursive inequalities:

$$\widehat{S}_1 \leqslant K,$$
$$\widehat{S}_n \leqslant K \sum_{s=1}^{\min(n-1,M)} \binom{M}{S} \sum_{\substack{(n_1,\dots,n_s):\\ n_1+\dots+n_s=n-1}} \widehat{S}_{n_1} \dots \widehat{S}_{n_s}, \quad n > 1, \tag{3}$$

with the sum $\sum_{(n_1,\dots,n_s)}$ being taken over all ordered collections $(n_1, \dots, n_s)$, $s > 0$, such that $n_1 + \dots + n_s = n-1$, $n_i \geqslant 0$. The inequality (3) is deduced in the following way. Let $n > 1$ and A_1 be the first (with respect to any ordering of the family $\mathfrak{A}$) of the sets of the collection $\Gamma = \{A_1, \dots, A_n\}$ which contains a fixed point $t \in T$. In addition, the collection $\{A_2, \dots, A_n\}$ splits up into s connected components $\Gamma_1, \dots, \Gamma_s$, $\widetilde{\Gamma}_i \cap \widetilde{\Gamma}_j \neq \emptyset$, $i \neq j$. Let t_i be the first (in the sense of some ordering in T) point in $\widetilde{\Gamma}_i \cap A_i$, $i = 1, \dots, s$.

Now we fix A_1, a number $s \leqslant |A_1|$, the points $t_1, \dots, t_s \in A_1$, and the number $n_i = |\Gamma_i|$, $i = 1, \dots, s$, such that $n_1 + \dots + n_s = n - 1$. Then the number of collections Γ of sets from $\mathfrak{A}$, having exactly s connected components $\Gamma_1, \dots, \Gamma_s$ consisting each of n_i different subsets and containing points $t_1, \dots, t_s$, respectively, is not larger than

$$\widehat{S}_{n_1} \dots \widehat{S}_{n_s}. \tag{4}$$

Summing further (4) over all collections $(n_1, \dots, n_s)$, over choices of points $t_s, \dots, t_s \in A_1$, over all numbers s, and also over all sets A_1 containing a point t, and considering the conditions 1) and 2), we get (3).

We further introduce the numbers $\overline{S}_n$, $n \geqslant 1$, defined recursively:

$$\overline{S}_1 = K,$$
$$\overline{S}_n = K \sum_{s=1}^{\min(n-1,M)} \binom{M}{S} \sum_{\substack{(n_1,\dots,n_s): \\ n_1+\dots+n_s=n-1}} \overline{S}_{n_1} \dots \overline{S}_{n_s}, \quad n > 1. \tag{5}$$

Obviously, $\widehat{S}_n \leqslant \bar{S}_n$. We define a function

$$f(z) = \sum_{n=1}^{\infty} \overline{S}_n z_n$$

(assuming that the series on the right-hand side converges for sufficiently small z). We find, according to (5), that $f(z)$ satisfies the equality

$$f(z) = Kz(f(z) + 1)^M,$$

which we rewrite in the form

$$z = W/(K(1 + W)^M) \equiv F(W),$$

with $W = f(z)$. Since the function $F(W)$ of the complex variable W is analytic in a neighbourhood of the point $W = 0$ and $F'(0) \neq 0$, the inverse function $W = f(z)$ is analytic and bounded in some disc $\{|z| < r\}$, $r > 0$. We get from it

$$\widehat{S}_n < B(1/r)^n, \tag{5'}$$

with $B = \max_{|z|=r} |f(z)|$.

Coming to the number S_n of arbitrary connected collections

$$\Gamma = (A_1, \dots, A_n), \quad t \in \widetilde{\Gamma},$$

we notice that each such a collection can be written down in the form $\Gamma = \{(A_1, k_1), \ldots, (A_s, k_s)\}$, $A_i \neq A_j$, $i \neq j$, with $(A_1, \ldots, A_s)$ being a connected collection of mutually different sets from $\mathfrak{A}$, and $k_i > 0$ the multiplicity, with which the set A_i, $i = 1, \ldots, s$, occurs in Γ. Obviously, $k_i + \cdots + k_s = n$, and therefore, the number of ordered collections $(k_1, \ldots, k_s)$ does not exceed $\binom{n-1}{s-1}$.

From this and (5′), we get (2).

We present an important corollary of the proved lemma.

Lemma 2. *Let the conditions of Lemma 1 be fulfilled. Then there are the constants $C_1 > 0, C_2 > 0$, $B_1 > 0$ and $\lambda_0 > 0$ (depending on M and K) such that for all $0 < \lambda < \lambda_0$, any finite set $Q \subset T$, and any n,*

$$\sum_{\Gamma} \lambda^{|\Gamma|} \leqslant B_1 (C_1 \lambda)^n C_2^{|Q|}, \tag{5''}$$

with the summation over all collections $\Gamma = \{A_1, \ldots, A_m\}$, $A_i \in \mathfrak{A}$, $i = 1, \ldots, m$, such that $|\Gamma| = m \geqslant n$, and the collection $(\Gamma, Q) \equiv \{A_1, \ldots, A_m, Q\}$ is connected.

Proof. We denote the connected components of the collection Γ by $\Gamma_1, \ldots, \Gamma_s$, $s = 1, 2, \ldots$. Obviously, $\tilde{\Gamma}_i \cap Q \neq \emptyset$, and let t_i be the first (with respect to some ordering in T) point in the set $\tilde{\Gamma}_i \cap Q$.

As follows from the preceding lemma, the sum $\sum \lambda^{|G|}$ over all such collections Γ, with a fixed number s of components Γ_i, cardinalities $|\Gamma_i| = n_i$, $i = 1, \ldots, s$, and points $t \in Q, i = 1, \ldots, s$, does not exceed $B^s (C\lambda)^{n_1} \ldots (C\lambda)^{n_s} = B^s (C\lambda)^N$, with $N = |\Gamma|$. The number of ordered collections $(n_1, \ldots, n_s)$ such that $n_1 + \cdots + n_s = N$ does not exceed 2^N, and the number of collections (no longer ordered) of mutually distinct sites $\{t_1, \ldots, t_s\} \subset Q$ is not greater than $C^s_{|Q|}$. From this we get (5″) with $\lambda_0 < 1/(2C)$, $C_1 = (2C)$, $C_2 = B + 1$, $B_1 = (1 - 2C\lambda_0)^{-1}$.

$\mathfrak{A}$-connectedness and d-connectedness. Let $\mathfrak{A}$ be some family of subsets. A finite set $R \subset T$ is called *$\mathfrak{A}$-composable* if there is a collection

$$\Gamma = \{B_1, \ldots, B_k\}, \quad B_i \in \mathfrak{A},$$

such that

$$\tilde{\Gamma} = R. \tag{6}$$

An $\mathfrak{A}$-composable set R is called *$\mathfrak{A}$-connected* if the collection

$$\Gamma = \{B_1, \ldots, B_k\}, \quad B_i \in \mathfrak{A},$$

from (6) can be chosen to the connected; every maximal $\mathfrak{A}$-connected subset $R \subset A$ of the set A is called the *$\mathfrak{A}$-connected component of* A (i.e., A has no $\mathfrak{A}$-connected subset $R' \subset A$ such that $R \subset R'$).

Lemma 3. *Each $\mathfrak{A}$-composable set R can be uniquely split up into its $\mathfrak{A}$-connected components:*

$$R'_1, \ldots, R'_p, \quad R'_i \cap R'_j = \emptyset, \quad i \neq j, \quad R = \bigcup_{i=1}^{p} R'_i.$$

The proof is simple and we omit it.

Now let $\{A_1, \ldots, A_k\}$ be a collection of finite subsets $A_i \subset T$; we say that an $\mathfrak{A}$-composable set R is $\mathfrak{A}$-connected with respect to the collection $\{A_1, \ldots, A_k\}$ if the collection $\{R'_1, \ldots, R'_p, A_1, \ldots, A_k\}$, with $R'_i \subset R$ being the $\mathfrak{A}$-connected components of R, is connected.

Let the family $\mathfrak{A}$ satisfy the conditions 1) and 2), and $L_{t,n}(\mathfrak{A})$ be the number of $\mathfrak{A}$-connected sets of the cardinality n that contain a fixed point $t \in T$, and

$$L_n \equiv L_n(\mathfrak{A}) = \max_{t \in T} L_{t,n}(\mathfrak{A}).$$

Lemma 4. *There are the constants $C_3 = C_3(\mathfrak{A})$ and $B_2 = B_2(\mathfrak{A})$ such that for any $n \geqslant 1$,*

$$L_n \leqslant B_2 (C_3)^n. \tag{7}$$

Proof. Obviously, the number of $\mathfrak{A}$-connected sets R of the cardinality n, containing t, is not greater than the number of connected collections $\Gamma = \{B_1, \ldots, B_s\}$, $B_i \in \mathfrak{A}$, such that $s \leqslant n$ and $t \in \widetilde{\Gamma}$. Estimate (7) follows from Lemma 1.

Now we introduce the following quantities.

Let $\mathfrak{A}$ be a family of sets and R an $\mathfrak{A}$-connected set. We put

$$d_R(\mathfrak{A}) = \min_{\Gamma : \widetilde{\Gamma} = R} |\Gamma|, \tag{8}$$

with the minimum taken over all connected collections $\Gamma = \{B_1, \ldots, B_s\}$, $B_i \in \mathfrak{A}$, such that $\widetilde{\Gamma} = R$.

Let $\mathcal{A} = \{A_1, \ldots, A_k\}$ be some collection of sets and let the set R be $\mathfrak{A}$-connected with respect to $\{A_1, \ldots, A_k\}$. We put

$$d_R(\mathfrak{A}, \mathcal{A}) = \min_{\Gamma : \widetilde{\Gamma} = R} |\Gamma|, \tag{9}$$

with the minimum taken over all collections $\Gamma = \{B_1, \ldots, B_s\}$, $B_i \in \mathfrak{A}$, such that $\{B_1, \ldots, B_s, A_1, \ldots, A_k\}$ is a connected collection and $\widetilde{\Gamma} = R$. It is easy to see that

$$d_R(\mathfrak{A}, \mathcal{A}) = \sum d_{R'_i}(\mathfrak{A}), \tag{10}$$

where $R'_1, \ldots, R'_p$ are the $\mathfrak{A}$-connected components of the set R.

We now define

$$\hat{d}_A(\mathfrak{A}) = \min_{R \supseteq A} d_R(\mathfrak{A}) \tag{11}$$

for any set A, where the minimum is taken over all $\mathfrak{A}$-connected sets R containing A ($\hat{d}_A(\mathfrak{A}) = \infty$ if there is no such set R). Further, we put

$$d(\mathfrak{A}, \mathcal{A}) = \min_{R} d_R(\mathfrak{A}, \mathcal{A}), \tag{12}$$

where $\mathcal{A} = \{A_1, \dots, A_k\}$ and the minimum is taken over all sets R that are $\mathfrak{A}$-connected with respect to the collection $\mathcal{A}$ (in case there does not exist such a set, we put $d(\mathfrak{A}, \mathcal{A}) = \infty$).

Let $H_{t,r}$, $t \in T$, $r \geqslant 1$, be the number of those finite subsets $A \subset T$ for which $\hat{d}_A(\mathfrak{A}) = r$, $t \in A$, and write $H_r = \max_{t \in T} H_{t,r}$.

Lemma 5. *Let a family $\mathfrak{A}$ satisfy the conditions 1) and 2). Then there are the constants $C_4 = C_4(\mathfrak{A})$ and $B_3 = B_3(\mathfrak{A})$ such that for any $r \geqslant 1$,*

$$H_r < B_3 C_4^r \tag{13}$$

The proof easily follows from Lemma 4.

Lemma 6. *Let a family $\mathfrak{A}$ satisfy the conditions 1) and 2). Then, for sufficiently small λ, the estimate*

$$\sum_R \lambda^{d_R(\mathfrak{A}, \{A_1, \dots, A_k\})} \leqslant \widehat{C}_5^{\sum_{i=1}^k |A_i|} (C_5 \lambda)^{d(\mathfrak{A}, \{A_1, \dots, A_k\})} \tag{13'}$$

is valid, with $\widehat{C}_5, C_5$ being constants depending on $\mathfrak{A}$, $\{A_1, \dots, A_k\}$ being an arbitrary collection of subsets of T, and the summation taken over all sets R that are $\mathfrak{A}$-connected with respect to the collection $\{A_1, \dots, A_k\}$.

Proof. Let $R_1', \dots, R_l'$ be the $\mathfrak{A}$-connected components of R. Then, with the help of (10) and definition (12), we find that the sum on the left-hand side of (13′) admits the estimate

$$\sum_R \lambda^{d_R(\mathfrak{A}, \{A_1, \dots, A_k\})} \leqslant (\kappa^{-1} \lambda)^{d(\mathfrak{A}, \{A_1, \dots, A_k\})} \sum_R \prod_{i=1}^{l} \kappa^{d_{R_i'}(\mathfrak{A})},$$

with $\kappa = \kappa(\mathfrak{A}) < 1$ to be specified later. Further, the sum $\sum_R$ in the last inequality does not exceed

$$\sum_{\{R_1', \dots, R_l'\}} \prod_{i=1}^{l} \kappa^{d_{R_1'}(\mathfrak{A})} \leqslant \prod_{t \in \bigcup_{i=1}^{k} A_i} \sum_{R : t \in R} \kappa^{d_R(\mathfrak{A})}, \tag{13''}$$

where the sum $\sum_{\{R'_1,\dots,R'_l\}}$ on the left-hand side denotes the sum over all possible unordered collections of mutually disjoint sets R'_i that are $\mathfrak{A}$-connected with respect to the set $\cup_{i=1}^k A_i$, and the sum $\sum_{R:t\in R}$ on the right-hand side denotes the sum over such sets containing a fixed point $t \in \cup A_i$. Further, by Lemma 1,

$$\sum_{R,t\in R} \kappa^{d_R(\mathfrak{A})} \leqslant B \sum_{n\geqslant 1}^{\infty} C^n \kappa^n = \widetilde{C} < \infty$$

for $\kappa < C^{-1}$, with C and B being defined in (2). From this, the right-hand side of (13″) does not exceed $\widetilde{C}^{|\cup A_i|} < \widehat{C}^{\Sigma|A_i|}$ (with $\widehat{C} = \max\{\widetilde{C}, 1\}$), and we get (13′). The lemma is proved.

Now let the set T be endowed with some metric ρ so that the cardinality of all balls of radius $r > 0$, i.e.,

$$B_t^{(r)} = \{t' \in T, \rho(t,t') \leqslant r\}$$

is uniformly bounded:

$$\left|B_t^{(r)}\right| \leqslant C(r) < \infty \tag{14}$$

(r arbitrary).

We shall often consider the case $T = Z^\nu$, $\nu \geqslant 1$, and the metric ρ on Z^ν (cf. §0, Chapter 1):

$$\rho(t,t') \equiv |t - t'| = \sum_{i=1}^{\nu} \left|t^{(i)} - t'^{(i)}\right|, \tag{14′}$$

with $t^{(i)}$ and $t'^{(i)}$ being the coordinates of the vectors t, $t' \in Z^\nu$.

In the role of the family $\mathfrak{A} = \mathfrak{A}_d$ we consider the class of subsets $B \subset T$ with $\operatorname{diam} B \leqslant d$. Notice that, for the metric ρ obeying condition (14), the family $\mathfrak{A}_d$ has the properties 1) and 2), and each subset $R \subset T$ is $\mathfrak{A}_d$-composable. Every $\mathfrak{A}_d$-connected set R is called simply d-connected (and, similarly, d-connected with respect to a collection $\{A_1, \dots, A_k\}$, $A_i \subset T$). In the case of a metric ρ fulfilling

$$\min_{t_1,t_2:t_1\neq t_2} \rho(t_1,t_2) = 1, \tag{14″}$$

the quantity $\hat{d}_A(\mathfrak{A}_{d=1})$ will be denoted by $\hat{d}_A$. Notice that, in the case of $T = Z^\nu$ and the metric (14′), the quantity $\hat{d}_A$ equals the minimal length of trees $\mathcal{T}$, with the set of vertices coinciding with A (the length of $\mathcal{T}$ equals the sum of the lengths of its edges). Let (14) and (14″) be fulfilled for the metric ρ, and $\mathfrak{A}$ be a family of subsets satisfying the conditions:

1′) the diameters of all set $B \in \mathfrak{A}$ are uniformly bounded:

$$\operatorname{diam} B < r, \quad B \in \mathfrak{A};$$

2′) for each pair of acjacent points t_1, $t_2 \in T$, $\rho(t_1,t_2) = 1$, the quantity $\hat{d}_{\{t_1,t_2\}}(\mathfrak{A})$ is bounded:

$$\hat{d}_{\{t_1,t_2\}}(\mathfrak{A}) < v;$$

here r and v are constants.

Lemma 7. *Let T be 1-connected and the conditions 1), 2) and (14), (14″) be fulfilled. Then there exist such constants C_6 and $\widehat{C}_6$ (depending only on r, v, and the metric ρ) that*

$$C_6 \hat{d}_A \leqslant \hat{d}_M(\mathfrak{A}) \leqslant \widehat{C}_6 \hat{d}_A \tag{15}$$

for any $A \subset T$.

Proof. By 2′), each 1-connected set A can be covered by a connected collection $\Gamma = \{B_1, \dots, B_s\}$, $B_i \in \mathfrak{A}$, such that $|\Gamma| \leqslant C_6|A|$. Hence, the second inequality from (15) follows. As T is 1-connected and the conditions 1′), (14), and (14″) are fulfilled, $\hat{d}_B \leqslant K$ for each $B \in \mathfrak{A}$, where $K = K(v, \rho)$. Hence, the first inequality from (15) follows.

Summation over trees. Let $\mathcal{N}$ be a finite set (the elements of which we enumerate for our convenience : $\mathcal{N} = (1, 2, \dots, n)$). We use $\mathcal{G}(\mathcal{N})$ to denote the set of connected (unoriented) graphs G with the set of vertices coinciding with $\mathcal{N}$, and we use $\mathcal{T} = (\mathcal{N}) \subset \mathcal{G}(\mathcal{N})$ to denote the corresponding set of trees. Sometimes, connected graphs with multiple edges, i.e., pairs $\gamma = \{G, k_G\}$ with $G \in \mathcal{G}(\mathcal{N})$ and $k_G = \{k(i,j), (i,j) \in G\}$ being a positive integer-valued function defined on the set of edges of the graph G (multiplicities of the edges), are considered. The set of such graphs is denoted by $\widehat{\mathcal{G}}(\mathcal{N})$.

Let $\varphi = \{\varphi(i,j),\ (i,j) \in \mathcal{N},\ i \neq j\}$ be a symmetric real function defined on the set of pairs of distinct elements of $\mathcal{N}$. The number

$$I(G, \varphi) = \prod_{(i,j) \in G} \varphi(i,j)$$

is called the contribution of the connected graph $G \in \mathcal{G}(\mathcal{N})$, and the number

$$I(\gamma, \varphi) = \prod_{(i,j) \in G} [\varphi(i,j)]^{k(i,j)}$$

is the contribution $I(\gamma, \varphi)$ of the graph $\gamma = (G, k_G)$ with multiple edges.

Lemma 8. *The inequality*

$$\left| \sum_{G \in \mathcal{G}(\mathcal{N})} I(G, \varphi) \right| \leqslant e^{S(\varphi)} \sum_{\tau \in \mathcal{T}(\mathcal{N})} I(\tau, |\varphi|) \tag{16}$$

holds true, with $S(\varphi) = \sum_{i<j} |\varphi(i,j)|$.

In the case $\max |\varphi(i,j)| < 1$, *the equality*

$$\sum_{\gamma \in \widehat{\mathcal{G}}(\mathcal{N})} I(\gamma, \varphi) = \sum_{G \in \mathcal{G}(\mathcal{N})} I(G, \psi) \tag{17}$$

holds true, with $\psi(i,j) = \varphi(i,j)/(1-\varphi(i,j))$.

Proof. Notice that each connected graph $G \in \mathcal{G}(\mathcal{N})$ can be represented, at least in one fashion, in the form $G = \tau \cup \widehat{\Gamma}$, where $\tau \in \mathcal{T}(\mathcal{N})$ is a tree with vertices $(1,\ldots,n)$ and $\widehat{\Gamma}$ is a graph (in general disconnected) with the set of vertices $V(\widehat{\Gamma}) \subset \mathcal{N}$ and with edges distinct from the edges of τ. Thus,

$$\left| \sum_{G \in \mathcal{G}(\mathcal{N})} I(G,\varphi) \right| \leqslant \sum_{\widehat{\Gamma}} \prod_{(i,j)\in\widehat{\Gamma}} |\varphi(i,j)| \sum_{\tau \in \mathcal{T}(\mathcal{N})} I(\tau, |\varphi|), \tag{18}$$

where $\sum_{\widehat{\Gamma}}$ denotes the summation over all those graphs $\widehat{\Gamma}$ (including the empty one) for which $V(\widehat{\Gamma}) \subset \mathcal{N}$.

On the other hand,

$$\sum_{\widehat{\Gamma}} \prod_{(i,j)\in\widehat{\Gamma}} |\varphi(i,j)| = \prod_{i<j} (1 + |\varphi(i,j)|) \leqslant e^{S(\varphi)}. \tag{19}$$

The inequality (16) follows from (18) and (19). The equality (17) is obvious.

Let the number $l_\tau(i)$ denote the degree of the vertex $i \in \mathcal{N}$ (i.e., the number of edges of τ incident with i) for each $\tau \in \mathcal{T}(\mathcal{N})$. Since the number of edges of any tree $\tau \in \mathcal{T}(\mathcal{N})$ equals $n-1$, we have

$$l_\tau(1) + \cdots + l_\tau(n) = 2(n-1). \tag{20}$$

Lemma 9. *For any collection* $(\bar{l}_1,\ldots,\bar{l}_n)$ *of positive integers satisfying condition (20), we use* $V_n(\bar{l}_1,\ldots,\bar{l}_n)$ *to denote the number of trees* $\tau \in \mathcal{T}(\mathcal{N})$ *with fixed degrees of vertices equal to*

$$l_\tau(i) = \bar{l}_i, \quad i \in \mathcal{N}.$$

Then

$$V_n(\bar{l}_1,\ldots,\bar{l}_n) \leqslant \frac{2^{n-2}(n-2)!}{\prod_{i=1}^{n} \bar{l}_i!}. \tag{21}$$

Proof. A vertex $i \in \mathcal{N}$ of a tree τ is called an end vertex if $l_\tau(i) = 1$. Let $M = \{i_1,\ldots,i_m\} \subset \mathcal{N}$ be a set of vertices of τ which are not end ones, and let $i \notin M$ be its end vertex with the least index.

Removing this vertex and the edge (i,i_s) incident with it, where $i_s \in M$, we get a tree $\tau' \in \mathcal{T}(\mathcal{N}')$, with $\mathcal{N}' = \mathcal{N} \setminus \{i\}$, such that the degree of i_s is decreased by one, and the degrees of the other vertices remain unchanged. Thus, we get the recursive equality

$$V_n(\bar{l}_1,\ldots,\bar{l}_n) = \sum_{i_s \in M} V_{n-1}(\bar{l}_1,\ldots,\check{\bar{l}}_i,\ldots,\bar{l}_{i_s}-1,\ldots,\bar{l}_n). \tag{22}$$

The inequality (21) can be easily deduced from (22) by the induction on the number n, considering that $\sum_{i_s \in M} l_{i_s} \leqslant 2(n-2)$ (any tree has at least two end vertices).

Let T be a countable set and $\varphi = \{\varphi(t_1, t_2); t_1, t_2 \in T, t_1 \neq t_2\}$ be a symmetric function of distinct variables $t_1, t_2 \in T$ such that

$$\sup_{t_1 \in T} \sum_{t_2 \in T} |\varphi(t_1, t_2)| = D < \infty. \tag{23}$$

Lemma 10. *For any $\bar{t} \in T$ and any integer $n \geqslant 2$, the estimate*

$$\left| \sum_{\substack{A \subset T: \\ \bar{t} \in A, |A| = n}} \sum_{\tau \in \mathcal{T}(A)} \prod_{t \in A} l_\tau(t)! \prod_{(t,t') \in \tau} \varphi(t, t') \right| \leqslant (8D)^{n-1} \tag{24}$$

holds true.

Proof. The equality

$$\sum_{\substack{A \subset T, \bar{t} \in A, \\ |A| = n}} = \frac{1}{(n-1)!} \sum_{(t_2, \ldots, t_n), t_i \neq \bar{t}}$$

is valid, with the second sum denoting the summation over all ordered collections of $(n-1)$ mutually distinct points from T which are distinct from the point t. Thus, the left-hand side of (24) can be rewritten in the form

$$\frac{1}{(n-1)!} \sum_{\tau \in \mathcal{T}(\mathcal{N})} \prod_{i \in \mathcal{N}} l_\tau(i)! \sum_{\substack{(t_2, \ldots, t_n): \\ t_i \neq \bar{t}, i \geqslant 2}} \prod_{(i,j) \in \tau} \varphi(t_i, t_j), \tag{24'}$$

with $\mathcal{N} = (1, \ldots, n)$ and $t_1 = \bar{t}$. Further, for a fixed tree $\tau \in \mathcal{T}(\mathcal{N})$, the estimate

$$\left| \sum_{\substack{\{t_2, \ldots, t_n\}: \\ t_i \neq \bar{t} = t_1}} \prod_{(i,j) \in \tau} \varphi(t_i, t_j) \right| \leqslant D^{n-1} \tag{25}$$

is valid.

Indeed, let i_1 be some end vertex of the tree τ, $i_2 (i_2 \neq i_1)$ be an end vertex of the tree $\tau' = \tau \backslash \{i\}$ obtained by cancelling the vertex i_1 together with the edge (i_1, j_1) emanating from it, i_3 be an end vertex of the tree $\tau'' = \tau' \backslash \{i_2\}$, and so on. Then

$$\left| \sum_{\substack{\{t_2, \ldots, t_n\}: \\ t_i \neq \bar{t} = t_1}} \prod_{(i,j) \in \tau} \varphi(t_i, t_j) \right| \leqslant \sum_{t_{i_{n-1}}} |\varphi(t_1, t_{i_{n-1}})| \sum_{t_{i_{n-2}}} \cdots \sum_{t_{i_1}} |\varphi(t_j, t_{i_1})| .$$

Now, using (23), we get (25). Since there are at most $2^{2(n-1)}$ sequences $(\bar{l}_1, \dots, \bar{l}_n)$ satisfying the condition (20), we get (24) from (24′), (25), and the estimate (21).

Corollaries of Lemma 10. *1) A weaker inequality*

$$\left| \sum_{\substack{A \subset T, \bar{t} \in A, \\ |A|=n}} \sum_{\tau \in \mathcal{T}(A)} \prod_{(t,t') \in \tau} \varphi(t,t') \right| < (8D)^{n-1} \tag{26}$$

follows from the inequality (24).

2) It follows from (26) and (16) that

$$\left| \sum_{\substack{A \subset T \\ \bar{t} \in A, |A|=n}} \sum_{G \in \mathcal{G}(A)} \prod_{(t,t') \in G} \varphi(t,t') \right| < (8D')^{n-1} \tag{27}$$

with $D' = e^{D/2} D$.

3) In case $\sup_{t,t' \in T} |\varphi(t,t')| = d < 1$, *we get*

$$\left| \sum_{\substack{A \subset T, \\ \bar{t} \in A, |A|=n}} \sum_{\gamma = (G, k_G) \in \widehat{\mathcal{G}}(A)} \prod_{(t,t') \in G} [\varphi(t,t')]^{k(t,t')} \right| \leqslant (8D'')^{n-1} \tag{28}$$

from (27) and (17), with $D'' = \exp\{1/2D(1-d)^{-1}\} d(1-d)^{-1}$.

§5 Estimates on Intersection Number

Let T be a countable set with a metric ρ satisfying the conditions (14.4) and (14″.4), and let $\alpha = (A_1, \dots, A_N)$ be a collection of finite subsets of T (not necessarily different). We define $v_i = v_i(\alpha), i = 1, \dots, N$, to be the number of the sets $A_j \in \alpha, j = 1, \dots, N$, such that $A_i \cap A_j \neq \emptyset$, and $n_i = n_i(\alpha)$ to be the number of sets $A_j \in \alpha, j = 1, \dots, N$, coinciding with A_i.

Theorem 1. *There is a constant* $\overline{C} = \overline{C}(\rho)$ *such that the inequality*

$$\overline{C} \sum_{i=1}^{N} |A_i| + \sum_{i=1}^{N} \ln n_i \geqslant \sum_{i=1}^{N} \ln v_i \tag{1}$$

is valid for any collection $\alpha = (A_1, \dots, A_N)$ *of 1-connected finite subsets* $A_i \subset T$.

Proof. Each collection $\alpha = (A_1, \dots, A_N)$ can be written in the form $\alpha = \{(\bar{A}_1, n_1), \dots, (\bar{A}_s, n_s)\}$, with $\{\bar{A}_1, \dots, \bar{A}_s\}$ being the collection of all mutually different sets from α, and with n_i, $i = 1, \ldots, s$, the multiplicities of the sets $\bar{A}_i$ in the collection α.

We will now partition the collection of sets $\{\bar{A}_1, \ldots, \bar{A}_s\}$ into blocks $\mathcal{B}_{kr}$: $\mathcal{B}_{kr} = \{\bar{A}_i : n_i = k, |\bar{A}_i| = r\}$, and let α_{kr} be the cardinality of the block $\mathcal{B}_{kr}$. Finally, the number of the pairs (i, j) for which $\bar{A}_i \in \mathcal{B}_{kr}$, $\bar{A}_j \in \mathcal{B}_{k'r'}$, and $\bar{A}_i \cap \bar{A}_j \neq \emptyset$, is denoted by $\alpha_{kr,k'r'}$.

Several obvious properties of the above introduced numbers are:

1) $\sum_{k,r} k\alpha_{kr} = N$ (normalization);

2) $\alpha_{kr,k'r'} = \alpha_{k'r',kr}$ (symmetry);

3) as follows from Lemma 4.4, for any A_i, the number of sets $\bar{A}_j$ of cardinality r that intersect A_i, does not exceed $|A_i|\, B_2(C_3)^r$, and, consequently,

$$\sum_k \alpha_{kr,k'r'} \leqslant B_2 C_3^r r' \alpha_{k'r'}. \tag{2}$$

Further,

$$\sum_{\substack{A_i:|A_i|=r,\\ n_i=k}} v_i = k \sum_{k'r'} k' a_{kr,k'r'}, \tag{3}$$

with the sum $\sum_{A_i}$ over all sets $A_i \in \alpha$ of cardinality $|A_i| = r$ and of multiplicity $n_i = k$. Hence, using the well-known inequality between the geometric and arithmetic means, $(a_1 a_2 \ldots a_l)^{1/l} \leqslant (a_1 + \cdots + a_l)/l$, with $l \geqslant 1$, $a_1 > 0, a_l > 0$, we get, by taking the logarithms,

$$\sum_{\substack{A_i:|A_i|=r,\\ n_i=k}} \ln v_i \leqslant k\alpha_{kr} \ln \frac{\sum_{k'r'} k' \alpha_{kr,k'r'}}{\alpha_{kr}}. \tag{4}$$

Thus it follows from (4) that the inequality (1) is implied by

$$\overline{C} N d + \sum_{kr} m_{kr} \ln m_{kr} > \sum_{kr} m_{kr} \ln \beta_{kr}, \tag{5}$$

with the notations $m_{kr} = k\alpha_{kr}$, $\beta_{kr} = \sum_{k'r'} k\alpha'_{kr,k'r'}$, and $d = 1/N \sum_i A_i = 1/N \sum_{kr} kr\alpha_{kr}$ (the average cardinality of the sets from the collection α).

It follows from (2) that

$$\sum_k \beta_{kr} = \sum_{k'r'} k' \sum_k a_{kr,k'r'} \leqslant B_2 C_3^r \sum k' r' \alpha_{k'r'} = BC_3^r N d. \tag{6}$$

Further, with the help of a standard computation of the constraint maximum of the function $\sum_{l=1}^q b_l \ln x_l$, $x_1 > 0, \ldots, x_q > 0$ under the constraint $\sum_{l=1}^q x_l = \text{const.}$, we find from (6) that, for any fixed r,

$$\sum_k m_{kr} \ln \beta_{kr} \leqslant \sum_k m_{kr} \ln m_{kr} - N_r \ln(N_r/N) + N_r(\ln B_2 + r \ln C_3 + \ln d), \tag{7}$$

where $N_r = \sum_k m_{kr}$ is the number of sets of cardinality r in the collection α. Summing further (7) over r and considering that $\sum_r N_r r = dN$, we find that it is enough to prove the inequality

$$-\sum_r p_r \ln p_r < \ln d + 1, \tag{8}$$

with $p_r = N_r/N$, to prove (5). Computing again the constraint maximum of the function $-\sum_r x_r \ln x_r$, $x_r \geqslant 0$, under the constraints $\sum_r x_r = 1$ and $\sum_r r x_r = d$, we find that this maximum is achieved for the values $x_r = \frac{1}{d-1}\left(\frac{d-1}{d}\right)^r$ and equals $\ln d + (d-1)\ln\left(\frac{d}{d-1}\right) < \ln d + 1$. The theorem is proved.

We shall now suppose that T is a 1-connected set, and $\{A_1, \ldots, A_N\}$ is an arbitrary collection of subsets A_i.

Theorem 2. *There exists a constant* $\widetilde{C} = \widetilde{C}(\rho)$ *such that the inequality*

$$\widetilde{C}\sum_1^N \hat{d}_{A_i} + \sum_1^N \ln n_i \geqslant \sum_1^N \ln v_i \tag{9}$$

is valid ($\hat{d}_A$ is the minimal cardinality of 1-connected subsets $B \subset T$ containing A; see (11.4)) for each collection α.

The proof of this theorem is carried out similarly to the proof of Theorem 1, with the only difference that the blocks $\mathcal{B}_{kr}$ are defined in the following way: $\mathcal{B}_{kr} = \{\bar{A}_i; n_i = k, \hat{d}_{\bar{A}_i} = r\}$. In addition, the property that corresponds to property 3) follows from Lemma 5.4.

Let a set $S = \{t_1, \ldots, t_m\} \subset T$ of points $t_i \neq t_j$, $i \neq j$, be given, and a collection $\alpha = (A_1, \ldots, A_N)$ of subsets A_i be partitioned into m subcollections $\delta_1, \ldots, \delta_m$, $\delta \subset \alpha$, $k = 1, \ldots, m$, so that the condition

$$t_k \in \bigcap_{A_i \in \delta_k} A_i, \quad k = 1, \ldots, m, \tag{9'}$$

holds true.

Lemma 3. *For any partitions $\delta_1, \ldots, \delta_m$ of a collection α satisfying (9'), the estimate*

$$\sum_{i=1}^N |A_i| + \sum_{k=1}^m |\delta_k| \ln |\delta_k| \geqslant \sum_{i=1}^N \ln n_i \tag{10}$$

is valid.

Proof. Let $\{\bar{A}_1, \ldots, \bar{A}_s\}$ be a set of mutually different sets from the collection α, and let $n_i^{(k)}$ be the number of sets from the collection δ_k coinciding with

$\bar{A}_i, k = 1, \ldots m$; $i = 1, \ldots, s$. Since $\sum_{i=1}^{s} n_i^{(k)} = |\delta_k|$, it follows from the obvious inequality $\sum_{i=1}^{s} n_i^{(k)} \ln(n_i^{(k)}/|\delta_k|) \leqslant 0$, $k = 1, \ldots, m$, that

$$\sum_{i,k} n_i^{(k)} \ln n_i^{(k)} \leqslant \sum_k |\delta_k| \ln |\delta_k| . \tag{11}$$

Let the set $\bar{A}_i$ be contained in exactly r subcollections $\delta_{k_1}, \delta_{k_2}, \ldots, \delta_{k_r}$. In this case, the set of points $\{t_{k_1}, t_{k_2}, \ldots, t_{k_r}\} \subset \bar{A}_i$, and $|\bar{A}_i| \geqslant r$.

Using the well-known entropy inequality:

$$\sum_{p=1}^{r} b_p \ln b_p \geqslant \left(\sum_{p=1}^{r} b_p\right) \ln \frac{1}{r} \left(\sum_{p=1}^{r} b_p\right)$$

for any $b_1 \geqslant 0, \ldots, b_r \geqslant 0$, and the equality $\sum_{p=1}^{r} n_i^{(k_p)} = n_i$ (n_i is the multiplicity of the set $\bar{A}_i$ in the collection α), we get $\sum_k n_i^{(k)} \ln n_i^{(k)} \geqslant n_i \ln(n_i/r)$ and, consequently,

$$n_i |\bar{A}_i| + \sum_k n_i^{(k)} \ln n_i^{(k)} \geqslant n_i \ln n_i . \tag{12}$$

Summing (12) over i and using (11), we obtain (10).

This lemma and Theorem 2 imply

Corollary 1. *There exists a constant $\widehat{C}$ so that, for any collection $\alpha = (A_1, \ldots, A_N)$, any set of points $\{t_1, \ldots, t_m\}$, $t_i \neq t_j$, $i \neq j$, and any partition $\delta_1, \ldots, \delta_m$ of the collection α satisfying condition (9′), the estimate*

$$\widehat{C} \sum_{i=1}^{N} \hat{d}_{A_i} + \sum_{k=1}^{m} |\delta_k| \ln |\delta_k| \geqslant \sum_{i=1}^{N} \ln v_i \tag{13}$$

is true.

Let now $\mathfrak{A}$ be a system of finite subsets T, satisfying the conditions 1) and 2) from §4, and let $\hat{d}_A(\mathfrak{A})$ be the minimal cardinality of connected collections of sets from $\mathfrak{A}$ that cover A (see (11.4)). Then we find, from (13) and Lemma 7.4, that

$$C^* \sum_{i=1}^{N} \hat{d}_{A_i}(\mathfrak{A}) + \sum_{k=1}^{m} |\delta_k| \ln |\delta_k| \geqslant \sum_{i=1}^{N} \ln v_i , \tag{14}$$

with $C^* = C^*(\mathfrak{A}, \rho)$ being an absolute constant.

§6 Lattices and Computations of Their Möbius Functions

Lattices. Let a partially ordered set P with an order relation $\leqslant$ be given. The upper bound of a set $X \subset P$ is defined to be an element $p \in P$ such that $x \leqslant p$ for all $x \in X$. The least upper bound, $\sup X$, is the upper bound that is less than, or equal to, all other upper bounds.

Similarly, also the greatest lower bound $\inf X$ of a set $X \subset P$ is defined. An ordered set P is called *a lattice* if any two elements $P_1, P_2 \in P$ have the smallest upper bound $p_1 \vee p_2 \equiv \sup\{p_1, p_2\}$ and the greatest lower bound $p_1 \wedge p_2 \equiv \inf\{p_1, p_2\}$. Later on, P is always supposed to be a lattice. The greatest and smallest elements of P are denoted $\underline{0}$ and $\underline{1}$, respectively.

Example 1. Let $P = \mathfrak{A}_{\mathcal{N}}$ be the set of all partitions $\alpha = \{A_1, \dots, A_k\}$ of the set $\mathcal{N} = (1, 2, \dots, n)$, and $\alpha \leqslant \beta = \{B_1, \dots, B_l\}$ if each element A_i of the partition α is contained in some element of β (α is "finer" than β) (see §1).

Example 2. Let G be a connected graph with the set of vertices $\mathcal{N}$, and let $\mathfrak{A}_{\mathcal{N}}^G \subset \mathfrak{A}_{\mathcal{N}}$ be the set of all partitions $\alpha = \{A_1, \dots, A_k\}$ such that each block A_i is a connected subgraph of G.

It is easy to convince oneself that $\mathfrak{A}_{\mathcal{N}}$ and $\mathfrak{A}_{\mathcal{N}}^G$ are lattices.

For each lattice P, we shall use P^* to denote the dual lattice with the same elements as P and with the order relation replaced by the opposite one:

$$x \leqslant y \quad \text{in} \quad P^* \quad \text{if} \quad y \leqslant x \quad \text{in} \quad P.$$

For any two elements $x, y \in P$, $x < y$, the set $[x, y] = \{z \in P; x \leqslant z \leqslant y\}$ is called *an interval*. It is evident that every interval is again a lattice.

Elementary facts on Möbius functions. Consider the set of all (complex or real) functions $f(x, y)$ of two variables $x, y \in P$ such that $f(x, y)$ may differ from zero only for $x \leqslant y$. This set of function forms an (incidence) algebra over the field of numbers with the usual pointwise addition, and multiplication by a scalar, and with the product defined by the convolution

$$(f \times g)(x, y) = \sum_{z \in [x,y]} f(x, z) g(z, y).$$

This algebra has a unit element, namely the Kronecker delta, $\delta(x, y)$. We define the ζ-function

$$\zeta(x, y) = \begin{cases} 1, & \text{if } x \leqslant y, \\ 0 & \text{otherwise}. \end{cases} \tag{1}$$

Lemma 1. *ζ-function is invertible in the incidence algebra.*

Proof. We define recursively the Möbius function $\mu_P = \mu(x, y)$ of the lattice P:

$$\mu(x, x) = 1, \quad \mu(x, y) = -\sum_{z : x \leqslant z < y} \mu(x, z). \tag{2}$$

It is easy to convince oneself that $\mu \times \zeta = \zeta \times \mu = \delta$.

Lemma 2 (Möbius inversion formula). *Let f be a function on P, and*

$$g(x) = \sum_{y \leqslant x} f(y). \tag{3}$$

Then

$$f(x) = \sum_{y \leqslant x} g(y)\mu(y, x). \tag{4}$$

Proof. Indeed,

$$\sum_{y \leqslant x} g(y)\mu(y,x) = \sum_{y \leqslant x} \sum_{z \leqslant y} f(z)\mu(y,x) = \sum_{z} f(z) \sum_{z \leqslant y \leqslant x} \zeta(z,y)\mu(y,x)$$
$$= \sum_{z} f(z)\delta(z,x) = f(x).$$

The following properties of the Möbius functions, to be needed later, follow easily from the definition (2):

1. Let $x < y$, z, $z' \in [x, y]$. Then

$$\mu_P(z, z') = \mu_{[x,y]}(z, z'). \tag{5}$$

2. Let P^* be the dual lattice to P. Then

$$\mu_P(x, y) = \mu_{P^*}(y, x). \tag{6}$$

3. Let $P \times Q$ be the Cartesian product of the lattices P and Q, i.e., the set of pairs (p, q), $p \in P$, $q \in Q$, with $(p_1, q_1) \leqslant (p_2, q_2)$ if and only if $p_1 \leqslant p_2$, and $q_1 \leqslant q_2$. In this case,

$$\mu_{P \times Q}[(p_1, q_1), (p_2, q_2)] = \mu_P(p_1, p_2)\mu_Q(q_1, q_2). \tag{7}$$

4. (The inclusion-exclusion principle.) Let P be a lattice of all subsets x of some set; in this case, $x \leqslant y$ signifies that $x \subset y$. Then

$$\mu_P(x, y) = (-1)^{|y|-|x|}, \quad x \leqslant y, \tag{8}$$

with $|x|$ being the number of elements of x.

This follows from the fact that P is isomorphic to the Cartesian power of the two-point lattices: $\{\underline{0}, \underline{1}\}$, with $\underline{0} < \underline{1}$ and $\mu_{\{\underline{0},\underline{1}\}}(\underline{0}, \underline{1}) = -1$.

It follows from (8) and (4) that

$$f(x) = \sum_{y \subseteq x} g(y)(-1)^{|x \setminus y|},$$

where $g(x) = \sum_{y \subset x} f(y)$. This is precisely the formula that is usually called *the inclusion-exclusion principle*.

Galois connections. Let P and Q be two lattices. We call a Galois connection any pair of mappings $\rho : P \to Q$ and $\pi : Q \to P$ such that

1) ρ and π change the order of elements: $\rho(a) \leqslant \rho(b)$ if $b \leqslant a$, $a, b \in P$, and similarly for π;

2) $\pi(\rho(p)) \geqslant p$ and $\rho(\pi(q)) \geqslant q$ for all $p \in P$ and $q \in Q$.

Lemma 3. *Let ρ, π define a Galois connection. Let, moreover, $\pi(\underline{1}) = \underline{0}$, and $\pi(q) = \underline{0}$ for $q \neq 1$. Then,*

$$\mu_Q(\underline{0}, \underline{1}) = \sum_{a:\rho(a)=\underline{0}} \mu_P(\underline{0}, a). \tag{9}$$

Proof. It follows from the definition of a Galois connection that $\pi(x) \geqslant b$ if and only if $x \leqslant \rho(b)$ (because $\pi(x) \geqslant b$ implies $x \leqslant \rho\ (\pi(x)) \leqslant \rho(b)$ and conversely). In other words, this means that

$$\sum_{a:a\geqslant b} \delta_P(\pi(x), a) = \zeta_Q(x, \rho(b)). \tag{10}$$

Fix x and consider $\zeta_Q(x, \rho(b))$ and $\delta_P(\pi(x), a)$ as functions on P (of b and a, respectively). We use the Möbius inversion formula in (10);

$$\delta_P(\pi(x), \underline{0}) = \sum_{a\geqslant\underline{0}} \zeta_Q(x, \rho(a))\mu_{P^*}(a, \underline{0}) = \sum_{a\geqslant\underline{0}} \zeta_Q(x, \rho(a))\mu_P(\underline{0}, a). \tag{11}$$

Put $n = \zeta - \delta$. The assumptions of the lemma imply that

$$\delta_P(\pi(x), \underline{0}) = 1 - n_Q(x, \underline{1}).$$

We rewrite (11) in the form

$$1 - n_Q(x, \underline{1}) = \zeta_Q(x, \rho(\underline{0})) + \sum_{\underline{0}<a} \mu_P(\underline{0}, a)\zeta_Q(x, \rho(a)).$$

Since $\rho(\underline{0}) = \rho(\pi(\underline{1})) = \underline{1}$, we have $\zeta_Q(x, \rho(\underline{0})) = 1$. Consequently, $-n_Q(x, \underline{1}) = \sum_{0<a} \mu_P(0, a)\zeta_Q(c, \rho(a))$. As $\mu = \delta - \mu n$, we get

$$\begin{aligned}\mu_Q(\underline{0}, \underline{1}) &= - \sum_{0\leqslant x\leqslant 1} \mu_Q(\underline{0}, x) n_Q(x, \underline{1}) = \sum_{\underline{0}\leqslant x \underline{1}} \sum_{a>\underline{0}} \mu_Q(\underline{0}, x)\mu_P(\underline{0}, a)\zeta_Q(x, \rho(a)) \\ &= \sum_{a>\underline{0}} \mu_P(\underline{0}, a)\delta_Q(\underline{0}, \rho(a)) = \sum_{a:\rho(a)=\underline{0}} \mu_P(\underline{0}, a).\end{aligned}$$

Lemma 4. *Let L be a lattice, and let a subset $R \subset L$ be given such that $\underline{1} \notin R, \underline{0} \notin R$, and that there is a $y \in R$ with $y \leqslant x$ for each $x \in L, x \neq \underline{0}$. Let*

$k \geqslant 2$ and q be the number of sets $X \subset R$ such that $|X| = k$ and $\sup X = \underline{1}$. *Then*

$$\mu_L(\underline{0}, \underline{1}) = q_2 - q_3 + q_4 - \cdots. \tag{12}$$

Proof. Let $Q = L^*$ and P be the lattice of all subsets of R in Lemma 3. We define $\pi : Q \to P$ as follows: $\pi(x)$, $x \in Q$, is the set of some elements $y \in R$ such that $y \leqslant x$. For $A \subset R$, we put $\rho(A) = \sup A$ in L. Then (12) follows from (8) and Lemma 3.

Explicit computation of the Möbius function. We define the rank $r(p)$ of the element $p \in P$ as the maximal number r such that there is a sequence $x_1, \ldots, x_{r+1} \in P$ for which $\underline{0} = x_1 < x_2 < \cdots < x_{r+1} = p$. The elements of rank 1 are called *atoms.*

Let R be a set of atoms of P. A lattice P is called a *geometric* lattice (or matroid lattice) if:

1. any element of P is the least upper bound of some set of atoms;

2. if p is an atom and a an arbitrary element, then either $p \leqslant a$ or $p \vee a$ covers a (i.e., there is no element x such that $a < x < p \vee a$).[3]

The set X of atoms is called *independent* if $r(\sup X) = |X|$, and *generating* if $\sup X = \underline{1}$.

Remark. It is easy to verify that the lattices $\mathfrak{A}_{\mathcal{N}}$ and $\mathfrak{A}_{\mathcal{N}}^G$, introduced in Examples 1 and 2, are geometric lattices. A partition $\alpha \in \mathfrak{A}_{\mathcal{N}}$ is an atom if and only if exactly one of its blocks consists of two elements, and the remaining ones contain one element each. In case this two-point element corresponds to an edge of the graph G, the partition $\alpha \in \mathfrak{A}_{\mathcal{N}}^G$ is, according to the foregoing, an atom in the lattice. Thus, the atoms of $\mathfrak{A}_{\mathcal{N}}^G$ may be identified with the edges of the graph G. Let $G_X \subset G$ denote the subgraph of the graph G spanned by the set of edges in X (atoms of the lattices $\mathfrak{A}_{\mathcal{N}}^G$) as above. Then the sets of vertices of connected components of G_X coincide with the block of the partition $\sup X \in \mathfrak{A}_{\mathcal{N}}^G$, containing more than one element.

The set of atoms $X \subset \mathfrak{A}_{\mathcal{N}}^G$ is independent if and only if G_X is a forest, i.e., G does not contain a closed path. A set X of atoms is generating in $\mathfrak{A}_{\mathcal{N}}^G$ if and only if G_X is connected, and the set of its vertices coincides with $\mathcal{N}$.

Lemma 5. *In a geometric lattice, the number of elements in each independent generating set of atoms equals r(1).*

We give a proof of this lemma only for the case of the lattice $\mathfrak{A}_{\mathcal{N}}^G$. Indeed, if X is an independent and generating set of atoms (edges of G), then G_X is a tree with the set of vertices $\mathcal{N} = \{1, \ldots, n\}$, and thus the number of its edges $|X| = n - 1$. On the other hand, in any chain of partitions

$$\underline{0} = \alpha_1 < \alpha_2 < \cdots < \alpha_{r+1} = \underline{1}, \tag{12'}$$

[3] Other equivalent definitions of geometric lattices are given in [7].

the number of blocks $|\alpha_i| \leqslant n-(i-1)$, $i = 1,\dots,r+1$, and, consequently, $r \leqslant n-1$. In addition, it is also easy to find a chain (12′) with $r = n-1$, i.e.,

$$r(\underline{1}) = n - 1. \tag{12''}$$

A subset T of atoms is called *a circuit* if T is not independent, but all its (proper) subsets are independent.

Remark 2. A subset T of atoms forms a circuit in $\mathfrak{A}_{\mathcal{N}}^{G}$ only if the corresponding set of edges of G (mentioned in Remark 1) is a closed self-avoiding path, i.e., each vertex of G belongs either to two edges of T or it does not belong to any edge of T. Indeed, in the graph $G_T \subset G$, constituted from the edges of T, there is a closed path. We denote the set of its edges by $C \subset T$. Since C is not an independent set, we have $C = T$.

We shall now assume that the set R of all atoms of P is enumerated in an arbitrary way: $a_1,\dots,a_k$. If $T = \{a_{i_1},\dots,a_{i_l}\}$ is a circuit, $i_1 < \dots < i_l$, then

$$T' = \varphi(T) = \{a_{i_1},\dots,a_{i_{l-1}}\}$$

is called an *open* circuit. We use $\varphi^{-1}(T')$ to denote some circuit T with $\varphi(T) = T'$ (under the hypothesis that T exists). Obviously, in the case of the lattice $\mathfrak{A}_{\mathcal{N}}^{G}$, such a circuit is unique.

Theorem 6. *Let P be a geometric lattice. Then*

$$\mu_P(\underline{0},\underline{1}) = (-1)^{r(\underline{1})} m_1,$$

where m_1 is the number of subsets $X \subset R$ of the set of atoms such that
1) $|X| = r(\underline{1})$;
2) X does not contain open circuits.

Proof. For a fixed enumeration of elements of the set R of all atoms, we can enumerate the set of all open circuits $T'_1,\dots,T'_h$ so that

$$f(T'_s) \leqslant f(T'_j) \quad \text{if} \quad s < j, \tag{13}$$

where $f(T') = i_k$ is the number (in the fixed enumeration of R) of the last atom in $T = \{a_{i_1},\dots,a_{i_k}\}$, $i_1 < \dots < i_k$.

We use R_j^i to denote the set of all generating subsets $X \subset R$ such that $|X| = j$ and X does not contain $T'_1,\dots,T'_i$; $R_j^{i,A}$ is the set of all generating subsets $X \subset R$ such that $|X| = j$, X does not contain $T'_1,\dots,T'_{i-1}$, but contains the circuit $\varphi^{-1}(T'_i)$; $R_j^{i,B}$ is the set of all generating subsets $X \subset R$ such that $|X| = j$, X does not contain $T'_1,\dots,T'_{i-1}$, contains T'_i, but does not contain $\varphi^{-1}(T'_i)$. Then, for $i = 0,1,\dots,$ we have

$$|R_j^i| = |R_j^{i+1}| + |R_j^{i+1,A}| + |R_j^{i+1,B}|. \tag{14}$$

We shall prove, by induction on i, that

$$\mu_P(\underline{0},\underline{1}) = |R_2^i| - |R_3^i| + |R_4^i| - \dots \tag{15}$$

for $i = 0, 1, \dots$. For $i = 0$, it is the statement of Lemma 4. For any $i > 0$, we get, by induction and by the formula (14), that

$$\begin{aligned}\mu_P(\underline{0},\underline{1}) = |R_2^{i-1}| - |R_3^{i-1}| + |R_4^{i-1}| - \dots = |R_2^i| - |R_3^i| + |R_4^i| - \dots \\ + |R_2^{i,A}| + \left(|R_2^{i,B}| - |R_3^{i,A}|\right) - \left(|R_3^{i,B}| - |R_4^{i,A}|\right) + \dots .\end{aligned}$$

But $R_2^{i,A} = 0$, since a circuit cannot contain exactly two elements (any two atoms form an independent set).

We shall define a one-to-one correspondence ζ between $R_j^{i,B}$ and $R_{j+1}^{i,A}$ in the following way: we put

$$\zeta(X) = X \cup \{\varphi^{-1}(T_i')\backslash T_i'\}$$

for $X \in R_j^{i,B}$; $\zeta(X) \in R_{j+1}^{i,A}$ because the elements $a_l = \varphi^{-1}(T_i')\backslash T_i'$ cannot belong to any circuit $T_1', \dots, T_{i-1}'$ according to the chosen enumeration of the system of open circuits; if $a_k \in T_j'$ for some $j < i$, then $l > f(T_i') \geqslant f(T_j') \geqslant k$. It follows from (15) for $i = h$ that $\mu_P(0,1) = (-1)^{r(\underline{1})}|R_{r(\underline{1})}^h|$, because if X does not contain open circuits (and, consequently, does not contain circuits at all), then X is independent. Since X is generating at the same time, it contains exactly $r(\underline{1})$ elements, and therefore $|R_j^h| = 0$ for $j = r(\underline{1})$. The theorem is proved.

Corollary. *In the case of the lattice $\mathfrak{A}_{\mathcal{N}}^G$, as follows from (12″),*

$$\mu_{\mathfrak{A}_{\mathcal{N}}^G}(\underline{0},\underline{1}) = (-1)^{n-1} m_1,$$

where m_1 is the number of subsets X of such edges of G that $|X| = n-1$ and X does not contain open circuits.

§7 Estimate of Semi-Invariants of Partially Dependent Random Variables

We recall that, in §1, a virtual field $\{f(T), T \subset \mathcal{N}\}$ on the set $\mathcal{N}$ was defined (as the set of its "moments"), and formal semi-invariants $\{g(T), T \subset \mathcal{N}\}$ were introduced with the help of the recursive formula (16.1), being rewritten in the form

$$\begin{aligned} f(T) &= \sum g(B_1)\dots g(B_k), & |T| > 1, \\ f(T) &= g(T), & |T| = 1, \end{aligned} \tag{1}$$

with the sum being taken over all partitions $\{B_1, \ldots, B_k\}$ of the set T.

Let now $\mathcal{N}$ be finite and $\alpha = \{A_1, \ldots, A_k\}$ be any of its partitions. We put

$$f(\alpha) = \prod_{i=1}^{k} f(A_i), \quad g(\alpha) = \prod_{i=1}^{k} g(A_i). \tag{2}$$

Let $\mathfrak{A}_{\mathcal{N}} = \mathfrak{A}_n$ be the lattice of all partitions of the set $\mathcal{N} = \{1, \ldots, n\}$ (cf. §6). We get

$$f(\alpha) = \sum_{\beta \leqslant \alpha} g(\beta) \tag{3}$$

from the relation (1). The Möbius inversion formula (cf. (4.6)) implies that

$$g(\alpha) = \sum_{\beta \leqslant \alpha} \mu_{\mathfrak{A}_{\mathcal{N}}}(\beta, \alpha) f(\beta). \tag{4}$$

Comparing (4) for $\alpha = \underline{1}$ with formula (1.16), we get

$$\mu_{\mathfrak{A}_{\mathcal{N}}}(\beta, \underline{1}) = (-1)^{|\beta|-1}(|\beta| - 1)!, \tag{5}$$

where $|\beta|$ is the number of blocks of the partition β.

Noticing that for any $\alpha \geqslant \beta$ the interval $[\beta, \alpha] \subset \mathfrak{A}_{\mathcal{N}}$ is isomorphic to the Cartesian product of the lattices $\mathfrak{A}_{n_1} \times \cdots \times \mathfrak{A}_{n_s}$, where $s = |\alpha|$ and n_i is the number of blocks of the partition α contained in the i-th block of α, we obtain

$$\mu_{\mathfrak{A}_{\mathcal{N}}}(\beta, \alpha) = \mu_{\mathfrak{A}_{n_1}}(\underline{0}, \underline{1}) \ldots \mu_{\mathfrak{A}_{n_s}}(\underline{0}, \underline{1}) = (-1)^{n-s} \prod_{i=1}^{s} (n_i - 1)!. \tag{6}$$

Let G be a graph with the set of vertices $\mathcal{N}$, and let a virtual field $\{f(T), T \subset \mathcal{N}\}$ be independent with respect to the graph G (see definition (1.17)). (For all graphs considered below, we shall assume that there is at most one edge connecting every two vertices.)

We denote, for any $A \subset \mathcal{N}$,

$$C_f(A) = \max \left| \prod_{i=1}^{k} f(B_1) \right|, \tag{7}$$

with the maximum taken over all partitions $\{B_1, \ldots, B_k\}$ of the set A.

Theorem 1. *Let a virtual field be independent with respect to a graph G, and let the subgraph G_A, spanned by the vertices of the set $A \subset \mathcal{N}$, be connected. Then*

$$|g(A)| \leqslant C_f(A) \frac{3}{2} \prod_{t \in A} (3v_t), \tag{8}$$

where $v_t = v_t^{(A)}$ *is the number of edges in* G_A *which are incident with* $t \in A$.

Corollary. *Let* μ_0 *be the distribution of an independent field* $\{x_t, t \in T\}$ *on a set* T *and* $F_{A_1}, \ldots, F_{A_N}$ *be a collection of bounded local functions (also coinciding ones may appear among them). Then*

$$\left|\langle F_{A_1}, \ldots, F_{A_N}\rangle_{\mu_0}\right| \leqslant \prod_{i=1}^{N} \sup|F_{A_i}| \prod_{i=1}^{N} \overline{C}^{\hat{d}_{A_i}} \prod_{i=1}^{N} n_i, \tag{8'}$$

where C *is an absolute constant, and* n_i, $i = 1, \ldots, N$, *is the number of sets* A_j *in the collection* $\{A_1, \ldots, A_N\}$ *coinciding with the set* A_i.

The estimate (8′) follows from (7), (8), and Theorem 2.5.

As to the proof of Theorem 1, we note that, with the help of Lemma 2.1, it is possible to rewrite formula (3) in the form

$$f(\alpha) = \sum_{\beta:\beta \leqslant \alpha} g(\beta), \quad \beta \in \mathfrak{A}_{\mathcal{N}}^{G} \tag{9}$$

(for the definition of the lattice $\mathfrak{A}_{\mathcal{N}}^{G}$, see the preceding section).

If we consider, in (9), only the α belonging to $\mathfrak{A}_{\mathcal{N}}^{G}$, then we can use the Möbius inversion formula for $\mathfrak{A}_{\mathcal{N}}^{G}$ and write

$$g(\alpha) = \sum_{\substack{\beta \leqslant \alpha \\ \beta \in \mathfrak{A}_{\mathcal{N}}^{G}}} \mu_{\mathfrak{A}_{\mathcal{N}}^{G}}(\beta, \alpha) f(\beta) \tag{10}$$

for $\alpha \in \mathfrak{A}_{\mathcal{N}}^{G}$.

Lemma 2. *The estimate*

$$\left|\mu_{\mathfrak{A}_{\mathcal{N}}^{G}}(\underline{0}, \underline{1})\right| \leqslant \prod_{t \in \mathcal{N}} v_t \tag{11}$$

is valid.

Proof. We shall use the corollary of Theorem 6.6. Hence $\left|\mu_{\mathfrak{A}_{\mathcal{N}}^{G}}(\underline{0}, \underline{1})\right| \leqslant m_1 \leqslant \widetilde{m}_1$, where $\widetilde{m}_1$ is the number of subsets X of the set G of edges such that $|X| = n - 1$ and X does not contain any circuit. To estimate $\widetilde{m}_1$, we notice first that X has to be a connected tree. We define a one-to-one mapping φ_X of the set X onto some $A \subset \mathcal{N}$, with $|A| = n - 1$, in the following way. Consider any edge $\gamma_1 \in X$ and choose $\varphi_X(\gamma_1)$ to be any of the two vertices incident with γ_1. We shall proceed by induction. Suppose that $\varphi_X(\gamma_1), \ldots, \varphi_X(\gamma_m)$ are defined for $\gamma_1, \ldots, \gamma_m \in X$, with $\{\gamma_1, \ldots, \gamma_m\}$ being a connected set of edges. Let V_m be a set of vertices incident with some of $\gamma_1, \ldots, \gamma_m$. We choose $\gamma_{m+1} \in X$ so that one of its vertices belongs to V_m; this is possible because

X is connected. Moreover, both vertices of γ_{m+1} cannot belong to V_m since $\{\gamma_1, \ldots, \gamma_{m+1}\}$ is connected by construction. We put $\varphi_X(\gamma_{m+1})$ to be equal to that of the two vertices of γ_{m+1} that belongs to V_m if it does not coincide with any from the vertices $\varphi_X(\gamma_1), \ldots, \varphi_X(\gamma_m)$ already defined, and to the other vertex in the opposite case.

We define a mapping ψ_X of the set $\mathcal{N}$ into the set of all edges of G writing $\psi_X(t) = \gamma$ if $\varphi_X(\gamma) = t$, $t \in \mathcal{N}$, and, in case no such γ exists, put $\psi_X(t)$ equal to any γ incident with t. The mappings are different for different X, thus the number of all possible X does not exceed the number of all mappings of the set $\mathcal{N}$ to the set of edges of G that assign to any $t \in \mathcal{N}$ some edge which is incident with t. The number of such mappings equals $\prod_{t \in \mathcal{N}} v_t$. The lemma is proved.

The theorem follows from (10) in an obvious way, for $G = G_A$, $\alpha = \underline{1}$, by proving

Lemma 3. *The estimate*

$$\sum_{\beta \in \mathfrak{A}_{\mathcal{N}}^G} \left| \mu_{\mathfrak{A}_{\mathcal{N}}^G}(\beta, \underline{1}) \right| \leqslant \frac{3}{2} \prod_{t \in \mathcal{N}} 3 v_t \tag{12}$$

is valid.

Notice that

$$\mu_{\mathfrak{A}_{\mathcal{N}}^G}(\beta, \underline{1}) = \mu_{\mathfrak{A}_{|\beta|}^{G(\beta)}}(\underline{0}, \underline{1}), \tag{13}$$

where the graph $G(\beta)$ is obtained from G by the identification of the vertices belonging to only one block of the partition β, i.e., the blocks of β form the vertices of $G(\beta)$. The vertices B_i and B_j are connected by an edge in $G(\beta)$ if there is an edge joining some $b_1 \in B_i$ and $b_2 \in B_j$ in G.

The equality (13) follows from property 1 of the Möbius function (cf. §6) and also from the fact that the interval $[\beta, \underline{1}] \subset \mathfrak{A}_{\mathcal{N}}^G$ is isomorphic to the lattice $\mathfrak{A}_{|\beta|}^{G(\beta)}$.

Proof of Lemma 3. The proof is based upon a special procedure of iterated "gluing" of the vertices of the graph G which produces all partitions $\beta \in \mathfrak{A}_{\mathcal{N}}^G$. It is suitable to interpret this procedure in the form of a construction of some tree $\mathcal{G} = \mathcal{G}(G)$ with vertices labelled by pairs (β, D), where $\beta \in \mathfrak{A}_{\mathcal{N}}^G$ and D is a subset of the blocks of the partition β. Moreover, each vertex $t \in \mathcal{G}$ is coloured; it is red or blue. The process of construction of the tree $\mathcal{G}$ consists of successive steps; the order k is designated to the vertices constructed in the k-th step. Further, each vertex $\bar{t}$ of the order $k+1$ is connected with the only vertex t of order k (we shall say that $\bar{t}$ lies below t). We pass to the construction of the tree $\mathcal{G}$.

1. In the first step, only one vertex of the order 1 labelled by the pair $(\beta = \underline{0}, D = \emptyset)$ of the red colour is constructed.

2. Second step. We enumerate all elements of the set $\mathcal{N}$ (the vertices of the graph G) so that

$$v_1 \leqslant v_2 \leqslant \ldots \leqslant v_n \tag{14}$$

(v_i is the number of the edges of the graph G that are incident with the i-th vertex). The vertices of order 2 are constructed as follows. The only blue vertex of order 2 is labelled by the pair ($\beta = \underline{0}, D = \{1\}$). Further, v_1 red vertices of order 2, labelled by pairs $\{B_t, D_t = \emptyset\}$, are introduced with β_t running over all the atoms of the lattice $\mathfrak{A}_{\mathcal{N}}^G$ corresponding to the edges of the graph G that are incident with the vertex 1.

3. Let all vertices of $\mathcal{G}$ up to the order k be constructed. Let $t = (\beta_t, D_t)$, $\beta_t \in \mathfrak{A}_{\mathcal{N}}^G$, be some vertex of order k, and $\xi_1, \ldots, \xi_m$ be the vertices of the graph $G(\beta_t)$ (blocks of β_t) that do not belong to D_t and are enumerated so that $v_{\xi_1}^t \leqslant v_{\xi_2}^t \leqslant \ldots \leqslant v_{\xi_m}^t$, where $v_{\xi_i}^t$ is the number of edges of the graph $G(\beta_t)$ that are incident with the vertex ξ_i, $i = 1, \ldots, m$. Each such vertex t is connected by an edge with at most one blue and no more than $v_{\xi_1}^t$ red vertices of order $k+1$. A blue vertex is constructed in case $m > 0$, and a pair $(\beta_t, D_t \cup \{\xi_1\})$ corresponds to it, and the red vertices are labelled by pairs of the form $(\bar{\beta}_t, D_t)$ where the partitions $\bar{\beta}_t$ are obtained from the partition β_t by gluing together the element ξ_1 with any of the elements $\xi_2, \ldots, \xi_m$ that are connected with ξ_1 by an edge of the graph $G(\beta_t)$ (in addition, the elements of the set D_t do not change, and they form the blocks of the partition $\bar{\beta}_t$).

Proposition 4. *For any partition $\beta \in \mathfrak{A}_{\mathcal{N}}^G$, there is a red vertex $t \in \mathcal{G}$ such that $\beta = \beta_t$.*

Proof. We proceed by induction on the number of elements of the set $\mathcal{N}$. It is evident from our construction that, for any vertex $t = (\beta_t, D_t)$ (of order k) and any vertex $\bar{t} = (\beta_{\bar{t}}, D_{\bar{t}})$ that lies below t (i.e., a vertex of an order greater than k that can be connected with t by passing through vertices of the order greater than k), the partition $\beta_{\bar{t}} \geqslant \beta$, and the blocks of B_t which are contained in the set D_t remain blocks of the partition $\beta_{\bar{t}}$ and belong to the set $D_{\bar{t}}$.

Let $\beta \in \mathfrak{A}_{\mathcal{N}}^G$ be a partition. We consider three cases.

1. $\beta = \underline{0}$. Then $\beta = \beta_t$ where t is the 1st vertex.

2. Let $\beta \neq \underline{0}$, and the vertex $1 \in \mathcal{N}$ of the graph G (the enumeration of the vertices being the same as in (14)) forms a separate block of the partition β. Then any blue vertex of the order 2 is of the form $t = (\beta = \underline{0}; D = \{1\})$, and the subtree $\mathcal{G}' \subset \mathcal{G}$ spanned by t and by the vertices of $\mathcal{G}$ lying below t is isomorphic to the tree $\mathcal{G}(G')$ (with the change of the colour of t to the red), where $G' \subset G$ is the subgraph of G spanned by the set of vertices $\mathcal{N}' = \mathcal{N} \backslash \{1\}$. Further, the induction hypothesis may be used.

3. The vertex 1 belongs to a block B of the partition β, $|\beta| \geqslant 2$. Now we choose a red vertex $t = (\beta_t, \emptyset) \in \mathcal{G}$ of the order 2 that is obtained by gluing together the vertex 1 with any other vertex from the block B. The partition

$\beta_t \leqslant \beta$ and the subtree $\mathcal{G}' \subset \mathcal{G}$ rooted in the vertex t is isomorphic to the tree $\mathcal{G}(G(\beta_t))$. Now we shall again use the induction hypothesis.

We put

$$\gamma_t = \prod_{\xi \in G(\beta_t)} v_\xi^t$$

for each $t \in \mathcal{G}$. Let t be a vertex of order k, and t' be a vertex of order $k+1$ lying below t. If t' is blue, then

$$\gamma_t = \gamma_{t'} \tag{15}$$

according to the construction. If t' is red, then

$$\frac{\gamma_{t'}}{\gamma_t} \leqslant \frac{v_{\xi_1}^t + v_{\xi_r}^t - 2}{v_{\xi_1}^t v_{\xi_r}^t} \leqslant \frac{2}{v_{\xi_1}^t}, \tag{16}$$

where ξ_r is the vertex glued together with ξ_1 in the construction of $G(\beta_{t'})$. We get $\sum_{t'} \gamma_{t'} \leqslant 2\gamma_t$ from (16), where the sum is taken over all red vertices (of order $k+1$) lying below t. Consequently,

$$\sum_{\substack{\text{red } t \\ \text{of order } k+1}} \gamma_t \leqslant 2 \sum_{\substack{\text{red } t \\ \text{of order } k}} \gamma_t + 2 \sum_{\substack{\text{blue } t \\ \text{of order } k}} \gamma_t \leqslant 2 \sum_{\substack{\text{red } t \\ \text{of order } k}} \gamma_t + 2 \sum_{\substack{\text{red } t \\ \text{of order } k-1}} \gamma_t$$
$$+ 2 \sum_{\substack{\text{blue } t \\ \text{of order } k-1}} \gamma_t \leqslant \ldots \leqslant 2 \sum_{\substack{\text{red } t \\ \text{of order } k}} \gamma_t.$$

Put $a_k = \sum_{\substack{\text{red } t \\ \text{of order } k}} \gamma_t$. We have $a_{k+1} \leqslant 2\sum_{j=1}^k a_j$ which implies that $a_k \leqslant 3^k a_1$.

Taking into account Proposition 4, Lemma 2, and (13), we have

$$\sum_{\mu \in \mathfrak{A}_{\mathcal{N}}^{\mathcal{G}}} |\mu(\beta, \underline{1})| \leqslant \sum_{\text{red } t} \gamma_t \leqslant a_1 \sum_{k=1}^n 3^k \cdot a_1 \leqslant \frac{3}{2} \prod_{t \in \mathcal{N}} 3v_t.$$

Lemma 3 is proved.

§8 Abstract Diagrams (Algebraic Approach)

In the theory of random point fields, the diagrams of the general form (and not only related to Gaussian variables) often turn out to be useful. We shall explain here these notions and the related algebraic constructions, limiting ourselves to the case of pure point fields.

Algebraic approach. Let Q be a (metric) space and λ a continuous measure on Q (i.e., the measure $\lambda(\{t\})$ of any singleton $\{t\} \subset Q$ equals zero). Let Ω_{fin} be the set of all finite subsets Q (with the σ-algebra $\mathfrak{B}$ defined in §1, Chapter 1), and $\mathcal{A}$ be the set of all measurable (complex) functions $\psi(x)$, $x \in \Omega_{\text{fin}}$, on Ω. The set $\mathcal{A}$ with the product

$$(\varphi_1 \times \varphi_2)(x) = \sum_{\substack{x_1, x_2: \\ x_1 \cap x_2 = \emptyset, x_1 \cup x_2 = x}} \varphi_1(x_1)\varphi_2(x_2), \tag{1}$$

where the summation is taken over ordered pairs of nonintersecting subsets (x_1, x_2), $x_i \subset x$, $i = 1, 2$, such that $x_1 \cup x_2 = x$ (including also the pairs $(\emptyset, x)$ and $(x, \emptyset)$), forms a commutative algebra with the unit

$$\mathbf{1}(x) = \begin{cases} 1, & x = \emptyset, \\ 0, & x \neq \emptyset. \end{cases} \tag{2}$$

Notice, that, for each element $\psi \in \mathcal{A}$, any of its powers $(\psi)^{\times n} \equiv \underbrace{\psi \times \cdots \times \psi}_{n \text{ times}}$ equals

$$\begin{aligned} (\psi)^{\times n}(x) &= \sum_{k=1}^{|x|} \psi^{n-k}(\emptyset) \frac{n!}{(n-k)!} \sum_{\{x_1,\ldots,x_k\}}^{(k)} \prod_{i=1}^{k} \psi(x_i), \\ (\psi)^{\times n}(\emptyset) &= \psi^n(\emptyset), \end{aligned} \tag{3}$$

where the sum $\sum^{(k)}_{\{x_1,\ldots,x_k\}}$ is taken over all partitions of the set x that consist of k elements (i.e., over unordered collections $\{x_1, \ldots, x_k\}$ of mutually disjoint nonempty subsets $x_i \subset x$ such that $\bigcup_{i=1}^{k} x_i = x$).

We use $\mathcal{A}_+ \subset \mathcal{A}$ to denote the ideal of the algebra $\mathcal{A}$ consisting of such functions that $\psi(\emptyset) = 0$. We introduce the mapping $\mathcal{A}_+ \to \mathbf{1} + \mathcal{A}_+$:

$$\psi \mapsto \operatorname{Exp} \psi = \mathbf{1} + \psi + \frac{1}{2!}(\psi)^{\times 2} + \cdots + \frac{1}{n!}(\psi)^{\times n} + \ldots. \tag{4}$$

We find, from formula (3), that

$$\begin{aligned} (\operatorname{Exp} \psi)(\emptyset) &= 1, \\ (\operatorname{Exp} \psi)(x) &= \sum_{\{x_1,\ldots,x_k\}} \prod_{i=1}^{k} \psi(x_i), \quad x \neq \emptyset, \end{aligned} \tag{5}$$

where the sum is taken over all partitions of the set x.

Lemma 1. *The inverse mapping to (4) defined on the set $\mathbf{1} + \mathcal{A}_+$ exists, and*

$$(\operatorname{Exp})^{-1}(\mathbf{1}+\psi) \equiv \operatorname{Log}(\mathbf{1}+\psi) = \psi - \frac{1}{2}(\psi)^{\times 2} + \frac{1}{3}(\psi)^{\times 3} - \cdots + \frac{(-1)^{n-1}}{n}(\psi)^{\times n} + \ldots \tag{6}$$

or, in accordance with formula (3),

$$(\mathrm{Log}(\mathbf{1}+\psi))(\emptyset) = 0,$$

$$(\mathrm{Log}(\mathbf{1}+\psi))(x) = \sum_{\{x_1,\ldots,x_k\}} (-1)^{k-1}(k-1)! \prod_{i=1}^{k} \psi(x_i), \quad x \neq \emptyset, \tag{7}$$

where the sum is taken over all partitions of the set x.

The proof is simple and we omit it. Notice that formula (5) is similar to the formula expressing the moments by means of semi-invariants and formula (7) to the formula expressing the semi-invariants by means of moments. (See §1, properties D and E.)

Let $\psi \in \mathcal{A}$ and $\xi = \{\xi(t), t \in Q\}$ be a bounded measurable (complex) function with compact support on the space Q ($\xi(t) = 0$ outside some bounded domain $\Lambda \subset Q$). We define a functional in ξ;

$$\psi(\xi) = \int \psi(x) \left(\prod_{t \in x} \xi(t) \right) d\nu_\lambda(x), \tag{8}$$

under the hypothesis that the integral on the right-hand side of (8) converges absolutely. Here ν_λ is a measure on Ω_{fin} defined with the help of the measure λ on Q (cf. §1, Chapter 1). The functional $\psi(\xi)$ is analytic on D_r if the integral (8) exists for all functions ξ, with compact supports, belonging to a ball

$$D_r = \{\xi; \max |\xi(t)| < r, \xi \quad \text{with compact support}\} \tag{9}$$

for some $r > 0$. Let $\tilde{\mathcal{A}} \subset \mathcal{A}$ be the set of those functions $\psi \in \mathcal{A}$ for which the functional $\psi(\xi)$ is defined in a domain D_r. It is easy to verify that $\tilde{\mathcal{A}} \subset \mathcal{A}$ is a sub-algebra of $\mathcal{A}$, and the mapping $\psi \mapsto \psi(\xi)$, $\psi \in \tilde{\mathcal{A}}$, is a homeomorphism of $\tilde{\mathcal{A}}$ to the set of germs of holomorphic functions in the point $\xi = 0$ (i.e., of functionals in ξ analytic in some domain D_r of the form (9), $r > 0$). In other words,

$$\psi_1(\xi) \cdot \psi_2(\xi) = (\psi_1 \times \psi_2)(\xi), \quad \psi_1, \psi_2 \in \tilde{\mathcal{A}}. \tag{10}$$

Diagrams. We call a pair $G = (x, \Gamma)$ a diagram (a hypergraph) if $x \in \Omega_{\mathrm{fin}}$, $x \neq \emptyset$, is the set of vertices of G and $\Gamma = \{x_1, \ldots, x_s\}$ is an arbitrary unordered collection of subsets $x_i \subset x$, $|x_i| \geqslant 2$, which are called the edges of G. For a diagram $G = \{x, \Gamma\}$, the connected components $\{\Gamma_1, \ldots, \Gamma_m\}$ of the collection $\Gamma = \{x_1, \ldots, x_s\}$ define connected diagrams $G_j = (y_j, \Gamma_j)$, where

$$y_j = \bigcup_{x_i \in \Gamma_j} x_i$$

are called connected components of G. For two diagrams $G_1 = (x_1, \Gamma_1)$ and

$G_2 = (x_2, \Gamma_2)$, for which $x_1 \cap x_2 = \emptyset$, we define their union $G_1 \cup G_2 = (x_1 \cup x_2, \Gamma_1 \cup \Gamma_2)$ ($\Gamma_1 \cup \Gamma_2$ is the collection obtained by the union of collections Γ_1 and Γ_2).

Let $\mathfrak{G}$ be a family of diagrams having the following properties:

1) for any diagram $G \in \mathfrak{G}$, all its connected components $G_j \in \mathfrak{G}$;

2) for any two diagram $G_1 = (x_1, \Gamma_1)$, $G_2 = (x_2, \Gamma_2)$ for which $x_1 \cap x_2 = \emptyset$, the diagram $G_1 \cup G_2 \in \mathfrak{G}$.

Any such a class of diagrams will be called admissible. Here are some examples of admissible families of diagrams:

1) all diagrams without multiple edges (i.e., all x_i are distinct in the collection $\Gamma = \{x_1, \ldots, x_s\}$);

2) all diagrams $G = (x, \Gamma)$ with the number of edges $x_i \in \Gamma$ incident, with any point $t \in X$ not exceeding a given number.

Let $\mathfrak{G}$ be an admissible class of diagrams. For any function $\psi \in \mathcal{A}_+$, we define two functions $Z \in \mathcal{A}_+$ and $F \in \mathcal{A}_+$ on Ω_{fin} by the formulas

$$Z(x) = \sum_{G=(x,\Gamma)\in\mathfrak{G}}^{(x)} \prod_{x_i\in\Gamma} \psi(x_i), \tag{11}$$

$$F(x) = \sum_{G=(x,\Gamma)\in\mathfrak{G}_{\text{conn}}}^{(x)} \prod_{x_i\in\Gamma} \psi(x_i), \tag{12}$$

where the sum $\sum_{G\in\mathfrak{G}}^{(x)}$ stands for the sum over the diagrams from $\mathfrak{G}$ with the set of vertices x, and $\sum_{G\in\mathfrak{G}_{\text{conn}}}^{(x)}$ for the analogous sum over connected diagrams from $\mathfrak{G}$.

As easily follows from the definitions (11) and (12), the functions Z and F are related by

$$Z = \operatorname{Exp} F. \tag{13}$$

We assume that Z and F belong to $\tilde{\mathcal{A}}$, and we define the functionals $Z(\xi)$ and $F(\xi)$ by the formula (8):

$$Z(\xi) = \int_{\Omega_{\text{fin}}} Z(x) \prod_{t\in x} \xi(t)\, d\nu_\lambda,$$

$$F(\xi) = \int_{\Omega_{\text{fin}}} F(x) \prod_{t\in x} \xi(t)\, d\nu_\lambda.$$

Lemma 2. *The functionals $Z(\xi)$ and $F(\xi)$ are related by*

$$Z(\xi) = \exp\{F(\xi)\}. \tag{14}$$

The proof follows directly from (10) and (13).

CHAPTER 3

GENERAL SCHEME OF CLUSTER EXPANSION

§1 Cluster Representation of Partition Functions and Ensembles of Subsets

The technique to be developed here is related to random fields in a countable set T. We note that to use this technique for point and functional fields, they first have to be reduced to fields in a countable set, as explained in Chapter 1. There is, however, a special technique for point fields.

Let T be a countable set, and for each finite $\Lambda \subset T$, let the Gibbs modifications in Λ and their partition functions be defined. The basic step in the construction of cluster expansions consists of a cluster representation of the partition functions Z_Λ.

We assume that a metric $\rho(\cdot,\cdot)$ is defined on T, where the cardinality of the d-neighbourhood of any point $t \in T$ does not exceed $v = v(d) < \infty$ for each $d > 0$. We fix $d > 0$ and call a partition $\alpha = \{A_1, \dots, A_n\}$ of a set Λ admissible (more exactly, d-admissible) if $\rho(A_i, A_j) \geqslant d$ for any two different elements A_i and A_j of α, $|A_i| \geqslant 2$, and $|A_j| \geqslant 2$.

Definition 1. We shall say that the partition functions admit a cluster representation, if a number k_A corresponds to each finite set $A \subset T$ so that the following conditions are fulfilled:

1) (cluster representation) for all finite $\Lambda \subset T$,

$$Z_\Lambda = \sum_\alpha k_{A_1} \dots k_{A_n}, \tag{1}$$

with the sum taken over all admissible partitions $\alpha = (A_1, \dots, A_n)$ of the set Λ;

2) (cluster estimate) for all $t \in T$ and all $n \geqslant 2$,

$$\sum_{A: t\in A, |A|=n} |k_A| \leqslant c\lambda^n, \quad n = 2, 3, \dots, \tag{2}$$

with $\lambda \geqslant 0$ called a cluster parameter and $c > 0$ small enough (see below);

3) for $|A| = 1$,

$$|k_A| \geqslant k > \lambda. \tag{3}$$

Remark 1. It is often suitable to use a slightly more general definition of the cluster representation of Z_Λ, $\Lambda \subset T$:

$$Z_\Lambda = \sum k_{\Gamma_1} \dots k_{\Gamma_n}, \tag{4}$$

where the "cluster" $\Gamma = (A, \vartheta)$ is a pair: A finite subset $A = A(\Gamma) \subset \widehat{T}$ of a countable set $\widehat{T}$ ($\widehat{T}$ often coincides with T), here $A(\Gamma)$ is called the *support* of the cluster, and an element ϑ of an abstract space, particular for each A; k_{Γ_i} are numbers assigned to the clusters. The summation in (4) is taken over all unordered collections of clusters such that the collection of their supports forms an admissible partition of the set $\hat{\Lambda} \subset \widehat{T}$, with $\hat{\Lambda}$ being uniquely defined by $\Lambda \subset T$. Here we need:

1) for any $\hat{t} \in \widehat{T}$ and any $n \geqslant 2$,

$$\sum_{\substack{\Gamma: \hat{t} \in A(\Gamma), \\ |A(\Gamma)| = n}} |k_\Gamma| < c\lambda^n; \tag{5}$$

2) for any $\hat{t} \in \widehat{T}$, there is exactly one cluster Γ with $A(\Gamma) = \{\hat{t}\}$ and, moreover,

$$|k_\Gamma| \geqslant k > \lambda \tag{6}$$

(for λ and c, the same conditions as above are fulfilled).

We note that, if $\widehat{T} = T$ and $k_A = \sum_{\Gamma: A(\Gamma) = A} k_\Gamma$ for any finite $A \subset T$, then we obtain (1) from (4). Here the ensemble of clusters is introduced similarly to the ensemble of subsets considered below.

It is a particular problem how to obtain a cluster representation of the partition function Z_Λ, and we shall examine it in the following chapters (cf. also §5 of this chapter). In some cases, this representation is found very easily (§5), while in others, it is obtained in a more difficult way (§7, Chapter 4; §8, Chapter 4). However, as soon as it is established, further construction of the cluster expansions is carried out by a unique scheme which we shall describe now.

Lemma 1. *A family of quantities $\{Z_\Lambda;\ \Lambda \subset T,\ |\Lambda| < \infty\}$ admits a cluster representation (1) (for all finite $\Lambda \subset T$) if and only if for any finite nonempty set $B \subset T$ and all $t \in B$, the following relations hold:*

$$Z_B = k_{\{t\}} Z_{B \setminus \{t\}} + \sum_{A \subseteq \Lambda: t \in A} k_A \left[\prod_{s \in \bar{A}} k_{\{s\}} \right] Z_{B \setminus \bar{A}}, \tag{7}$$

with $Z_\emptyset = 1$, $\bar{A} = \bar{A}_d = \{t, \rho(t, A) < d\}$, $|A| \geqslant 2$, *and* $\bar{A} = A$ *for* $|A| = 1$.

Proof. In (1), we put $\Lambda = B$, and let $t \in A_1$ for a fixed point $t \in \Lambda = B$. Writing $A = A_1$ and summing, in (1), over $A_2, \dots, A_n$, we obtain formula (7). Conversely, by virtue of the recursive character of the relations (7) and of the equality $Z_\emptyset = 1$, the system of quantities Z_Λ satisfying (7) is uniquely determined, and consequently, it is of the form (1).

Lemma 2. *Let the family of quantities* $\{Z_\Lambda, \Lambda \subset T\}$ *admit a cluster representation (1) (or, which is equivalent, let it satisfy (7)) such that the conditions (2) and (3) are fulfilled, with the constant c in (2) sufficiently small. Then*

$$Z_\Lambda \neq 0 \tag{8}$$

for all finite $\Lambda \subset T$.

The proof of this lemma, and also its essential refinements, will be obtained as corollaries of further constructions. In this and subsequent sections, we shall assume, not mentioning it in particular cases, that quantities k_A are such that condition (8) is fulfilled. We notice that this condition is fulfilled in an obvious way.

Ensemble of subsets. It is useful to interpret the representation (1) and the notion, introduced further, with the help of the "ensemble of subsets," that we shall now describe. For any fixed Λ, we introduce the space of elementary events $\mathfrak{A}(\Lambda)$ as the set of all admissible partitions of Λ. The number

$$p^{(\Lambda)}(\alpha) = Z_\Lambda^{-1} k_{A_1} \dots k_{A_n}, \tag{9}$$

for any admissible partition $\alpha = \{A_1, \dots, A_n\}$, is called its "probability." We note that, in the case of negative (or complex) k_A, the numbers $p^{(\Lambda)}(\alpha)$ do not actually define a usual probability.

Let $\Gamma = \{A_1, \dots, A_n\}$ be an arbitrary admissible (i.e., such that $\rho(A_i, A_j) \geqslant d$ for $|A_i| \geqslant 2$, $|A_j| \geqslant 2$) unordered collection of mutually disjoint subsets of Λ. We write

$$\widetilde{\Gamma} = \bigcup_{i=1}^{n} A_i, \quad \overline{\widetilde{\Gamma}} = \bigcup_{i=1}^{n} \bar{A}_i.$$

Define a "probability" of Γ as

$$\rho^{(\Lambda)}(\Gamma) = \sum_{\alpha : \Gamma \subseteq \alpha} \rho^{(\Lambda)}(\alpha) = k_{A_1} \dots k_{A_n} \prod_{t \in \overline{\widetilde{\Gamma}} \setminus \widetilde{\Gamma}} k_{\{t\}} \frac{Z_{\Lambda \setminus \overline{\widetilde{\Gamma}}}}{Z_\Lambda}, \tag{10}$$

with the sum taken over all admissible partitions α such that each $A_i \in \Gamma$ is a block of the partition α. The function $\rho^{(\Lambda)}(\Gamma)$, defined on the set of admissible collections Γ, is called the correlation function of the ensemble of subsets of Λ.

Theorem 3. *Let the partition function Z_Λ, $\Lambda \subset T$, admit a cluster representation (1) fulfilling the conditions (2) and (3), with the constant c, in (2), satisfying the condition $c < c_0(\lambda/k)$ with c_0 from (23). Then there is a limit correlation function $\rho(\Gamma)$, with Γ being a finite admissible collection of sets, such that*

$$\lim_{\Lambda \nearrow T} \rho^{(\Lambda)}(\Gamma) = \rho(\Gamma). \tag{11}$$

This theorem will be proved in §3.

A particular role is played by the relations

$$f_A^{(\Lambda)} = \frac{Z_{\Lambda \setminus A}}{Z_\Lambda} \prod_{t \in A} k_{\{t\}}, \quad f_\emptyset = 1. \tag{12}$$

If $A = \{t_1, \dots, t_k\}$, then it is obvious that

$$f_A^{(\Lambda)} = \rho^{(\Lambda)}(\Gamma), \tag{13}$$

with $\Gamma = \{\{t_1\}, \dots, \{t_k\}\}$ consisting of singletons. Hence, in case the assumptions of Theorem 1 are fulfilled, the limit

$$\lim_{\Lambda \nearrow T} f_A^{(\Lambda)} \stackrel{\text{def}}{=} f_A = \rho(\Gamma) \tag{14}$$

exists. We find, from (10) and (12), that

$$\rho^{(\Lambda)}(\Gamma) = g_{A_1} \cdots g_{A_n} f^{(\Lambda)}_{\underset{\Gamma}{=}}, \quad \Gamma = \{A_1, \dots, A_n\}, \tag{15}$$

using the notation

$$g_A = k_A \Big/ \prod_{t \in A} k_{\{t\}}. \tag{15'}$$

It is evident from (15) that it suffices to prove the existence of the limit (11) in the particular case (14).

As is evident from (12), the quantities $f_A^{(\Lambda)}$ are rational functions of the variables $\{g_B; |B| \geqslant 2, B \subset \Lambda\}$, and $f_A^{(\Lambda)}(g_B \equiv 0) = 1$. Hence $f_A^{(\Lambda)}$ expands in a power series in these variables converging in a neighbourhood of zero that, in general, depends on A and Λ. However, as we shall show later (and this is the main purpose of cluster expansions), this neighbourhood may be chosen uniquely for all $A \subset \Lambda$, and its dimensions do not depend on Λ.

We use $\eta = \{A_1, \dots, A_n\}$ to denote an arbitrary unordered collection of subsets, $|A_i| \geqslant 2$, $i = 1, \dots, n$, among which the same set may repeat several times, and let $g_\eta = \prod_{A \in \eta} g_A$. Then the desired series are of the form

$$f_A^{(\Lambda)} = \sum_\eta D^{(\Lambda)}(A; \eta) g_\eta. \tag{16}$$

The coefficients $D^{(\Lambda)}(A;\eta)$ can be easily expressed in an explicit form by means of functions similar to the Möbius functions for some lattices (cf. §6, Chapter 2). However, such a representation of these coefficients does not help too much in the investigation of the convergence of the series (16). Therefore, we shall employ another expression for these series.

We fix a point $t = t_D \in D$ for each finite $D \subset T$.

Lemma 4. *Under any choice of points $t_D \in D$, the system of quantities $\{f_D^{(\Lambda)}, D \subset \Lambda\}$ satisfies the relations*

$$f_D^{(\Lambda)} - f_{D\setminus\{t_D\}}^{(\Lambda)} + \sum\nolimits^{(D)} g_A f_{D\cup(\bar{A}\cap\Lambda)}^{(\Lambda)} = 0, \tag{17}$$

with $|D| \geqslant 2$, and for $|D| = 1$,

$$f_D^{(\Lambda)} + \sum\nolimits^{(D)} g_A f_{D\cup(\bar{A}\cap\Lambda)}^{(\Lambda)} = 1, \tag{17'}$$

with the summation in $\sum^{(D)}$ taken over sets $A \subset \Lambda$ such that $A \cap D = \{t_D\}$ and $|A| \geqslant 2$.

Proof. Formula (7), with the set $B \subset \Lambda$ replaced by $\Lambda\setminus B$, after division by Z_Λ and multiplication by $\prod_{t'\in B} k_{\{t'\}}$, can be rewritten as

$$f_B^{(\Lambda)} = f_{B\cup\{t\}}^{(\Lambda)} + \sum g_A f_{B\cup(\bar{A}\cap\Lambda)}^{(\Lambda)}.$$

Letting here $t = t_D$ and $B = D\setminus\{t_D\}$, we obtain (17) and (17′).

It is convenient to write the system (17), (17′) in an operator form. To do this, we consider the vector space $\mathcal{B}^{(\Lambda)}$ of the complex vectors $\varphi^{(\Lambda)} = \{\varphi_A^{(\Lambda)}, A \subset \Lambda\}$, with the coordinates indexed by all nonempty subsets $A \subset \Lambda$ (the superscript will often be dropped for the sake of brevity).

We define an operator R (a generalized "shift") on $\mathcal{B}^{(\Lambda)}$:

$$(R\varphi)_A = \begin{cases} \varphi_{A\setminus\{t_A\}}, & |A| \geqslant 2, \\ 0, & |A| = 1, \end{cases}$$

and an operator $K = K^{(\Lambda)}$ on $\mathcal{B}^{(\Lambda)}$:

$$(K\varphi)_A = -\sum\nolimits^{(A)} g_B \varphi_{A\cup(\bar{B}\cap\Lambda)}. \tag{18}$$

We use the notation $\delta = \delta^{(\Lambda)}$ for the vector with coordinates

$$\delta_A = \begin{cases} 1, & |A| = 1, \\ 0, & |A| > 1. \end{cases}$$

Under these notations, the system of equations (17), (17′) can be rewritten in the form

$$f^{(\Lambda)} - (R+K)f^{(\Lambda)} = \delta. \tag{19}$$

To investigate it, we introduce, in $\mathcal{B}^{(\Lambda)}$, a norm

$$\|\varphi\|_M = \sup_{A\subseteq\Lambda} |\varphi_A| \frac{1}{M^{|A|}}, \tag{20}$$

with a constant $M > 0$ that will be specified below.

Notice that $\|R\| < 1/M$, and it follows from (2) and (3) that if $\frac{\lambda}{k}M^v < 1$ (recall that v denotes the maximal cardinality of a d-neighbourhood of any point $t \in T$), then

$$\|K\| < \frac{1}{M} \sup \sum\nolimits^{(A)} |g_B| M^{|\bar{B}\cap\Lambda|} \leqslant \frac{c}{M} \sum_{n\geqslant 2} (M^v \lambda/k)^n. \tag{21}$$

Hence, it follows from (21) that

$$\|R+K\| \leqslant \left(\frac{\lambda}{k}\right)^{1/v} \frac{1}{x^{1/v}} \left(1 + c\frac{x^2}{1-x}\right),$$

with $x = (\lambda/k)M^v$.

We use $h(c)$ to denote the quantity

$$h(c) = \min_{0<x<1} \frac{1}{x^{1/v}} \left(1 + c\frac{x^2}{1-x}\right)$$

and $x_0 = x_0(c) \in (0,1)$ to denote the point at which this maximum is attained. Choosing

$$M = [(\lambda/k)^{-1} x_0(c)]^{1/v}, \tag{22}$$

we get

$$\varepsilon_0 \equiv \|R+K\| \leqslant (\lambda/k)^{1/v} h(c). \tag{22′}$$

We shall always suppose that

$$(\lambda/k)^{1/v} h(c) < 1 \tag{23}$$

and refer to this condition as to the cluster condition later on.

As $h(0) = 1$, the function $h(c)$ is monotone for $c > 0$ and increases as $c > 0$; for $\lambda/k < 1$, there is only one solution $c_0 = c_0(\lambda/k)$ of the equation

$$(\lambda/k)^{1/v} h(c_0) = 1, \tag{23′}$$

with $c_0(\lambda/k) \to \infty$ as $\lambda/k \to 0$.

Obviously, the cluster condition (23) is fulfilled for $\lambda/k < 1$ and $c < c_0(\lambda/k)$. It follows from (22′) and (23) that $\varepsilon_0 < 1$ and that the solution of (19) exists, is unique, and satisfies the estimate

$$\left|f_A^{(\Lambda)}\right| < \frac{\|\delta\|}{1-\varepsilon_0} M^{|A|} < \frac{1}{M(1-\varepsilon_0)} \left[\left(\frac{\lambda}{k}\right)^{-1} x_0(c)\right]^{|A|/v}. \tag{24}$$

Moreover, this solution can be expressed in the form of a series

$$f^{(\Lambda)} = \sum_{n=0}^{\infty}(R+K)^n\delta = \sum_{m_1=0}^{\infty} R^{m_1}\delta + \sum_{p=1}^{\infty} \sum_{(m_1,\dots,m_{p+1})} R^{m_1}KR^{m_2}K\dots R^{m_p}KR^{m_{p+1}}\delta, \tag{25}$$

with the last sum taken over all (ordered) collections of nonnegative integers $(m_1,\dots,m_{p+1})$; $m_i \geqslant 0$, $i = 1,\dots,p+1$, and the series converges in the norm (20) in $\mathcal{B}^{(\Lambda)}$.

§2 Cluster Expansion of Correlation Functions

It is easy to deduce the representation of $f_A^{(\Lambda)}$ in the form (16) in §1 from the expansion (25) in §1. Namely, we express the solution of (19) in §1 in a geometrically intuitive form that is convenient for further application (in the form of a series over "connected groups of clusters").

Therefore, we introduce an oriented graph (a connected tree) with the set of vertices to be the set $\mathcal{F} = \mathcal{F}(T)$ of all finite subsets of T. An edge is connecting a vertex $A \in \mathcal{F}$ and a vertex $B \in \mathcal{F}$ if and only if $B = A\backslash\{t_A\}$. We say that a vertex B lies below a vertex A, if there is a path through the graph $\mathcal{F}$ beginning at A and ending at B. The lowest vertex of the tree $\mathcal{F}$ is $\emptyset$. There are exactly $|A|$ vertices below the vertex A.

We use the notation γ for the collection of nonempty subsets

$$\gamma = (B_1, A_1; \dots; B_p, A_p), \quad p = p(\gamma) \geqslant 1, \tag{1}$$

such that

1) for all $i = 2,\dots,p$, either B_i lies below $C_i = B_{i-1} \cup \bar{A}_{i-1}$, or $B_i = C_i$; we write $m_i = |C_i\backslash B_i|$;

2) $A_i \cap B_i = \{t_{B_i}\}$, $|A_i| \geqslant 2$ for all $i = 1,\dots,p$; we denote, for each collection (1), the ordered collection $(A_1,\dots,A_p)$ by $\eta = \eta(\gamma)$, and let

$$g_\gamma = g_{\eta(\gamma)} = \prod_{i=1}^{p} g_{A_i}. \tag{2}$$

The set $\operatorname{supp}\gamma = \tilde{\eta}(\gamma) = \bigcup_{i=1}^{p} A_i$ is called the support of γ.

Lemma 1. *If the condition (23) in §1 is fulfilled, then*

$$f_A^{(\Lambda)} = 1 + \sum\nolimits^{(\Lambda,A)} (-1)^{p(\gamma)} g_\gamma, \tag{3}$$

with the summation taken over all collections γ of the form (1) for which 1) $\operatorname{supp}\gamma \subset A$; *2) $B_i \subset \Lambda$; 3) either $B_1 = \Lambda$ or B_1 lies below A. Here, the series on the right-hand side converges absolutely.*

Remark 1. In case $d = 1$, i.e., $\bar{A} = A$, each collection of the form (1) that enters the sum (3) can be uniquely expressed in the form of a path through the graph $\mathcal{F}$:

$$(C_1, B_1; C_2, B_2; \ldots; C_p, B_p), \quad C_1 = A,$$

where the path to $B_i \subset C_i$ leads downwards in $\mathcal{F}$, and the path to $C_{i+1} \supset B_i$ leads upwards.

Proof of Lemma 1. We choose a basis $\{\chi^{(A)}; A \subset \Lambda, A \neq \emptyset\}$ in $\mathcal{B}^{(\Lambda)}$, with $\chi^A = \{\chi_B^A\}$, $\chi_B^A = \delta_{AB}$ (the Kronecker delta).

It is easily seen that, with respect to this basis, the matrix elements $(R^m)_{C,B}$ or the operator R^m equal

$$(R^m)_{C,B} = \begin{cases} 1, & \text{if } B \text{ lies below } C \text{ in } \mathcal{F} \text{ and } |C \backslash B| = m, \\ 0 & \text{otherwise,} \end{cases} \tag{4}$$

and the matrix elements of the operator K equal

$$(K)_{B,C} = \begin{cases} -g_A, & \text{if there is such an } A \subset \Lambda \text{ that} \\ & |A| \geqslant 2, B \cap A = \{t_B\}, C = B \cup (\bar{A} \cap \Lambda) \\ & \text{(if such an } A \text{ exists, it is unique),} \\ 0 & \text{otherwise.} \end{cases} \tag{5}$$

Noticing that $\delta = \sum_{t \in \Lambda} \chi^{\{t\}}$, we find, from (25) in §1, that

$$f_A^{(\Lambda)} = \sum_{t \in \Lambda} \Bigg[\sum_{m_1} (R^{m_1})_{A,\{t\}} + \sum_{p \geqslant 1} \sum_{(m_1, \ldots, m_{p+1})} \sum_{B_1, \ldots, B_p} (R^{m_1})_{A,B_1} \\ \times (K)_{B_1,C_1} (R^{m_2})_{C_1,B_2} \cdots (K)_{B_p,C_p} (R^{m_{p+1}})_{C_p,\{t\}} \Bigg]. \tag{6}$$

Obviously, for any $A \subset \Lambda$, it is

$$\sum_{t \in \Lambda} \sum_m (R^m)_{A,\{t\}} = 1,$$

and it is evident from (4) and (5) that each nonzero summand of the second sum corresponds to a collection γ of the form (1) (and all such collections enter this sum) and equals $(-1)^{p(\gamma)}g_\gamma$.

Thus expansion (3) is proved.

Notice that, in case $-g_A \geqslant 0$ for all A, each summand from (6) is positive and the convergence of the series (3) follows from (24) and (25) in §1. To prove the absolute convergence of the series (3) for arbitrary g_A, the term $-g_A$ has to be replaced by $|g_A|$ in the above considerations. This does not change the norm of the operator K.

Corollary 1. *The expansion (16) in §1 follows immediately from the expansion (3) by putting the coefficient $D(A,\eta)$ equal to $(-1)^{|\eta|}$, for any nonempty collection $\eta = \{A_1, \ldots, A_p\}$, multiplied by the number of collections γ from the representation (3), for which $\eta = \eta(\gamma)$ and $D(A,\emptyset) = 1$.*

Remark 2. The representation (16) in §1 and expressions for the coefficients $D(A,\eta)$ can be obtained by a direct substitution of the series (16) in §1 to the equations (17) and (17′), §1. Hence the following recursive system of equations for $D(A,\eta)$ arises:

$$D(A,\eta) - D(A\backslash\{t_\Lambda\},\eta) + \sum_{\substack{B\in\eta \\ B\cap A=\{t_A\}}} D(A\cup(\bar{B}\cap\Lambda),\eta\backslash\{B\}) = 0,$$

$$|A| \geqslant 2, \quad \eta \neq \emptyset$$

$$D(A,\eta) + \sum_{B\in\eta} D(A\cup(\bar{B}\cap\Lambda),\eta\backslash\{B\}) = 0,$$

$$A = \{t\}, \quad \eta \neq \emptyset, \quad t = t_A \in B, \quad D(A,\emptyset) = 1.$$

It is useful to sum partially the series (3). We use the notation

$$b_R^{(\Lambda)}(A) = \sum_{\operatorname{supp}\gamma=R}^{(\Lambda,A)} (-1)^{p(\gamma)} g_\gamma, \tag{7}$$

with the summation taken over all γ defined in (3) and such that $\operatorname{supp}\gamma = R$. Then (3) can be rewritten in the form

$$f_A^{(\Lambda)} = 1 + \sum_{R\subseteq\Lambda} b_R^{(\Lambda)}(A). \tag{8}$$

§3 Limit Correlation Function and Cluster Expansion of Measures

We shall prove Theorem 1 of §1.

Lemma 1. *Provided the cluster condition (23) in §1 is satisfied, the limit (14) in §1 exists, and f_A obey the estimate (24) in §1, and*

$$f_A = 1 + \sum\nolimits^{(T,A)} (-1)^{p(\gamma)} g_\gamma = 1 + \sum_{R \subset T} b_R(A), \tag{1}$$

where

$$b_R(A) = \sum_{\gamma : \operatorname{supp} \gamma = R}^{(T,A)} (-1)^{p(\gamma)} g_\gamma = \lim_{\Lambda \nearrow T} b_R^{(\Lambda)}(A), \tag{2}$$

and in $\sum^{(T,A)}$, the sum is taken over all collections γ of the form (1) in §2 such that either $B_1 = A$ or B_1 lies below A in the graph $\mathcal{F}$. The series (1) converges absolutely.

Proof. Any summand of the series $\sum^{(T,A)} |g_\gamma|$ enters $\sum^{(\Lambda,A)} |g_A|$ for all sufficiently large Λ. Since these sums are uniformly bounded in Λ, the series $\sum^{(T,A)} |g_\gamma|$ converges. Hence (1) follows. It is obvious from (2) and (7) in §2 that

$$b_R(A) = b_R^{(\Lambda)}(A) \tag{3}$$

for all $A \subset \Lambda$ and R with $\overline{R} \subset \Lambda$. (2) follows from (3). The lemma is proved.

The above statements and (15) in §1 imply the statement of Theorem 1 of §1, with

$$\rho(\Gamma) = \left(\prod_{A \in \Gamma} g_A \right) f_{\overline{\overline{\Gamma}}}. \tag{4}$$

Lemma 2. *Let $\{\mu_\Lambda, \Lambda \subset T\}$ be a family of Gibbs modifications of measures μ_Λ, $\Lambda \subset T$, obtained by means of the energies $\{U_\Lambda, \Lambda \subset T\}$. Suppose that*

1) the set $\{\mu_\Lambda, \Lambda \subset T\}$ of measures is weakly locally compact;

2) the partition functions Z_Λ admit a cluster representation (1) in §1, with the estimates (2) and (3) in §1, and the cluster condition (23) in §1 fulfilled;

3) there is a set $G \subset C_0(\Omega)$ of local functions, with the linear span being everywhere dense in the space $C(\Omega)$ of all bounded continuous functions such that, for any function $F \equiv F_A \in G$, its mean $\langle F_A \rangle_{\mu_\Lambda} \equiv \langle F_A \rangle_\Lambda$, $A \subset \Lambda$, is of the form

$$\langle F_A \rangle_\Lambda = k_\emptyset(F_A) f_A^{(\Lambda)} + \sum_{R \subseteq \Lambda} k_R(F_A) f_{A \cup R}^{(\Lambda)}, \tag{5}$$

with $f_B^{(\Lambda)}$ being a correlation function and $k_R(F_A)$ some quantities obeying the estimate

$$|k_R(F_A)| < m(F_A) \sum_{\substack{\Gamma = \{B_1, \ldots, B_n\} \\ \overline{\overline{\Gamma}} = R}} \prod_{j=1}^{n} r_{B_i}^{(A)}. \tag{6}$$

Here, $m(F_A)$ is a constant depending only on A and F_A and not depending on R, and the summation is taken over all d-admissible collections $\Gamma = \{B_1, \ldots, B_n\}$ of finite subsets B_i, $|B_i| \geqslant 2$, $B_i \cap A \neq \emptyset$, $i = 1, \ldots, n$. Moreover, the numbers $r_B^{(A)} \geqslant 0$ satisfy the condition: for any $t \in T$ and any $n \geqslant 2$,

$$\sum_{\substack{B:t\in B\\|B|=n}} r_B^{(A)} < b_A \left(\frac{\lambda}{k}\right)^n, \tag{7}$$

with a constant $b_A > 0$ (depending on A) and the same parameters $0 < \lambda < k < \infty$ as in the estimates (2) and (3) in §1. Then the set $\{\mu_\Lambda, \Lambda \subset T\}$ admits a cluster expansion.

Proof. Define

$$b_R^{(\Lambda)}(F_A) = k_R(F_A) f_{R\cup A}^{(\Lambda)} \tag{8}$$

for $F_A \in G$.

We find, from (6) and (24) in §1, that the estimate (18, §1, Chapter 1) is fulfilled, with

$$C_R(F_A) = m(F_A)\frac{M^{|A|}}{M(1-\varepsilon_0)} \sum_{\Gamma=\{B_1,\ldots,B_n\}} \prod_{i=1}^{n} r_{B_i}^{(A)} M^{v|B_i|}. \tag{9}$$

Hence, with the help of (22) in §1 and (7), we obtain

$$\sum_R C_R(F_A) \leqslant m(F_A)\frac{M^{|A|}}{M(1-\varepsilon_0)} \left[\frac{b_A x_0^2(c)}{1-x_0(c)}\right]^{|A|} < \infty.$$

Further, (14) in §1 implies that

$$\lim_{\Lambda \nearrow T} b_R^{(\Lambda)}(F_A) = k_R(F_A) f_{A\cup R} \equiv b_R(F_A).$$

Thus the conditions (17), (18), and (19), §1, Chapter 1, are fulfilled. The lemma is proved.

Besides the cluster expansion of the means $\langle F_A\rangle_\Lambda$ with the quantities $b_R^{(\Lambda)}(F_A)$ of the form (8), the representation (5) and expansion (8) in §2 imply another expansion for the means:

$$\langle F_A\rangle_\Lambda \equiv \sum_{R\subseteq\Lambda} \tilde{b}_R^{(\Lambda)}(F_A), \tag{10}$$

with the quantities $\tilde{b}_R^{(\Lambda)}(F_A)$ being equal to

$$\tilde{b}_R^{(\Lambda)}(F_A) = \sum_{\substack{R',R''\subseteq\Lambda,\\ R'\cup R''=R}} k_{R'}(F_A) b_{R''}^{(\Lambda)}(A\cup R'). \tag{11}$$

Lemma 3. *Let the conditions 2) and 3) of the previous lemma be fulfilled. Then the quantities $\tilde{b}_R^{(\Lambda)}(F_A)$ fulfill the conditions (19) from §1 in Chapter 1.*

Proof. The existence of the limit

$$\lim_{\Lambda \nearrow T} \tilde{b}_R^{(\Lambda)}(F_A) = \tilde{b}_R(F_A) = \sum_{\substack{R',R'' \subset T \\ R' \cup R'' = R}} k_{R'}(F_A) b_{R''}(A \cup R')$$

follows from Lemma 1. Further,

$$\left|\tilde{b}_R^{(\Lambda)}(F_A)\right| \leqslant \sum_{R',R'',R' \cup R'' = R} |k_{R'}(F_A)| \, \bar{b}_{R''}^{(\Lambda)}(A \cup R')$$
$$\leqslant \sum_{R',R''} |k_{R'}(F_A)| \, \bar{b}_R(A \cup R') = \widetilde{C}_R(F_A),$$

where we have used the notation $\bar{b}_R^{(\Lambda)}(B) \geqslant b_R^{(\Lambda)}(B)$ for the quantities in the expansion (8) in §2 of the correlation function $\bar{f}_B^{(\Lambda)} > 0$ that is defined by the quantities $\{-|g_B|\}$ (compare with the proof of Lemma 1 of §2). The quantities $\bar{b}_R(B) = \lim_{\Lambda \nearrow T} \bar{b}_R^{(\Lambda)}(B)$ have an analogous sense.

Obviously,

$$\sum_R \widetilde{C}_R(F_A) = \sum_R |k_R(F_A)| \, \bar{f}_{A \cup R} < \infty,$$

as follows from the preceding lemma.

As we shall see below, the cluster expansion of the form (10) with the quantities (11) in some cases possesses a remarkable property. We shall need it only in §3, Chapter 4.

Definition. The cluster expansion of the means $\langle F_A \rangle_\mu$, $F_A \in G$, with μ being a measure on the space S^T,

$$\langle F_A \rangle_\mu = \sum_{R \subset T} b_R(F_A) \tag{12}$$

is called *regular* if the following conditions are fulfilled:

1) There is a family of finite subsets $\mathfrak{A}$ satisfying the conditions 1) and 2) from §4, Chapter 2, and such that in the expansion (12),

$$b_R(F_A) \neq 0 \tag{13}$$

only in case the set R is $\mathfrak{A}$-connected with respect to A (cf. §4, Chapter 2).

2) In case $F_A = F_{A_1} \cdot F_{A_2}, F_A, F_{A_1}, F_{A_2} \in G$, and $A_1 \cap A_2 = \emptyset$, $A_1 \cup A_2 = A$, and the set R is not $\mathfrak{A}$-connected with respect to $\{A_1, A_2\}$, and $b_R(F_A) \neq 0$, the equality

$$b_R(F_A) = b_{R_1}(F_{A_1}) \cdot b_{R_2}(F_{A_2}) \tag{14}$$

holds true, with

$$R_1 = \bigcup_{j:R'_j \cap A_1 \neq \emptyset} R'_j, \quad R_2 = \bigcup_{j:R'_j \cap A_2 \neq \emptyset} R'_j$$

and $R'_1, \dots, R'_p$ being the $\mathfrak{A}$-connected components of the set R (cf. §4, Chapter 2). Notice that $R_1 \cup R_2 = R$ because R is $\mathfrak{A}$-connected with respect to A (and, in addition, $R_1 \cap R_2 = \emptyset$).

In case, besides 1) and 2), condition 3) is fulfilled, the expansion (12) is called *exponentially-regular*:

3) For all $F_A \in G$ and all R that are $\mathfrak{A}$-connected with respect to A, the estimate

$$|b_R(F_A)| < C(F_A) \cdot \lambda^{d_R(\mathfrak{A}, \{A\})} \tag{15}$$

holds true, with $C(F_A)$ being a constant depending only on F_A, $0 < \lambda < 1$. Recall that $d_R(\mathfrak{A}, \{A\})$ (cf. (9), §4, Chapter 2) is the minimal cardinality of a collection $\Gamma = \{B_1, \dots, B_s\}$, $b_i \in \mathfrak{A}$, $i = 1, \dots, s$, such that $\{B_1, \dots, B_s, A\}$ is connected and $\widetilde{\Gamma} = R$.

A regular (or exponentially-regular) cluster expansion of the means $\langle F_A \rangle_\Lambda$, for a family of measures $\{\mu_\Lambda, \Lambda \subset T\}$, is defined similarly, assuming that the constants $C(F_A)$ and parameter λ in the estimate (15) do not depend on $\Lambda \subset T$.

Let a family of Gibbs modification $\{\mu_\Lambda, \Lambda \subset T\}$ be given, and let the partition function Z_Λ admit a cluster representation (11) with $d = 1$, where

$$k_A \neq 0, \quad A \subset T, \quad |A| \geqslant 2, \tag{16}$$

only in case A is $\mathfrak{A}$-connected ($\mathfrak{A}$ is a family of subsets satisfying the conditions 1) and 2) of §4, Chapter 2), and the estimate

$$\begin{aligned} |k_A| &\leqslant \bar{c}\bar{\lambda}^{d_A(\mathfrak{A})}, \quad |A| > 2, \\ |k_{\{t\}}| &\geqslant 1 \end{aligned} \tag{17}$$

is fulfilled, with $\bar{c} > 0$ and a sufficiently small constant $\bar{\lambda} < 1$. Then, as follows from Lemma 2 in §4, Chapter 2, the conditions (2) in §1 (with the parameter $\lambda = c\bar{\lambda}$ where $c = c(\mathfrak{A})$ is a constant) and (3) in §1, and also the cluster condition (23) in §1, are fulfilled, and consequently, there is a correlation function $f_A^{(\Lambda)}$ that admits an expansion (8) in §2. In addition, let for any function $F_A \in G$ (cf. Lemma 2), its mean $\langle F_A \rangle_\Lambda$ admit the expansion (5), with the quantities $k_R(F_A)$ satisfying the conditions 1), 2), and 3) from the definition of the exponentially-regular cluster expansion, where

$$|k_R(F_A)| < \bar{C}(F_A)(\bar{\lambda})^{d_R(\mathfrak{A}, \{A\})}, \tag{18}$$

with a constant $\bar{C}(F_A)$, depending only on F_A, and the same $\bar{\lambda}$ as in (17).

Lemma 4. *Under the above assumptions, the conditions (6) and (7) of Lemma 2 are fulfilled, and the expansion (10) is exponentially-regular.*

Proof. The estimate (6) is a consequence of (18), with

$$r_B^{(A)} = \begin{cases} (\bar{\lambda})^{d_B(\mathfrak{A})}, & \text{if } B \in \mathfrak{A} \text{ connected and } B \cap A \neq \emptyset, \\ 0 & \text{otherwise.} \end{cases}$$

Here, the estimate (7), with $\lambda = c\bar{\lambda}$, follows again from Lemma 2 in §4, Chapter 2. The condition 1) from the definition of a regular expansion follows immediately from formula (11) if we notice that it holds for the quantities $b_R^{(\Lambda)}(A)$ by (7) in §2. The condition 2) follows from (11), the analogous condition 2) for the quantities $k_R(F_A)$, and from the fact that the quantities $b_R^{(\Lambda)}(A)$ possess a property analogous to (14):

$$b_R^{(\Lambda)}(A_1 \cup A_2) = b_{R_1}^{(\Lambda)}(A_1) \cdot b_{R_2}^{(\Lambda)}(A_2),$$

where $A_1 \cap A_2 = \emptyset$ and the set R is not $\mathfrak{A}$-connected with respect to $\{A_1, A_2\}$ (but $\mathfrak{A}$-connected with respect to $A_1 \cup A_2$), R_1 and R_2 are defined in (14). In order to verify the condition (15) of the exponential regularity, we put

$$k_A = (\kappa^{-1}\bar{\lambda})^{d_A(\mathfrak{A})}\bar{k}_A$$

and

$$k_R(F_A) = (\kappa^{-1}\bar{\lambda})^{d_R(\mathfrak{A},\{A\})}\bar{k}_R(F_A),$$

where κ, $0 < \kappa < 1$, is a constant that will be defined below. Notice that $\bar{k}_A$ and $\bar{k}_R(F_A)$ satisfy the estimates (17) and (18), respectively, in which λ is replaced by κ. Choosing now $\kappa \ll 1$ sufficiently small, but such that $\kappa^{-1}\lambda$ is also sufficiently small, we verify the cluster condition for the quantities $\bar{k}_A$, and the quantities $\bar{k}_R(F_A)$ satisfy the conditions (7) and (8). Moreover, it follows from (7) in §2 that

$$|b_R(A)| < (\kappa^{-1}\lambda)^{d_R(\mathfrak{A},\{A\})}\bar{b}_R(A), \tag{19}$$

where the quantities $\bar{b}_R(A) > 0$ define an expansion of the correlation function $\bar{f}_A$ defined with the help of $\bar{g}_B = -\left|\bar{k}_B\right|$. Further, we find from (11) and (19) that

$$\left|\tilde{b}_R(F_A)\right| < (\kappa^{-1}\lambda)^{d_R(\mathfrak{A},\{A\})}\bar{b}_R(F_A),$$

where $\bar{b}_R(F_A)$ are defined by (11), in which the quantities $k_{R''}(F_A)$ and $b_{R'}(A)$ are replaced by $\bar{k}_{R''}(F_A)$ and $\bar{b}_{R'}(A)$, respectively. It remains to use the considerations from Lemma 3. The lemma is proved.

Remark 1. In some cases, the quantities k_A which enter the cluster representation (1) in §1 of the partition function and the quantities $k_R(F_A)$ in the expansion (5) for the means $\langle F_A\rangle_\Lambda$ may depend on the set $\Lambda : k_A = k_A^{(\Lambda)}$ and

$k_R(F_A) = k_R^{(\Lambda)}(F_A)$. This is, for example, the case of the cluster expansion of the measures $\{\mu_\Lambda, y^{\partial\Lambda}\}$, where $y^{\partial\Lambda}$ is a boundary configuration; see §2, Chapter 1. However, if the quantities $k_A^{(\Lambda)}$ satisfy, uniformly in Λ, the cluster estimate (2) and the condition (3) in §1, and the quantities $k_A^{(\Lambda)}(F_A)$ the estimates (6) and (7), and if there exist limits

$$k_A = \lim_{\Lambda\nearrow T} k_A^{(\Lambda)}, \quad k_R(F_A) = \lim_{\Lambda\nearrow T} k_R^{(\Lambda)}(F_A), \tag{20}$$

then all the constructions of this chapter remain valid.

§4 Cluster Expansion and Asymptotics of Free Energy. Analyticity of Correlation Functions

In the preceding sections, we assumed that the partition functions Z_Λ do not vanish. Now we prove Lemma 2 of §1, i.e., we shall prove that the condition $Z_\Lambda \neq 0$ is a consequence of the cluster representation (1) in §1.

In a finite-dimensional complex space $\widetilde{\mathcal{B}}^{(\Lambda)} \subset \mathcal{B}^{(\Lambda)}$ of the vectors $\{g_B;\ B \subset \Lambda, |B| \geqslant 2\}$, we consider a polydisc

$$\prod_{\kappa,c}^{(\Lambda)} = \left\{ \sum_{B:t\in B,|B|=n} |g_B| < c\kappa^n; t \in \Lambda, n \geqslant 2 \right\}. \tag{1}$$

As shown in §2, the series (3) converges uniformly in Λ and absolutely in any domain $\prod_{\kappa,c}^{(\Lambda)}$ with $\kappa < 1$, $c < c_0(\kappa)$ (cf. 23′ in §1) and defines an analytic function $f_A^{(\Lambda)}(\{g_B\})$ in this domain.

We introduce a polynomial in $\{g_B, B \subset \Lambda\}$:

$$Q_\Lambda = Z_\Lambda \Big/ \prod_{t\in\Lambda} k_{\{t\}} = \sum_{\Gamma=\{B_1,\ldots,B_n\}} \prod_{i\in 1}^{n} g_{B_i}, \tag{2}$$

where the sum is taken over all d-admissible collections Γ of the sets $B_i \subset \Lambda$. For all values $\{g_B\}$ for which $Q_\Lambda \neq 0$, the equality

$$Q_\Lambda \cdot f_\Lambda^{(\Lambda)} = 1, \tag{3}$$

following from (12), §1, holds true. In particular, the equality (3) is fulfilled for all positive real collections $\{g_B\}$ from the domain $\prod_{\kappa,c}^{(\Lambda)}$. The set $\widetilde{\prod}_{\kappa,c}^{(\Lambda)} \subset \prod_{\kappa,c}^{(\Lambda)}$ of such collections is the set of unicity for the domain $\prod_{\kappa,c}^{(\Lambda)}$.

Recall that the set D_0, contained in a domain $D \subset \mathbf{C}^n$, is called a set of unicity for D if any two analytic functions on D coinciding on D_0 coincide on all D.

Lemma 1. *Each domain D_0 in the real space $R^n=\{(z_1,\ldots,z_n);\mathrm{Im}\, z_i=0, i=1,\ldots,n\}\subset \mathbf{C}^b$ is a set of unicity for any domain $C\subset\mathbf{C}^n$ containing D_0.*

Proof. We proceed by induction on the number n of variables. In the case $n=1$, the statement is well known. For an arbitrary n, it is enough to consider the case when D is a convex neighbourhood of D_0 that is entirely contained in the tube $\{(z_1,\ldots,z_n),(\mathrm{Re}\, z_1,\ldots,\mathrm{Re}\, z_n)\in D_0\}$. Considering the cross sections of the domain D by hyperplanes $z_n=$ const., we reduce our statement for the case of n variables to the case of $n-1$ variables. The lemma is proved.

By Lemma 1, the equality (3) holds true in the whole domain $\prod_{\kappa,c}^{(\Lambda)}$. This implies Lemma 2 of §1.

The proved lemma, and the fact that $Q_\Lambda(g_B\equiv 0)=1$, imply that in the domain $\prod_{\kappa,c}^{(\Lambda)}$, an analytic function F_Λ is defined:

$$F_\Lambda(\{g_B\})=\ln Q_\Lambda(\{g_B\}), \tag{4}$$

$$F_\Lambda(\{g_B\equiv 0\})=0. \tag{5}$$

The function F_Λ admits an expansion in a power series, in the variables $\{g_B\}$, which converges in the polydisc $\prod_{\kappa,c}^{(\Lambda)}$.

Lemma 2. *This power series is of a form analogous to (3), §2:*

$$F_\Lambda=\sum_{t\in\Lambda}\sum_{\gamma}^{(\Lambda,\{t\})}((-1)^{p(\gamma)}g_\gamma)\Big/\sum_{A\in\eta(\gamma)}|A|, \tag{6}$$

with the sum $\sum_\gamma^{(\Lambda,\{t\})}$ taken over all collections γ of the form (1) in §2 and defined in Lemma 1 of §2.

Corollary 1. *If the expansion (6) is rewritten in the form*

$$F_\Lambda=\sum_\eta d_\eta^{(\Lambda)}g_\eta, \tag{7}$$

by analogy with (16) in §1, then (6) implies that the coefficients $d_\eta^{(\Lambda)}$ are of the form

$$d_\eta^{(\Lambda)}=\sum_{t\in\Lambda}D^{(\Lambda)}(\eta,\{t\})\Big/\sum_{A\in\eta}|A|, \tag{8}$$

with $D^{(\Lambda)}$ being the coefficients from the expansion (16), §1.

Corollary 2. *By virtue of (2) and (6), we find that*

$$\ln Z_\Lambda=\sum_{t\in\Lambda}\left[\ln k_{\{t\}}+\sum_\gamma^{(\Lambda,\{t\})}\left((-1)^{p(\gamma)}g_\gamma\Big/\sum_{A\in\eta(\gamma)}|A|\right)\right]. \tag{9}$$

Proof of Lemma 2. Fix a collection of parameters $\{g_A^0 : A \subset \Lambda\} \in \prod_{\kappa,c}^{(\Lambda)}$ and consider the one-parameter family of collections $\{g_A^s, A \subset \Lambda\} \in \prod_{\kappa,c}^{(\Lambda)}$:

$$g_A^s = g_A^0 e^{-s|A|}, \quad 0 \leqslant s < \infty. \tag{10}$$

Letting

$$F_\Lambda^s = F_\Lambda(\{g_A^s\}), \quad Q_\Lambda^s = (\{g_A^s\}), \quad g_\eta^s = \prod_{A\in\eta} g_A^s,$$

and using (10) and (14) of §1, and (3) of §2, we get

$$\begin{aligned}\frac{dF_\Lambda^s}{ds} &= \frac{1}{Q_\Lambda^s}\sum_\eta g_\eta^s\left(-\sum_{A\in\eta}|A|\right) = -\sum_{A\subset\, L, |A|\geqslant 2} |A|\rho^{(\Lambda)}(\{A\}) \\ &= \sum_{A\subset\Lambda} |A|(-g_A^s)\cdot f_{\bar{A}\cap\Lambda}^{(\Lambda)}(\{g_B^s\}) = \sum_{t\in\Lambda}\cdot\sum_\gamma^{(\Lambda,\{t\})}(-1)^{p(\gamma)} g_\gamma^s.\end{aligned} \tag{11}$$

Integrating (11) over s from 0 to ∞, and using the fact that $F_\Lambda^s \to 0$ as $s \to \infty$ and $\int_0^\infty g_\eta^s\, ds = g_\mu^0/\sum_{A\in\eta}|A|$, we obtain (6); Lemma 2 is proved.

We introduce the quantities

$$\begin{aligned}\hat{b}_R^{(\Lambda)}(t) &= \sum_{\gamma:\operatorname{supp}\gamma=R}^{(\lambda,\{t\})}(-1)^{p(\gamma)}\left(g_\gamma \Big/ \sum_{A\in\eta(\gamma)}|A|\right), \\ \hat{b}_R(t) &= \sum_{\gamma:\operatorname{supp}\gamma=R}^{(T,\{t\})}(-1)^{p(\gamma)}\left(g_\gamma \Big/ \sum_{A\in\eta(\gamma)}|A|\right)\end{aligned} \tag{12}$$

(the sum $\sum^{(T,\{t\})}$ is defined in Lemma 1 of §3).

It is obvious that

$$\hat{b}_R^{(\Lambda)}(t) = \hat{b}_R(t) \tag{13}$$

for any t and any R such that $t \in R$, $\bar{R} \subset \Lambda$. Here,

$$F_\Lambda = \sum_{t\in\Lambda}\sum_{R:t\in R,\bar{R}\subseteq\Lambda}\hat{b}_R^{(\Lambda)}(t) = \sum_{t\in\Lambda}\sum_{R:t\in R}\hat{b}_R(t) + \Delta F_\Lambda, \tag{14}$$

where the "boundary term" is

$$\Delta F = \sum_{t\in\Lambda}\left[\sum_{\substack{R:t\in R\subset\Lambda,\\ \bar{R}\cap(T\setminus\Lambda)\neq\emptyset}}(\hat{b}_R^{(\Lambda)}(t) - \hat{b}_R(t)) - \sum_{R:t\in R\not\subset\Lambda}\hat{b}_R(t)\right]. \tag{15}$$

The first sum on the right-hand side of (14) is called the bulk part of F_Λ. Such a notion is justified by the following theorem, which we formulate in a general form.

Let G be a transitive group of transformations of a set T that preserve the metric ρ, and let the small subgroup of each point be finite. These conditions are in particular fulfilled if $T = Z^\nu$, and $G = Z^\nu$ is the group of translations in Z^ν.

We use r to denote any class of congruent (with respect to G) sets from T, and for any $\Lambda \subset T$, let $S_\Lambda(r)$ denote the number of such sets $R \in r$ that $R \cap \Lambda \neq \emptyset$, $\bar{R} \cap (T \backslash \Lambda) \neq \emptyset$.

We say that an increasing sequence of finite subsets $\Lambda_n \subset T$ tends to T as $n \to \infty$ in the sense of van Hove if $\bigcup \Lambda_n = T$, and for any class r,

$$S_{\Lambda_n}(r)/\left|\Lambda_n\right| \to 0, \quad n \to \infty. \tag{16}$$

Theorem 3. *In the cluster representation (1) in §1, let the quantities k_A be invariant under G, i.e.,*

$$k_{gA} = k_A \tag{17}$$

for all $g \in G$.

Then for any sequence Λ_n tending to T in the sense of van Hove, there is a limit

$$f = \lim \frac{\ln Z_{\Lambda_n}}{|\Lambda_n|} = \sum_{R:t\in R} \hat{b}_R(t) + \ln k, \tag{18}$$

where $t \in T$ is an arbitrary fixed point and $k \equiv k_{\{t\}}$.

Proof. Notice that, under condition (17), the sums $\sum_{R:t\in R} \hat{b}_R(t)$ are identical for all t, and at the same time, (14) implies that

$$\frac{F_\Lambda}{|\Lambda|} = \sum_{R:t\in R} \hat{b}_R(t) + \frac{\Delta F}{|\Lambda|}. \tag{19}$$

The second sum in (15) admits the estimate

$$\begin{aligned} &\left| \sum_{t\in\Lambda_n} \sum_{R:t\in R\not\subset\Lambda_n} \hat{b}_R(t) \right| \\ &\qquad \leqslant \sum_{\substack{R\not\subset\Lambda,\\ R\cap\Lambda\neq 0}} \sum \frac{1}{|R|} \sum_{\gamma:\operatorname{supp}\gamma=R}^{(T,\{t\})} |g_\gamma| < \sum_r S_{\Lambda_n}(r) C_r, \end{aligned} \tag{20}$$

with

$$C_r = C_R = \sum_{t\in R} \frac{1}{|R|} \sum_{\gamma:\operatorname{supp}\gamma=R}^{(T,\{t\})} |g_\gamma|$$

for any $R \in r$. Since

$$\sum_r \kappa_r C_r = \sum_{R: t_0 \in R} C_R \leqslant \sum_{\gamma'}^{(T,\{t_0\})} |g_\gamma| < \infty, \tag{21}$$

where κ_r is the number of sets from the class r containing an arbitrary fixed point $t_0 \in T$ ($\kappa_r \leqslant |R| \cdot |G_{t_0}|$, G_{t_0} is the small group of the point t_0), we obtain, from (21), the estimate $S_{\Lambda_n}(r) \leqslant |\Lambda_n| \, \kappa_r$, and from (16) that the sum on the left-hand side of (20) is of the order $o(|\Lambda_n|)$. Similarly, the first sum in (15) can be estimated. The theorem is proved.

Let now the quantities $k_A = k_A(z_1, \ldots, z_m)$ be analytic functions of the complex parameters $z_1, \ldots, z_m$ in a domain $D \subset \mathbf{C}^m$, and let the conditions (1), (2), (3), and (23) in §1 be fulfilled uniformly in D.

Theorem 4. *Under the assumptions formulated above, the correlation functions $\rho^{(\Lambda)}(\Gamma)$, $\rho(\Gamma)$, and also $\ln Z_\Lambda$, and (if the conditions of Theorem 3 are fulfilled) the quantity f are analytic functions on D.*

The proof follows from the representations (3), §2; (15), §1; (1), §3; (3), §3; (9), (12), and (18) of this section.

§5 Regions of Cluster Expansions for the Ising Model

This section is a continuation of §0, Chapter 1, and at the same time, serves as an introduction to the following chapters. Here, we illustrate, on a simple example, how the cluster expansion of partition functions is obtained. We consider the Ising model and use the notations from §0, Chapter 1.

We distinguish four regions in the plane of parameters (β, h) with different types of a corresponding cluster representation of the partition function for the Ising Model. It turns out to be suitable to consider a cluster expansion not of the original partition function but of a quantity which differs from it by a simple factor.

I. High-Temperature Region (Small β)

We obtain here a cluster representation of the partition function Z_Λ with empty boundary conditions (see §2, Chapter 1).

For $\beta = 0$, the limit measure $\mu_{h,\beta=0}$ is a measure with independent values in the case of the Ising model (see §0, Chapter 1). It is natural to suppose that, for each small β, the limit Gibbs measure defines a field that is close to the independent one, and the cluster representation of the partition function can be obtained by expanding $\exp(-U_\Lambda)$ in a series in β. Joining the terms with the same collections of potential supports in this representation and averaging with the help of the measure μ_0, we obtain the required representation. Technically, this is to be carried out in the following way: Notice that ($\langle\cdot\rangle_{\mu_h,\beta=0} = \langle\cdot\rangle_h$)

$$Z_\Lambda = \left\langle \exp\left(\beta \sum_{t,t'}^{(\Lambda)} \sigma_t\sigma_{t'}\right)\right\rangle_h (e^h + e^{-h})^{|\Lambda|},$$

and deduce the cluster representation of

$$\left\langle \exp\left(\beta \sum_{t,t'}^{(\Lambda)} \sigma_t\sigma_{t'}\right)\right\rangle_h ,$$

with $\sum_{t,t'}^{(\Lambda)}$ denoting the sum over the pairs of nearest neighbours in Λ:

$$\begin{aligned}\left\langle \exp\left(\beta \sum_{t,t'}^{(\Lambda)} \sigma_t\sigma_{t'}\right)\right\rangle_h &= \left\langle \prod_{t,t'}^{(\Lambda)} [\exp(\beta\sigma_t\sigma_{t'}) - 1 + 1]\right\rangle_h \\ &= 1 + \sum_{\Gamma} \left\langle \prod_{\{t,t'\}\in\Gamma} [\exp(\beta\sigma_t\sigma_{t'}) - 1]\right\rangle_h , \end{aligned} \tag{1}$$

and with the sum $\sum_\Gamma$ taken over unordered nonempty collections Γ of different pairs $\{t,t'\}$ of nearest neighbours.

By splitting Γ up into connected components $\Gamma_1, \ldots, \Gamma_s$ and by using the independence of the random variables $\prod_{\{t,t'\}\in\Gamma_i}[\exp(\beta\sigma_t\sigma_{t'}) - 1]$, $i = 1, \ldots, s$, we obtain the required cluster representation (4) of §1, in which

$$k_\Gamma = \left\langle \prod_{\{t,t'\}\in\Gamma} [\exp(\beta\sigma_t\sigma_{t'}) - 1]\right\rangle_h . \tag{2}$$

Thus the cluster here is interpreted as a connected collection of pairs of nearest neighbours with the support $\widetilde{\Gamma}$, and the cluster estimate (5) in §1 follows from (2) and Lemma 1, §4, Chapter 2.

II. Region of Large Magnetization (Large $|h|$)

We shall investigate the case $h > 0$, for large h, and we get a cluster representation of the partition functions $Z_{\Lambda,+}$ for the (+)-boundary conditions. Similarly, the symmetric case $h < 0$ and the representation of the partition function Z_Λ with (−)-boundary conditions can be examined.

In our region, the values $\sigma_t = -1$ are attained with a small probability (more exactly, each value $\sigma_t = -1$ produces a small factor $z = e^{-h}$ in the corresponding term in the partition function). In this connection, we define a cluster to be the maximal 1-connected set of such points $t \in Z^\nu$ that $\sigma_t = -1$. More exactly, for each configuration σ^Λ, let

$$B_-(\sigma^\Lambda) = \{t \in \Lambda : \sigma_t = -1\}, \quad B(\sigma^\Lambda) = \overline{B_-(\sigma^\Lambda)},$$

with

$$\bar{B} = \{t \in Z^\nu : \rho(t, B) \leqslant 1\}$$

for any $B \subset Z^\nu$.

For any $B \subset Z^\nu$, we use the notation $\gamma(B)$ to denote the set of bonds (i.e., of pairs $\{t, t'\}$ of nearest neighbours) such that $t' \in \partial_i B$, $t \in \operatorname{Int} B$, with $\partial_i B = \{t \in B, \rho(t, Z^\nu \backslash B) = 1\}$, $\operatorname{Int} B = B \backslash \partial_i B$. Then the partition function $Z_{\Lambda,+}$ may be written in the form ($\langle \cdot \rangle_0 = \langle \cdot \rangle_{\beta=0,h=0}$)

$$Z_{\Lambda,+} = \exp\left[h|\Lambda| + \beta\left|\Lambda^{(2)}\right|\right] \left\langle \exp\left\{-2\beta\left|\gamma(B(\sigma^\Lambda))\right| - 2h\left|\operatorname{Int} B(\sigma^\Lambda)\right|\right\}\right\rangle_0, \tag{3}$$

with $\Lambda^{(2)}$ being the set of all bonds $\{t, t'\}$ such that $t \in \Lambda$, $t' \in \bar{\Lambda}$.

Now we get a cluster expansion for

$$2^{|\Lambda|} Z_{\Lambda,+} \exp\left\{-h|\Lambda| - \beta\left|\Lambda^{(2)}\right|\right\} = 1 + \sum_{B \subset \bar{\Lambda}} \exp\{-2\beta\left|\gamma(B)\right| - 2h\left|\operatorname{Int} B\right|\}, \tag{4}$$

with the summation taken over all nonempty $B \subset \bar{\Lambda}$ such that each of their 1-connected components has a nonempty interior. Let

$$k_B = \exp\{-2\beta\left|\gamma(B)\right| - 2h\left|\operatorname{Int} B\right|\} \tag{5}$$

for each 1-connected set B with a nonempty interior. Let $k_{\{t\}} = 1$ and $k_B = 0$ for other B. Then we obtain the cluster representation (1) of §1 with $d = 2$ from (4).

Noticing that $|\operatorname{Int} B| \geqslant |B|/(2\nu + 1)$, $|\gamma(B)| \leqslant 2\nu\,|\operatorname{Int} B|$ and that the number of 1-connected sets of the cardinality n that contain a fixed point t does not exceed const $(\overline{C})^n$, $\overline{C} = \overline{C}(\nu)$ (cf. Lemma 4, §4, Chapter 2), we obtain the estimate (2) in §1 with

$$c = 1, \quad \gamma = \overline{C} \exp\{-2(h - 2\nu\beta)/(2\nu + 1)\}. \tag{6}$$

For sufficiently large $h > 0$, the cluster condition (23) in §1 is fulfilled.

III. Low-Temperature Ferromagnetic Region of the First-Order Phase Transitions ($h = 0$, $\beta > 0$ and Large)

We use again the notations of §0, Chapter 1, and assume $\nu = 2$ for simplicity.

In the capacity of the set $\widehat{T}$ (see Remark 1, §1), we take the set of all bonds of the dual lattice $\tilde{Z}_l$ and put $\hat{\Lambda} = \tilde{\Lambda}$, with $\tilde{\Lambda}$ being the set of bonds of the dual lattice that intersect bonds from $\Lambda^{(2)}$ (i.e., that lie completely inside $\bar{\Lambda} = \Lambda \cup \partial\Lambda$). We call a cluster an arbitrary contour $\Gamma \subset \hat{\Lambda}$ (i.e., a closed connected polygon consisting of the bonds from $\hat{\Lambda}$). Then the cluster representation of the quantity $Z_{\Lambda,+}e^{-\beta|\hat{\Lambda}|}$ takes the form

$$Z_{\Lambda,+}e^{-\beta|\hat{\Lambda}|} = \sum_{\sigma^\Lambda} \exp\left\{-2\beta\left|\gamma(\sigma^\lambda\right|\right\} = \sum_{\{\Gamma_1,\dots,\Gamma_n\}} \prod_{i=1}^{n} \exp\{-2\beta\,|\Gamma_i|\}, \qquad (7)$$

with $\gamma(\sigma^\Lambda) \subset \hat{\Lambda}$ being the boundary of the configuration σ^Λ, and the last sum in (7) taken over all collections of mutually nonintersecting contours. Thus, (7) yields the desired cluster expansion, in which

$$k_\Gamma = \begin{cases} \exp\{-2b|\Gamma|\}, & \text{if } \Gamma \subset \hat{\Lambda} \text{ is a contour,} \\ 1, & |\Gamma| = 1,\ \Gamma \subset \hat{\Lambda}, \\ 0 & \text{for other subsets } \Gamma \subset \hat{\Lambda}. \end{cases}$$

The cluster estimate (2) of §1 for these quantities k_Γ may be proved by analogy with the preceding cases.

IV. Low-Temperature Ferromagnetic Region of Uniqueness ($h \neq 0$, $\beta > 0$ and Large)

We examine the case $\nu = 2$, as before, and get here a cluster expansion for the quantity

$$\begin{aligned} \tilde{Z}_{\Lambda,+} &= Z_{\Lambda,+} \exp\left\{-\beta\left|\Lambda^{(2)}\right| - h|\Lambda|\right\} \\ &= \sum_{\sigma^\Lambda \in \Omega_\Lambda^{(+)}} \exp\left\{-2\beta\left|\gamma(\sigma^\Lambda)\right| - 2h\left|B_-(\sigma^\Lambda)\right|\right\} \end{aligned} \qquad (8)$$

for $h > 0$. Here, as above, $B_-(\sigma^\Lambda) \subset \Lambda$ is the set of sites in Λ, in which the configuration σ^Λ takes the values $\sigma = -1$, $\gamma(\sigma^\Lambda) \subset \tilde{\Lambda}$ is the boundary of the configuration σ^Λ, $\Omega_\Lambda^{(+)} \subset \Omega$ is the set of the configurations in Λ with the boundary consisting only of contours, and the set $B_-(\sigma^\Lambda)$ lies inside $\gamma(\sigma^\Lambda)$ (i.e., any site $t \in B_-(\sigma^\Lambda)$ is separated from ∂A by a contour $\Gamma \subset \gamma(\sigma^\Lambda)$). The set $\Omega_\Lambda^{(-)} \subset \Omega_\Lambda$ is defined similarly. We use, for any contour Γ, the notation

$$\mathring{\Gamma} = \operatorname{Int}\Gamma \backslash \partial_i(\operatorname{Int}\Gamma),$$

with $\operatorname{Int}\Gamma \subset Z^2$ being the set of sites encircled by the contour Γ. Obviously, $\partial\mathring{\Gamma} = \partial_i(\operatorname{Int}\Gamma)$. Now we choose $\widehat{T}$ and $\hat{\Lambda}$ just as in the preceding case. A contour $\Gamma' \subset \widehat{T}$ is called internal with respect to a contour $\Gamma \subset \widehat{T}$ if Γ' lies in $\operatorname{Int}\Gamma$ and there are no sites of $\partial\Lambda$ inside Γ'. Then it is simple to see that

$$\tilde{Z}_{\Lambda,+} = \sum_{\{\Gamma_1,\dots,\Gamma_n\}}' \exp\left(-2\beta\sum_{i=1}^{n}|\Gamma_i| - 2h\sum_{i=1}^{n}\left|\partial\mathring{\Gamma}_i\right|\right)\prod \tilde{Z}_{\mathring{\Gamma}_i,-}, \tag{9}$$

with the sum taken over all possible collections of mutually nonintersecting contours $\Gamma_i \subset \hat{\Lambda}$, none of which is internal with respect to the remaining ones. Here, we use $\tilde{Z}_{\Lambda,-}$, $\Lambda \subset Z^2$, to denote the sum analogous to (8) taken only over the set $\Omega_\Lambda^{(-)} \subset \Omega_\Lambda$ of configurations. Defining for each contour $\Gamma \subset \widehat{T}$,

$$k_\Gamma = \exp\left\{-2\beta|\Gamma| - 2h\left|\partial\mathring{\Gamma}\right|\right\}\frac{\tilde{Z}_{\mathring{\Gamma},-}}{\tilde{Z}_{\mathring{\Gamma},+}}, \tag{10}$$

we get from (9) that

$$\tilde{Z}_{\Lambda,+} = \sum_{\{\Gamma_1,\dots,\Gamma_n\}}' \prod_{i=1}^{n} k_{\Gamma_i}\tilde{Z}_{\mathring{\Gamma}_i,+}. \tag{11}$$

Using the expansion (11) repeatedly for each sum $\tilde{Z}_{\mathring{\Gamma}_i,+}$, we finally obtain

$$\tilde{Z}_{\Lambda,+} = \sum_{\gamma=\{\Gamma_1,\dots,\Gamma_n\}} \prod k_\Gamma, \tag{12}$$

with the sum taken over all collections γ of mutually disjoint contours.

Lemma 1. *For any finite set Λ and $h > 0$, the inequality*

$$\tilde{Z}_{\Lambda,-}/\tilde{Z}_{\Lambda,+} \leqslant 1 \tag{13}$$

holds true.

Proof. It is $\tilde{Z}_{\Lambda,-} = \tilde{Z}_{\Lambda,+}$ for $h = 0$. Further,

$$\frac{d}{dh}\left(\frac{\tilde{Z}_{\Lambda,-}}{\tilde{Z}_{\Lambda,+}}\right) = \frac{\tilde{Z}_{\Lambda,-}}{\tilde{Z}_{\Lambda,+}}\left(\frac{\left(\tilde{Z}_{\Lambda,-}\right)_h'}{\tilde{Z}_{\Lambda,-}} - \frac{\left(\tilde{Z}_{\Lambda,+}\right)_h'}{\tilde{Z}_{\Lambda,+}}\right). \tag{14}$$

As easily follows from (8),

$$\left(\tilde{Z}_{\Lambda,+}\right)_h'/\tilde{Z}_{\Lambda,+} = -2\langle B_-(\sigma^\Lambda)\rangle_{\Lambda,+} = \sum_{t\in\Lambda}\langle\sigma_t\rangle_{\Lambda,+} - |\Lambda|, \tag{15}$$

with $\langle\cdot\rangle_{\Lambda,+}$ being the mean value under the distribution with (+)-boundary conditions in Λ. Since, under the mapping $\sigma^\Lambda \mapsto -\sigma^\Lambda$, it is $\Omega_{\Lambda,-} \to \Omega_{\Lambda,+}$ and $B_-(\sigma^\Lambda) = \Lambda \backslash B_-(-\sigma^\Lambda)$, we get

$$\left(\tilde{Z}_{\Lambda,-}\right)'_h / \tilde{Z}_{\Lambda,-} = -2|\Lambda| + 2\langle B_-(\sigma^\Lambda)\rangle_{\Lambda,+} = -\sum \langle \sigma_t \rangle_{\Lambda,+} - |\Lambda|. \tag{16}$$

As $\langle \sigma_t \rangle_{\Lambda,+} > 0$ for $h > 0$ (the Griffith inequality, cf. (13), §0, Chapter 1), we find from (14), (15), and (16) that for $h > 0$,

$$\frac{d}{dh}\left(\frac{\tilde{Z}_{\Lambda,-}}{\tilde{Z}_{\Lambda,+}}\right) \leqslant 0.$$

Hence (13) follows.

The cluster estimates (2) in §1 for the quantities k_Γ follow easily from (13) and (10).

§6 Point Ensembles

We shall preserve here the notations of §1, Chapter 1, and consider a pure point field in the space Q in which a continuous measure is given (i.e., such that the measure $\lambda(\{t\})$ of any singleton, $t \in Q$, equals zero). The generalization of the constructions occurring here to the case of labelled fields does not cause any trouble. The case of a purely discrete measure λ reduces to an examination of fields in a countable set.

On the set $\Omega_{\text{fin}} \subset \Omega$ of finite configurations (subsets) $x \subset Q$, let a function $k(x)$ be given so that, for any bounded domain $\Lambda \subset Q$,

$$\iint\limits_{y \in Q(\Lambda), x \in Q_{\text{fin}}} k(x \cup y)\, d\nu(x)\, d\nu(y) < \infty. \tag{1}$$

Recall that $\Omega(\Lambda)$ denotes the set of configurations in the domain Λ, and the measure ν on the space Ω_{fin} (or its restriction $\nu|_\Lambda$ to the space $\Omega(\Lambda) \subset \Omega_{\text{fin}}$) is defined by formula (8), §1, Chapter 1.

It follows from (1) that, for any bounded domain Λ, it is

$$\int\limits_{\Omega(\Lambda)} |k(y)|\, d\nu(y) < \infty, \tag{1a}$$

and, for ν-almost all $y \in \Omega_{\text{fin}}$, the integral fulfills

$$\int\limits_{\Omega_{\text{fin}}} |k(y \cup x)|\, d\nu(x) < \infty. \tag{1b}$$

Assume that the densities $p_\Lambda = \frac{d\mu_\Lambda}{d\nu}$ of the Gibbs modifications $\{\mu_\Lambda;$ $\Lambda \subset Q$ is a bounded domain$\}$ with respect to the measure $\nu = \nu|_\Lambda$ on the subspace $\Omega(\Lambda)$ may be expressed in the form

$$p_\Lambda(x) = Z_\Lambda^{-1} \sum\nolimits^{(x)} k(x_1)\dots k(x_n), \quad x \neq \emptyset,$$

$$p_\Lambda(\emptyset) = Z_\Lambda^{-1}, \tag{2}$$

with $\sum^{(x)}$ denoting the sum over all partitions $x = x_1 \cup x_2 \cup \dots \cup x_n$ to nonempty subsets $x_i \subset x$, $i = 1, \dots, n$; $n = 1, \dots, |x|$, and the partition function equals

$$Z_\Lambda = \int\limits_{\Omega(\Lambda)} \left(\sum\nolimits^{(x)} k(x_1)\dots k(x_n) \right) d\nu(x). \tag{3}$$

This representation of Z_Λ is also called the cluster representation.

Lemma 1. *The following formula*

$$\int\limits_{\Omega_{\text{fin}}} F(x) \left(\sum_{(x_1,\dots,x_p)} \varphi_1(x_1)\dots\varphi_p(x_p) \right) d\nu$$

$$= \overbrace{\int\limits_{\Omega_{\text{fin}}} \int\limits_{\Omega_{\text{fin}}}}^{p \text{ times}} F(x_1 \cup \dots \cup x_p)[\varphi_1(x_1)\dots\varphi_p(x_p)]\, d\nu(x_1)\dots d\nu(x_p) \tag{4}$$

is valid under the assumption that one of the integrals in (4) converges absolutely. Here, $F, \varphi_1, \dots, \varphi_p$ are arbitrary functions defined on the space Ω_{fin}, and $\sum_{(x_1,\dots,x_p)}$ is taken over all ordered collections of nonempty mutually disjoint subsets $x_i \subset x$, $i = 1, \dots, p$, such that $\cup x_i = x$.

The proof of this lemma follows easily from the definition of the measure ν; see §1, Chapter 1 (for more details, see [45]).

Using formula (4), inequality (1a), and the representation (3), we easily find that

$$Z_\Lambda = \sum_{n \geqslant 1} \frac{1}{n!} \int\limits_{\Omega(\Lambda)} \left(\sum_{\substack{(x_1,\dots,x_n) \\ \bigcup_{i=1}^n x_i = x}} k(x_1)\dots k(x_n) \right) d\nu(x)$$

$$= 1 + \sum_{n \geqslant 1} \frac{1}{n!} \left[\int\limits_{\Omega(\Lambda)} k(x)\, d\nu(x) \right]^n = \exp\left\{ \int\limits_{\Omega(\Lambda)} k(x)\, d\nu(x) \right\}. \tag{5}$$

Theorem 2. *Let the assumption (1) be fulfilled. Then for all bounded domains $A \subset \Lambda \subset Q$, the density $p_\Lambda^{(A)}(y) = d\mu_\Lambda|_A/d\nu$ of the restriction $\mu_\Lambda|_A$ of the measure μ_Λ to the σ-algebra $\sum_A$ is of the form*

$$p_\Lambda^{(A)}(y) = (\tilde{Z}_\Lambda^{(A)})^{-1} \sum\nolimits^{(y)} r_\Lambda^{(A)}(y_1) \ldots r_\Lambda^{(A)}(y_n), \tag{6}$$

with

$$r_\Lambda^{(A)}(y) = \int\limits_{\Omega(\Lambda \setminus A)} k(y \cup x)\, d\nu(x), \tag{7}$$

$$\tilde{Z}_\Lambda^{(A)} = \exp\left\{ \int\limits_{\Omega(A)} r_\Lambda^{(A)}(y)\, d\nu(y) \right\}. \tag{8}$$

There exists a limit cylindric measure (or measure) μ on the space Ω:

$$\mu = \lim_{\Lambda \nearrow Q} \mu_\Lambda$$

(in the sense of the weak local convergence), where, for any $A \subset Q$, the density $\rho^{(A)}(y) = d\mu|_A/d\nu$ of its restriction to $\sum_A$ equals

$$p^{(A)}(y) = \lim_{\Lambda \nearrow Q} p_\Lambda^{(A)}(y) = (\tilde{Z}^{(A)})^{-1} \sum\nolimits^{(y)} \prod_{i=1}^{n} r^{(A)}(y_1), \tag{9}$$

with

$$r^{(A)}(y) = \lim_{\Lambda \nearrow Q} r_\Lambda^{(A)}(y) = \int\limits_{\Omega_{\text{fin}}(Q \setminus A)} k(x \cup y)\, d\nu(x) \tag{10}$$

and

$$\tilde{Z}^{(A)} = \lim_{\Lambda \nearrow Q} Z_\Lambda^{(A)} = \exp\left\{ \int\limits_{\Omega(A)} r^{(A)}(y)\, d\nu(y) \right\}. \tag{11}$$

The expansions (6) and (9) are usually called the cluster expansions of the Gibbs modifications μ_Λ and the limit measure μ, respectively.

Proof. From an obvious property of the measure $\nu : d\nu(x \cup y) = d\nu(x)\, d\nu(y)$, $x \cap y = \emptyset$, we find that

$$p_\Lambda^{(A)}(y) = \int\limits_{\Omega(\Lambda \setminus A)} p_\Lambda(y \cup x)\, d\nu(x).$$

Using the equality

$$\sum^{(y\cup x)} k(x_1)\dots k(x_n) = \sum_{\substack{z_1,z_2:\\ z_1\cap z_2=\emptyset,\\ z_1\cup z_2=x}} \left[\widetilde{\sum}^{(y\cup z_1)} k(y_1\cup x_1')\dots k(y_n\cup x_{n'}')\right]$$

$$\times\left[\sum^{(z_2)} k(x_1'')\dots k(x_{n''}'')\right],$$

with $\widetilde{\sum}^{(y\cup z_1)}$ denoting the summation over all partitions $y\cup z_1 = (y_1\cup x_1')\cup\dots\cup(y_n\cup x_{n'}')$, $y_i\subset y$, $x_i'\subset z_1$, in which the y_i are nonempty subsets of y, and after that successively applying Lemma 1 several times, we arrive at the decomposition (6). From the conditions (1) and (1a), we deduce the existence of the limit (9) (and also of the limits (10) in (11)), and hence, we derive the existence of a limit measure μ on Ω.

CHAPTER 4

SMALL PARAMETERS IN INTERACTIONS

§1 Gibbs Modifications of Independent Fields with Bounded Potential

Let S be a complete separable metric space and μ_0 a probability measure on S^T that is the product of probability measures ν_t on S (in general, different for distinct $t \in T$); write $\langle\cdot\rangle_0 = \langle\cdot\rangle\mu$.

Let a potential $\{\Phi_A; A \subset T, |A| < \infty\}$ be given, and let μ_Λ be the Gibbs modification of the measure μ_0 with the density

$$\frac{d\mu_\Lambda}{d\mu_0} = Z_\Lambda^{-1} \exp(-U_\Lambda), \quad U_\Lambda = \sum_{A \subseteq \Lambda} \Phi_A. \tag{1}$$

In addition, we shall assume that the one-point potential $\Phi_{\{t\}} = 0$, replacing the measures ν_t on S, in the opposite case, by the measures $\widetilde{\nu}_t$ with the densities $d\widetilde{\nu}_t/d\nu_t = C_t^{-1} \exp(-\Phi_{\{t\}})$ where $C = \int\limits_S \exp\{-\Phi_{\{t\}}\} d\nu_t$.

In this section, the boundedness of the potential $\{\Phi_A\}$ is supposed:

$$\sup_x |\Phi_A(x)| \equiv b_A < \infty \tag{2}$$

for all finite $A \subset T$, with

$$\sum_{\substack{A:t\in A,\\ |A|=n}} b_A < c\lambda^n \tag{2'}$$

for all $t \in T$ and $n \geqslant 2$, with $c > 0$ and λ, $0 < \lambda < 1$, being some parameters.

Introducing the quantities $\kappa_A(x) = e^{-\Phi_A(x)} - 1$, the conditions (2) and (2′) become equivalent to $\widehat{\kappa}_A = \sup_x |\kappa_A(x)| < \infty$, and for all $t \in T$ and $n \geqslant 2$,

$$\sum_{\substack{A:t\in A,\\ |A|=n}} \widehat{\kappa}_A < \bar{c}\lambda^n \text{ with } ce^{-c} \leqslant \bar{c} \leqslant ce^c. \tag{3}$$

The stability conditions (2), §1, Chapter 1, are obviously fulfilled.

Theorem 1. *If the conditions (2) and (2′) are fulfilled, and the constant c is sufficiently small ($c < c_o(\lambda)$), the family of measures $\{\mu_\Lambda, \Lambda \subset T\}$ admits a cluster expansion.*

Proof. For any finite set $B \subset T$ and any bounded $\sum_A$-measurable function F_A on $\Omega = S^T$, we use the notations

$$k_B(F_A) = \sum_\zeta \left\langle F_A \prod_{A_i \in \zeta} \kappa_{A_i} \right\rangle,$$
$$k_B = k_B(\mathbf{1}), \quad |B| \geqslant 2, \quad k_{\{t\}} = 1, \quad t \in T, \tag{4}$$

with the sum taken over all unordered collections $\zeta = \{A_1, \ldots, A_n\}$ of finite sets such that $\tilde{\zeta} = B$ and $\{A, A_1, \ldots, A_n\}$ is connected. In the case of $F_A = \mathbf{1}$, the very collection ζ is supposed to be connected.

Lemma 2. *The partition functions admit a cluster expansion with the terms k_B defined by the formulas (4), where for the case $\bar{c} < (1-\lambda)^2/4$,*

$$\sum_{\substack{B: t \in B, \\ |B| = n}} |k_B| < \frac{4\bar{c}}{1-\lambda} (\bar{\lambda})^n, \quad \bar{\lambda} = \lambda(2-\lambda) < 1 \tag{5}$$

for all $t \in T$ and $n \geqslant 2$.

Moreover, for any bounded local function F_A, the decomposition

$$\langle F_A \rangle_\Lambda = \langle F_A \rangle_0 f_A^{(\Lambda)} + \sum_{\substack{R \subset \Lambda, \\ R \cap A \neq \emptyset}} k_R(F_A) \cdot f_{R \cup A}^{(\Lambda)} \tag{6}$$

takes place.

Proof. To derive a cluster representation of Z_Λ, we proceed in the same way as in §5 of Chapter 3 in the high-temperature region. Notice that

$$Z_\Lambda = \left\langle \prod_{A \subset \Lambda} (1 + \kappa_A) \right\rangle_0 = \sum_\zeta \left\langle \prod_{A \subseteq \zeta} \kappa_A \right\rangle_0, \tag{7}$$

with the sum taken over all collections ζ of mutually different sets $A_1, \ldots, A_n \subset \Lambda$. For each collection $\zeta = (A_1, \ldots, A_n)$, we use the notation $\kappa_\zeta = \prod_{A \in \zeta} \kappa_A$. Let $\zeta_1, \ldots, \zeta_k$ be the connected components of the collection ζ. The random variables $\kappa_{\zeta_1}, \ldots, \kappa_{\zeta_k}$ are independent, and consequently, $\langle \kappa_\zeta \rangle_0 = \prod_{i=1}^k \langle \kappa_{\zeta_i} \rangle_0$. Moreover, (7) may be rewritten as

$$Z_\Lambda = \sum_{\{\zeta_1, \ldots, \zeta_k\}} \prod_{i=1}^k \langle \kappa_{\zeta_i} \rangle_0. \tag{8}$$

With the help of definition (4), we obtain from (8) the cluster expansion (1), §1, Chapter 3. We pass to the proof of estimate (5). Obviously, for all $\xi \geqslant 1$ and $t \in T$,

$$h_t(\xi) \stackrel{\text{def}}{=} \sum_{B:t\in B} |k_B| \xi^{|B|} \leqslant \sum_{\zeta:t\in\tilde{\zeta}} \prod_{A\in\zeta} \widehat{\kappa}_A \xi^{|A|} \leqslant \sum_{k=1}^{\infty} S_k(\xi), \tag{9}$$

with

$$S_k^{(t)}(\xi) = \sum_{\substack{\zeta:t\in\tilde{\zeta} \\ |\zeta|=k}} \prod_{A\in\zeta} \widehat{\kappa}_A \xi^{|A|}, \quad S_k(\xi) = \sup_{t\in T} S_k^{(t)}(\xi).$$

Here, the collections ζ are supposed to be connected in all sums. Notice that the quantities $S_k(\xi)$ satisfy the following recursive inequalities:

$$S_k(\xi) \leqslant \bar{c} \sum_{m=2}^{\infty} (\lambda\xi)^m \sum_{s=0}^{\min(k-1,m)} \binom{m}{s} \sum_{(k_1,\ldots,k_s)} S_{k_1}(\xi)\ldots S_{k_s}(\xi), \tag{10}$$

with the summation in the last sum taken over the ordered collections $(k_1, \ldots, k_s)$ of integers such that $k_1 + \cdots + k_s = k-1$ and $k_i \geqslant 1$, $i = 1, \ldots, s$. The inequalities (10) can be deduced similarly to the inequalities (3), §4, Chapter 2.

Further, we introduce the numbers $\overline{S}_k = \overline{S}_k(\xi)$, defined recursively by the formulas

$$\overline{S}_k = \bar{c} \sum_{m=2}^{\infty} (\lambda\xi)^m \sum_{s=0}^{\min(k-1,m)} \binom{m}{s} \sum_{(k_1,\ldots,k_s)} \overline{S}_{k_1} \ldots \overline{S}_{k_s}, \quad k > 1,$$

$$\overline{S}_1 = S_1 = \bar{c} \sum_{m=2}^{\infty} (\lambda\xi)^m. \tag{11}$$

It is obvious that $S_k(\xi) \leqslant \overline{S}_k$. We introduce the function

$$f(z) = \sum_{k=1}^{\infty} \overline{S}_k z^k \tag{12}$$

under the hypothesis that the sum on the right-hand side converges for sufficiently small z.

We obtain from (11), under the condition

$$|\lambda\xi(1 + f(z))| < 1, \tag{13}$$

that

$$f(z) = \bar{c}z \frac{[\lambda\xi(1 + f(z))]^2}{1 - \lambda\xi(1 + f(z))}. \tag{14}$$

With the help of an explicit solution of this equation (which reduces to a quadratic equation), we can easily convince ourselves that there is an analytic function $f(z)$ on a neighbourhood of $z = 0$ that satisfies the conditions (13) and (14). Under the conditions

$$\lambda\xi < 1/(2-\lambda) \quad \text{and} \quad \bar{c} < (1-\lambda)^2/4, \tag{15}$$

the function is analytical on the disc $|z| < 1$, satisfies condition (13) in this disc, and is bounded by $4\bar{c}/(1-\lambda)$. Thus,

$$\sum_{k=1}^{\infty} S_k(\xi) \leqslant \sum_{k=1}^{\infty} \overline{S}_k(\xi) = f(1) \leqslant 4\bar{c}/(1-\lambda).$$

It follows from here that for $|\xi| < 1/[\lambda(2-\lambda)]$ and any $t \in T$, the function $h_t(\xi)$ is analytic in ξ and does not exceed $4\bar{c}/(1-\lambda)$. Hence we obtain estimate (5) with the help of the Cauchy formula. The decomposition (6) can be deduced by analogy with the cluster representation of partition functions. The lemma is proved.

It follows from (4) that

$$|k_R(F_A)| < \sup_{x} |F_A(x)| \sum_{\substack{\Gamma=\{B_1,\ldots,B_n\},\widetilde{\Gamma}=R,\\ B_i\cap A\neq\emptyset}} \prod |k_{B_i}|.$$

Thus, for $4\bar{c}/(1-\lambda) < c_0(\bar{\lambda})$ (see (23′), §1, Chapter 3), the cluster representation of Z_Λ satisfies the cluster condition (23) in §1 of Chapter 3, and the expansions of the means $\langle F_A\rangle_\Lambda$ are of the form (5) in §3, Chapter 3, and satisfy the conditions (6) and (7) in §3, Chapter 3.

Further, as is easily seen from (1), (2), and (2′), the density $d\mu_{\Lambda|A}/d\mu_0$ of the restriction of the measure μ_Λ to the σ-algebra $\sum_A$, $A \subset \Lambda$, admits the estimate

$$\left|\frac{d\mu_{\Lambda|A}}{d\mu_0}\right| < e^{m|A|}\left|\frac{Z_{\Lambda\setminus A}}{Z_\Lambda}\right| = e^{m|A|}\left|f_A^{(\Lambda)}\right|$$

with $m = \sup_t \sum_{B:t\in B} b_B < \infty$. This and Lemma 3 of §1 in Chapter 1 imply the weak local compactness of the family of measures $\{\mu_\Lambda, \Lambda \subset T\}$. The theorem is proved.

Remark. In the case of a finite-range potential $\{\Phi_A\}$, i.e., $\Phi_A \equiv 0$ for $\operatorname{diam} A > R$, that is sufficiently small, $\sup_x |\Phi_A(x)| < \beta$ for all supports A ($\beta \ll 1$ a small number), all the conditions of Theorem 1 are fulfilled. Moreover, the expansion of the form (10) in §3 of Chapter 3 for $\langle F_A\rangle_\Lambda$ turns out to be exponentially-cluster (see §3, Chapter 3) if the role of the family $\mathfrak{A}$ is played by the set of the supports of $\{\Phi_A\}$, i.e., of sets $A \subset T$ with $\Phi_A \not\equiv 0$. This fact follows easily from the proof of Theorem 1 and Lemma 4, §3, Chapter 3.

§2 Unbounded Interactions in the Finite-Range Part of a Potential

In this section, we generalize Theorem 1 of §1 to the case when the potential Φ_A is unbounded for some $A \subset T$. Namely, we suppose that the potential $\{\Phi_A\}$ that defines the Gibbs modification (1), §1, is expressed as the sum of two potentials

$$\Phi_A = \widehat{\Phi}_A + \Psi_A, \quad A \subset T, \tag{1}$$

with $\widehat{\Phi}_A$ satisfying the conditions (2) and (2′) of §1 and Ψ_A the following conditions:

1) Ψ_A is finite-range, i.e., its supports have uniformly bounded diameters: $\operatorname{diam} A \leqslant R$; hence $|A| \leqslant D = D(R)$ and each point $t \in T$ belongs to at most M supports ($M = M(R)$);

2) $\Psi_A \geqslant 0$ for all A; (2)

3) for any support A of Ψ_A,

$$\left\langle \left| e^{-\Psi_A} - 1 \right| \right\rangle_0 < \delta \tag{2'}$$

with δ being sufficiently small.

As before, we assume that the one-particle potential $\Psi_{\{t\}} \equiv 0$, $t \in T$.

Theorem 1. *Let $\widehat{\Phi}_A$ from (1) satisfy the conditions of Theorem 1 in §1 and the constant $\delta > 0$ in (2) be sufficiently small ($\delta < \delta_0(c, \lambda)$). Then the family of Gibbs modification μ_Λ, defined by the formula (1) in §1, admits a cluster expansion.*

Remark. The potential of the form

$$\Phi_A = \widehat{\Phi}_A + \beta \Psi_A, \tag{3}$$

with $\hat{\Phi}_A$ fulfilling the conditions (2) and (2′) of §1 with a sufficiently small constant c, and Ψ_A, a finite-range potential uniformly bounded from below, satisfies, for sufficiently small β, the conditions of Theorem 1.

Proof of Theorem 1. The partition function Z_A, just as in the case of a bounded potential, admits a cluster expansion with the numbers k_B defined by (4) of §1. We prove for them the cluster estimate

$$\sum_{\substack{B: t \in B, \\ |B| = n}} |k_B| < \frac{8d}{1 - \lambda} (\bar{\lambda})^n, \quad \bar{\lambda} = \lambda(2 - \lambda) < 1, \tag{4}$$

with

$$d = \max \left\{ \bar{c}, \frac{(1 + \bar{c})M}{\lambda^D} \delta^{1/(1+MD)} \right\}.$$

We have

$$e^{-(\widehat{\Phi}_A + \Psi_A)} - 1 = (e^{-\widehat{\Phi}_A} - 1)(e^{-\Psi_A} - 1) + (e^{-\widehat{\Phi}_A} - 1) + (e^{-\Psi_A} - 1).$$

Using (3) of §1, we obtain that

$$\left|e^{-(\widehat{\Phi}_A+\Psi_A)}-1\right| \leqslant \kappa_A^{(1)}+\kappa_A^{(2)}(\bar{c}+1), \tag{5}$$

with

$$\kappa_A^{(1)}=\left|e^{-\widehat{\Phi}_A}-1\right|, \quad \kappa_A^{(2)}=\left|e^{-\Psi_A}-1\right|.$$

Thus, we get from (5) and (4) of §1 that

$$|k_B|<\sum_{\Gamma,\Gamma'}(1+\bar{c})^{k'}\left\langle\prod_{A_i\in\Gamma}\kappa_{A_i}^{(1)}\cdot\prod_{A'_j\in\Gamma'}\kappa_{A'_j}^{(2)}\right\rangle_0, \tag{6}$$

with the summation over all pairs of collections $\Gamma=\{A_1,\dots,A_k\}$, $\Gamma'=\{A'_1,\dots,A'_k\}$ of mutually different sets, where the collection $\Gamma\cup\Gamma'$ is connected, and $B=\widetilde{\Gamma}\cup\widetilde{\Gamma}'$, and the sets $A'_j\in\Gamma'$ are supports of the potential Ψ. Notice that for each collection $\Gamma'=\{A'_1,\dots,A'_{k'}\}$ a sub-collection $\Gamma''\subset\Gamma'$ consisting of at least $[k'/(1+MD)]+1$ mutually disjoint sets may be chosen. Since $\kappa_A^{(2)}\leqslant 1$,

$$\left\langle\prod_{A\in\Gamma}\kappa_A^{(1)}\prod_{A'\in\Gamma'}\kappa_{A'}^{(2)}\right\rangle_0\leqslant\prod_{A\in\Gamma}\widehat{\kappa}_A^{(1)}\delta^{|\Gamma'|/(1+MD)}.$$

Thus, for any $\xi>1$,

$$\begin{aligned}h_t(\xi)&\stackrel{\text{def}}{=}\sum_{B:t\in G}k_B\xi^{|B|}\leqslant\sum_{\Gamma,\Gamma'}a^{|\Gamma'|}\left(\prod_{A'\in\Gamma'}\xi^{|A'|}\right)\left(\prod_{A\in\Gamma}\widehat{\kappa}_A^{(1)}\xi^{|A|}\right)\\&\leqslant\sum_{\substack{k,k'\\k+k'>0}}S_{kk'}(\xi),\end{aligned}$$

with

$$S_{kk'}=\sup_{t\in\Gamma}\sum_{\substack{\Gamma,\Gamma':\\t\in\widetilde{\Gamma}\cup\widetilde{\Gamma}',\\|\Gamma|=k,|\Gamma'|=k'}}a^{k'}\left(\prod_{A'\in\Gamma'}\xi^{|A'|}\right)\left(\prod_{A\in\Gamma}\widehat{\kappa}_A^{(1)}\xi^{|A|}\right),$$

$$a=(1+\bar{c})\delta^{1/(1+MD)}.$$

As before, we obtain the following recursive inequalities for $S_{kk'}$:

$$\begin{aligned}S_{kk'}(\xi)\leqslant{}&\bar{c}\sum_{m\geqslant 2}(\lambda\xi)^m\sum_{s=0}^{\min(m,k+k'-1)}\binom{m}{s}\sum_{\substack{k_1+\dots+k_s=k-1,\\k'_1+\dots+k'_s=k'}}S_{k_1k'_1}\dots S_{k_sk'_s}\\&+a\sum_{m=1}^{D}N_m\xi^m\sum_{s=0}^{\min(m,k+k'-1)}\binom{m}{s}\sum_{\substack{k_1+\dots+k_s=k,\\k'_1+\dots+k'_s=k'-1}}S_{k_1k'_1}\dots S_{k_sk'_s},\end{aligned}$$

which may be derived similarly to the inequalities (10) in §1.

Here, $N_m < M$ is the maximal number of supports of the potential Ψ of the cardinality m, which contain a fixed site. Since $N_m \leqslant M\lambda^m/\lambda^D$, $m = 2, \ldots, D$, the numbers $S_{kk'}$ are bounded, $S_{kk'} \leqslant \overline{S}_{kk'}$, with $\overline{S}_{kk'}$ recursively defined by the equalities

$$\overline{S}_{kk'} = d\sum_{m=2}^{\infty}(\lambda\xi)^m \sum_{s=0}^{\min(m,k+k'-1)} \binom{m}{s} \times \left[\sum_{\substack{k_1+\cdots+k_s=k-1,\\ k_1+\cdots+k'_s=k'}} \overline{S}_{k_1k'_1}\ldots\overline{S}_{k_sk'_s} + \sum_{\substack{k_1+\cdots+k_s=k,\\ k'_1+\cdots+k'_s=k'-1}} \overline{S}_{k_1k'_1}\ldots\overline{S}_{k_sk'_s} \right], \tag{7}$$

$$d = \max(\bar{c}, aM/\lambda^D).$$

We introduce the function

$$\varphi(z_1, z_2) = \sum_{k,k'} \overline{S}_{kk'} z_1^k z_2^{k'},$$

and under the condition that $\lambda\xi\,|1 + \varphi(z_1, z_2)| \leqslant 1$, we find from (7) that $\varphi(z_1 + z_2)$ satisfies the equality

$$\varphi = 2d\frac{z_1 + z_2}{2}\frac{[\lambda\xi(1+\varphi)]^2}{1 - \lambda\xi(1+\varphi)}.$$

Thus $\varphi = f_{2d}((z_1 + z_2)/2)$, with $f_{\bar{c}}(z)$ denoting the solution of equation (14) in §1. Using the results of the preceding section, we find that for $\xi = 1/[\lambda(2-\lambda)]$ and $d < (1-\lambda)^2/8$, the function φ is analytic in the polydisc $|z_1| \leqslant 1$, $|z_2| \leqslant 1$ and bounded by the constant $8d/(1-\lambda)$ on it. Hence, similarly as before, we obtain estimate (4). The remaining part of the proof is completely analogous to the proof of Theorem 1 of §1.

§3 Gibbs Modifications of d-Dependent Fields

A random field $\{\xi_t, t \in T\}$ in an arbitrary set T (not necessarily countable) with a metric $\rho(\cdot,\cdot)$ (for example, $T = Z^\nu$ or $T = R^\nu$) is called d-dependent if, for any finite set $\{t_1, \ldots, t_n\}$, $t_i \in T$ and $\rho(t_i, t_j) > d$, the random variables $\xi_{t_1}, \ldots, \xi_{t_n}$ are mutually independent.

Now let T be countable and the measure μ_0 on S^T define a d-dependent field with values in S.

Theorem 1. *Theorems 1 of §1 and 1 of §2 remain valid if the free measure μ_0 used therein is d-dependent. (Under the corresponding change of the admissible boundaries for the small constants c and δ.)*

The proof follows the proofs of Theorems 1 of §1 and 1 of §2 with the difference that d-admissible partitions have to be considered in cluster representations of the partition functions. Here, k_B is defined by (1), §1, as before, with the sum taken over d-connected collections ζ. Recall that this means that ζ cannot be split up into two collection ζ' and ζ'' such that the distance between $\widetilde{\zeta'}$ and $\widetilde{\zeta''}$ is greater than d. The necessary changes in these considerations are obvious.

Remark. A lot of examples of d-dependent translation invariant random fields in R^ν can be shown. With the help of the coarse graining device (discretization) described in §1 of Chapter 1, we may construct and examine the Gibbs modifications of such fields.

§4 Gibbs Point Field in R^ν

We consider the Gibbs modifications $\{\mu_\Lambda, \Lambda \subset R^\nu\}$ (Λ being a bounded domain) of the Poisson pure point field μ_0 in R^ν given by the Lebesgue measure $d\lambda$ in R^ν. The density $p_\Lambda(x) = \frac{d\mu_\Lambda}{d\nu}(x)$, $x \in \Omega(\Lambda)$, of the measure μ_Λ with respect to the measure ν on $\Omega(\Lambda)$ (which differs only by the factor $\exp\{\lambda(\Lambda)\}$ from the measure $\mu_0|_\Lambda$; see (9), §1, Chapter 1) equals

$$p_\Lambda(x) = Z_\Lambda^{-1} \exp\{-\beta U_\Lambda(x)\}, \tag{1}$$

with a Hamiltonian of the form (11), §1, Chapter 1:

$$U_\Lambda(x) = \hat{\mu}|x| + \sum_{\{t,t'\}\subseteq x} \varphi(t-t'), \tag{2}$$

where the function φ is bounded from below and satisfies

1) the stability condition: for any $x \in \Omega_{\text{fin}}$,

$$\sum_{\{t,t'\}\subseteq x} \varphi(t-t') > -B|x|, \tag{3}$$

with $B \geqslant 0$ being a constant that does not depend on x;

2) the regularity condition:

$$C(\beta) \equiv \int_{R^\nu} \left|e^{-\beta\varphi(t)} - 1\right| d^\nu t < \infty \tag{4}$$

for some $\beta > 0$ (and, consequently, for all $\beta > 0$).

Theorem 1. *Let the conditions (3) and (4) be fulfilled and the quantity* $z = e^{-\beta\hat{\mu}}$ *(activity) be sufficiently small. Then there is a function* $k(x)$, $x \in \Omega_{\text{fin}}$, *defined on the space* Ω_{fin}, *satisfying the condition (1), §6, Chapter 3, and such that for the density* $p_\Lambda(x)$ *and the partition function* Z_Λ, *the representation (2) and (3) of §6, Chapter 3, containing the function* $k(x)$, *are valid.*

It follows from this theorem and Theorem 2, §6, Chapter 3, that the Gibbs measures $\{\mu_\Lambda, \Lambda \subset R^\nu\}$ admit the cluster expansion (6) in §6, Chapter 3, and that there is a limit measure (or cylinder measure) μ on the space Ω admitting also the cluster expansion (9), §6, Chapter 3.

Proof. For $x \in \Omega_{\text{fin}}$, $|x| \geqslant 2$, we write

$$
\begin{aligned}
e^{-\beta U(x)} &= z^{|x|} \prod_{\{t,t'\} \subseteq x} \left(1 + (e^{-\beta\varphi(t-t')} - 1)\right) \\
&= z^{|x|} \left(1 + \sum_{\Gamma} \prod_{\{t,t'\} \in \Gamma} (e^{-\beta\varphi(t-t')} - 1)\right),
\end{aligned}
\tag{5}
$$

with $\sum_\Gamma$ taken over all possible graphs Γ constructed on the subsets $\tilde{x}(\Gamma) \subset x$. We use k_Γ to denote the expression

$$
k_\Gamma = z^{|\tilde{x}|} \prod_{\{t,t'\} \in \Gamma} (e^{-\beta\varphi(t-t')} - 1). \tag{6}
$$

Each graph Γ splits up into its connected components $\gamma_1, \ldots, \gamma_s$, and

$$
k_\Gamma = k_{\gamma_1} \ldots k_{\gamma_s}. \tag{7}
$$

We introduce, for each set $x \in \Omega_{\text{fin}}$, the function

$$
k(x) = \begin{cases} z, & |x| = 1, \\ \sum_\gamma^{(x)} k_\gamma, & |x| \geqslant 2, \end{cases} \tag{8}
$$

with the sum $\sum_\gamma^{(x)}$ taken over all possible connected graphs with the set of vertices coinciding with x. We get from (5), (6), (7), and (8) the expansion (2), §6, Chapter 3:

$$
p_\Lambda(x) = Z_\Lambda^{-1} \sum^{(x)} k(x_1) \ldots k(x_n). \tag{9}
$$

We shall show that $k(x)$ satisfies condition (1), §6, Chapter 3.

Lemma 2. *For the function $k(x)$, the estimate*

$$|k(x)| \leqslant (|z|e^{2\beta B})^{|x|} \sum_{T}^{(x)} \prod_{(t,t')\in T} \left| \left(e^{-\beta\varphi(t-t')} - 1 \right) \right| \tag{10}$$

is valid, where the sum $\sum_T^{(x)}$ is taken over all trees T, with the set of vertices coinciding with x.

Proof. We introduce a function $\bar{k}(x,y)$, $x, y \in \Omega_{\text{fin}}$, $x \cap y = \emptyset$, $x \neq \emptyset$:

$$\bar{k}(x,y) = \sum_{\Gamma}^{(x,y)} k_\Gamma,$$

where the sum is taken over all graphs Γ, with the set of vertices $x \cup y$, and simultaneously, each connected component $\gamma \subset \Gamma$ of the graph Γ contains at least one vertex from the set x. It is easy to verify that $\bar{k}(x,y)$ satisfies the following relations. Let $t_1 \in x$ be an element of $x \in \Omega_{\text{fin}}$ that fulfils the condition

$$\sum_{t\in x'} \varphi(t_1 - t) > -2B \tag{11}$$

with $x' = x \backslash \{t_1\}$. The stability condition (3) implies that there is at least one such point. The equalities

$$\begin{aligned} \bar{k}(x,y) &= ze^{-\beta \sum_{t\in x'} \varphi(t_1-t)} \\ &\quad \times \left[k(x',y) + \sum_{\substack{u \subseteq y, \\ u \neq \emptyset}} \bar{k}(x' \cup u, y\backslash u) \prod_{t\in u} (e^{-\beta\varphi(t_1-t)} - 1) \right], \quad y \neq \emptyset, \\ \bar{k}(x,\emptyset) &= e^{-\beta U(x)} \end{aligned} \tag{12}$$

hold true.

Notice that the equalities (12) are of a recursive nature: the value $\bar{k}(x,y)$, for $|x|+|y| = n$, may be expressed by the values of $\bar{k}(\tilde{x},\tilde{y})$ for $|\tilde{x}|+|\tilde{y}| = n-1$. Hence, from this and (11), it follows that $\bar{k}(x,y)$ is majorized $\left(\left|\bar{k}(x,y)\right| \leqslant Q(x,y)\right)$ by a solution $Q(x,y) \geqslant 0$ of the system of relations

$$\begin{aligned} Q(x,y) &= |z|e^{2\beta B} \\ &\quad \times \left[Q(x',y) + \sum_{\substack{u \subseteq y, \\ u \neq \emptyset}} Q(x' \cup y, y\backslash u) \prod_{t\in u} \left| e^{-\beta\varphi(t_1-t)} - 1 \right| \right], \quad y \neq 0, \\ Q(x,\emptyset) &= |z|e^{2\beta B|x|}, \quad Q(\emptyset, y) = 0. \end{aligned} \tag{13}$$

With the help of the induction on the number $n = |x| + |y|$, we can easily convince ourselves that the solution of (13) is of the form

$$Q(x,y) = \sum_{(y_1,\dots,y_s)} Q(\{t_1\},y_1)Q(\{t_2\},y_2)\dots Q(\{t_s\},y_s), \tag{14}$$

with $x = \{t_1,\dots,t_s\}$, and the sum taken over ordered collections $(y_1,\dots,y_s)$ of mutually disjoint sets $y_i \subset y$ (possibly empty) such that $\bigcup_{i=1}^{s} y_i = y$ and

$$Q(\{t\},y) = (|z|e^{2\beta B})^{|y|+1} \sum_{\mathcal{T}} \prod_{\{t',t''\}\in\mathcal{T}} \left|e^{\beta\varphi(t'-t'')} - 1\right|, \tag{15}$$

with the sum taken over all trees $\mathcal{T}$, with the set of vertices $\{t\} \cup y$. Since $k(x) = \bar{k}(\{t\},x')$, (10) follows from (15). The lemma is proved.

Further, we have from (10) that

$$\int_{y\in\Omega(\Lambda)} \int_{x\in\Omega_{\text{fin}}} |k(x\cup y)|\, d\nu(x)\, d\nu(y) \leqslant \sum_{m,n,m>0} \frac{(|z|e^{2\beta B})^{m+n}}{m!n!}$$

$$\times \sum_{\mathcal{T}} \overbrace{\int_\Lambda \cdots \int_\Lambda}^{m \text{ times}} \overbrace{\int_{R^\nu} \cdots \int_{R^\nu}}^{n \text{ times}} \prod_{(i,j)\in\mathcal{T}} \left|e^{-\beta\varphi(t_i-t_j)} - 1\right| \prod_{i=1}^{m+n} dt_i, \tag{16}$$

with the sum $\sum_{\mathcal{T}}$ taken over all trees with the set of vertices $\mathcal{N}_{m+n} = \{1,\dots,m+n\}$, and the integration with respect to the first m variables $(t_1,\dots,t_m)$ is carried out over the set Λ and with respect to the other variables $(t_{m+1},\dots,t_{m+n})$ over the whole space R^ν. It is easy to verify that for any tree $\mathcal{T}$,

$$\overbrace{\int_\Lambda\int_\Lambda \cdots \int_\Lambda}^{m \text{ times}} \overbrace{\int_{R^\nu} \cdots \int_{R^\nu}}^{n \text{ times}} \prod_{(i,j)\in\mathcal{T}} \left|e^{-\beta\varphi(t_i-t_j)} - 1\right| \prod_{i=1}^{m+n} dt_i \leqslant |\Lambda|(C(\beta))^{m+n-1}. \tag{17}$$

Since (as follows, e.g., from (21), §4, Chapter 2), the number of trees with k vertices does not exceed $c^k k!$ with a constant c (the exact value of this number is k^{k-2} (cf. [52])), we find from (16) and (17) that the integral (16) is finite if

$$|z| < \frac{1}{2c}[e^{2\beta B}C(\beta)]^{-1}. \tag{18}$$

The theorem is proved.

§5 Models with Continuous Time

We consider here Gibbs fields $\{\xi_{t,s}, t \in T, s \in R^1\}$ taking values in a finite set $Q = (q_1, \ldots, q_l)$ and defined in the "space time" $T \times R^1$, where the "space" T is countable and the "time" R^1 is continuous.

First we define the free measure μ_0. Let a stationary ergodic Markov chain $\xi(s)$ with continuous time and with the set Q of values be given. To it corresponds the measure ν_0 on the space Ω_0 of piece-wise constant (and continuous from the right) trajectories (functions) $\{\xi(s), s \in R^1\}$, $\xi(s) \in Q$, defined on R^1.

Consider the space $\Omega = \Omega_0^T$ with the measure $\mu_0 = \nu_0^T$, i.e., the collection $\{\xi_t, t \in T\}$ of mutually independent chains $\{\xi_t(s) = \xi_{t,s}\}$ associated with each point $t \in T$.

We now define the modification $\mu_{\Lambda \times L}$ of the measure μ_0 ($\Lambda \subset T$ is a finite subset of T, $L = [l_1, l_2] \subset R^1$ a finite segment):

$$\frac{d\mu_{\Lambda \times L}}{d\mu_0} = Z_{\Lambda \times L}^{-1} \exp\{U_{\Lambda \times L}\}, \tag{1}$$

where

$$U_{\Lambda \times L} = \beta \sum_{(t,t')}^{(\Lambda)} \int_L \Phi(\xi_t(s), \xi_{t'}(s))\, ds; \tag{2}$$

$\sum_{(t,t')}^{(\Lambda)}$ is the sum over $t, t' \in \Lambda$, $\rho(t,t') = 1$, and Φ is a real symmetric function on $Q \times Q$.

Theorem 1. *For sufficiently small β, the partition functions $Z_{\Lambda \times L}$ admit a cluster representation and the measures $\mu_{\Lambda \times L}$ a cluster expansion (and thus the limit measure $\mu = \lim_{\Lambda \nearrow T, L \nearrow R^1} \mu_{\Lambda \times L}$ on the space Ω exists).*

Since the random field $\xi_{t,s}$ is not defined in a countable set, the assertions of the theorem require, strictly speaking, some refinement. Namely, let us consider the set $T \times Z_a^1$, where $Z_a^1 = \{an, n = 0, \pm 1, \pm 2, \ldots\}$ is the lattice with the lattice spacing $a > 0$ to be chosen sufficiently large. Let $S_{t,n}$ be the space of trajectories $S_{t,s}, an \leqslant s \leqslant a(n+1)$, on the interval $[an, a(n+1)]$ associated with the site $t \in T$. Clearly, $\Omega = \prod_{t,n} S_{t,n}$, and in the case of a segment $L = [n_1 a, n_2 a]$ where $n_1, n_2 (n_1 < n_2)$ are integers, the Hamiltonian (2) may be represented in the form

$$U_{\Lambda \times L} = \sum_{\delta \subset \Lambda \times L_a} \Phi_\delta,$$

where

$$\delta = \delta(t, t', n) = \{t, t'\} \times \{na, (n+1)a\} \subset T \times Z_a^1, \tag{3}$$

$\{t,t'\} \subset T$ is a pair of nearest neighbours, $L_a = L \cap Z_a^1 = \{na, n_1 \leqslant n \leqslant n_2-1\}$ and

$$\Phi_\delta = \beta \int_{na}^{(n+1)a} \Phi(\xi_t(s), \xi_{t'}(s))\, ds. \tag{4}$$

Thus, the measure $\mu_{\Lambda\times L}$ (with $L = [n_1 a, n_2 a]$) may be regarded as a modification of a random field in the countable set $T \times Z_a^1$ taking values in the spaces $S_{t,n}$; namely, regarding the measures $\mu_{\Lambda\times L}$ in that way, we shall establish their cluster expansion.

Proof. We use $\hat{q} = \hat{q}(\Lambda, L_a)$ to denote the set of values taken by the random trajectories $\xi_t(s)$ at points in $\Lambda \times L_a$, and let $\mu_0^{\hat{q}}(\cdot) = \mu_0(\cdot|\hat{q})$ be the conditional distribution on Ω under a fixed set $\hat{q}$. The distribution of values $\hat{q}$ (with respect to the measure μ_0) will be denoted by $p_0(\hat{q})$. The values $\xi_t(s)$ taken by trajectories in different segments $\varepsilon_{t,n} = \{t\} \times [na, (n+1)a] \subset \Lambda \times L$ are mutually independent under the conditional measure $\mu_0^{\hat{q}}$.

We put $V_\delta = e^{\Phi_\delta} - 1$ with δ being a set of the form (3), and for any unordered collection $\Delta = \{\delta_\Lambda, \dots, \delta_k\}$ of such mutually different sets, we use the notation $V_\Delta = \prod_{\delta\in\Delta} V_\delta$. Hence,

$$\begin{aligned} Z_{\Lambda\times L} = \langle \exp U_{\Lambda\times L}\rangle_{\mu_0} &= \sum_{\hat{q}} \Big\langle \prod_{\delta\subset\Lambda\times L_a} e^{\Phi_\delta} \Big\rangle_{\mu_0^{\hat{q}}} p_0(\hat{q}) \\ &= \sum_{\Delta}^{(\Lambda\times L_a)} \sum_{\hat{q}} \langle V_\Delta\rangle_{\mu_0^q} p_0(q), \end{aligned} \tag{5}$$

where $\sum_\Delta^{(\Lambda\times L_a)}$ is taken over all collections Δ (including the empty one) lying in $\Lambda \times L_a$.

A collection $\Delta = \{\delta(t_1, t_1', n_1), \dots, \delta(t_k, t_k', n_k)\}$ will be called connected if $n_1 = n_2 = \dots = n_k$ and the collection $\{\{t_1, t_1'\}, \dots, \{t_k, t_k'\}\}$ is connected. Let $\Delta_1, \dots, \Delta_p$ be the connected components of the collection Δ in (5). From the independence of V_{Δ_i} for different Δ_i, we find that

$$Z_{\Lambda\times L} = \sum_{\Gamma=(\Delta_1,\dots,\Delta_p)}^{(\Lambda\times L_a)} \sum_{\hat{q}} \prod_i \langle V_{\Delta_i}\rangle_{\mu_0^q} p_0(\hat{q}), \tag{6}$$

where $\sum_\Gamma^{(\Lambda\times L_a)}$ is taken over sets Γ of connected collections Δ_i (each of which is maximal, i.e., any union $\Delta_1 \cup \Delta_2$, where Δ_1, Δ_2 are two arbitrary collections from Γ, is not connected any more).

Furthermore,

$$p_0(\hat{q}) = \prod_{(t,s)\subset\Lambda\times L_a} p_0(q_{t,s}) \prod_{\substack{t\in\Lambda,\\ n_1\leqslant n<n_2}} \frac{p_a(q_{t,a(n+1)}, q_{t,an})}{p_0(q_{t,a(n+1)})}, \tag{7}$$

where $p_0(q)$ are stationary probabilities of the process $\xi(s)$ and $p_a(q_2, q_1) = \Pr(\xi(a) = q_2, \xi(0) = q_1)$ are transition probabilities of the process $\xi(s)$ over the period of time a. In view of the ergodicity of the process $\xi(s)$, we have

$$\min_q p_0(q) = p_0 > 0 \tag{8}$$

and

$$|p_a(q_2, q_1) - p_0(q_2)| < Ce^{-\tau a}, \tag{9}$$

where $C > 0$ and $\tau > 0$ are some constants.

For sets of the form

$$\alpha = \alpha_{t,n} = \{t\} \times \{na, (n+1)a\} \subset T \times Z_a, \tag{10}$$

we use r_α to denote the quantity

$$r_\alpha = \frac{p_a(q_{t,(n+1)a}, q_{t,na})}{p_0(q_{t,(n+1)a})} - 1,$$

and for each unordered collection $\eta = \{\alpha_1, \dots, \alpha_l\}$ of mutually different sets α of the form (10), we set $r_\eta \doteq \prod_{\alpha \in \eta} r_\alpha$.

Thus, if we use $\hat{\mu}_0^a$ to denote the probability distribution on the space $\widehat{Q}^{T \times Z_a}$ that equals the infinite product of measures p_0 on Q at each point $(t, na) \in T \times Z_a$, it follows from (6) and (7) that

$$Z_{\Lambda \times L} = \sum_{\Gamma = \{\Delta_1, \dots, \Delta_p\}}^{(\Lambda \times L_a)} \sum_{\eta = \{\alpha_1, \dots, \alpha_l\}}^{(\Lambda \times L_a)} \left\langle \prod_{\Delta \in \Gamma} \langle V_\Delta \rangle_{\mu_0^q} r_\eta \right\rangle_{\hat{\mu}_0^a}. \tag{11}$$

A pair (Γ, η), $\Gamma = \{\Delta_i\}$, $\eta = \{\alpha_j\}$, is called connected if the collection of sets $\{\alpha, \dots, \alpha_l, \tilde{\Delta}_1, \dots, \tilde{\Delta}_p\}$ is connected in $\Gamma \times Z_a$. Let $(\Gamma_1, \eta_1), \dots, (\Gamma_m, \eta_m)$ be connected components of the pair (Γ, η) in the sum (11). Since the quantities

$$g_{(\Gamma_i, \eta_i)} = \prod_{\Delta \in \Gamma_i} \langle V_\Delta \rangle_{\mu_0^q} r_{\eta_i}$$

are independent for different components (Γ_i, η_i), we get from (11) the final cluster representation of the partition function $Z_{\Lambda \times L}$ of the form (1), §1, Chapter 3, where $k_B = 1$ for $B \subset T \times Z_a$, $|B| = 1$, and

$$k_B = \sum_{(\Gamma, \eta)}^{(B)} \langle g_{(\Gamma, \eta)} \rangle_{\hat{\mu}_0^a}, \quad |B| \leqslant 2, \tag{12}$$

where the sum $\sum_{(\Gamma, \eta)}^{(B)}$ is taken over all connected pairs (Γ, η) such that $(\bigcup_{\Delta \in \Gamma} \tilde{\Delta}) \cup \tilde{\eta} = B$ (if there is no such a pair, then $k_B = 0$).

It follows from (12), (8), (9), (4), and Lemma 1 in §4, Chapter 2, that, with a suitable choice of $a > 0$, we get

$$\sum_{\substack{B: t \in B, \\ |B|=n}} |k_B| < c\lambda^n,$$

where $\lambda = \lambda(\beta) = \min C \left[\frac{1}{p_0} e^{-\tau a} + ka\beta e^{ka\beta}\right] \to 0$ as $\beta \to 0$, C and c are absolute constants, and $k = \max |\Phi|$.

A cluster expansion of the means $\langle F_A \rangle_{\Lambda \times L}$ of the form (5), (6), and (7) of §3, Chapter 3, for the bounded local function F_A may be obtained in a similar manner. Finally, with the help of the preceding arguments, it is easy to verify that the densities $(d\mu_{\Lambda \times L|A \times l})/(d\mu_{0|A \times l})$, where $\mu_{\Lambda \times L|A \times l}$ and $\mu_{0|A \times l}$ are the restrictions of the measures $\mu_{\Lambda \times L}$ and μ_0 onto the σ-algebra $\sum_{A \times l}$, $A \subset T$, $l \subset R^1$, are bounded by a constant that does not depend on Λ and L. Thus the collection of measures $\mu_{\Lambda \times L}$ is locally weakly compact (see Lemma 3, §1, Chapter 1). The theorem is proved.

§6 Expansion in Semi-Invariants. Perturbation of a Gaussian Field

Let μ_0 be a probability measure on S^T and $\{\mu_\Lambda, \Lambda \subset T\}$ be a family of its Gibbs modifications by means of a one-point potential. More precisely, let the Hamiltonian be of the form

$$U_\Lambda = \beta \sum_{t \in \Lambda} \Phi_{\{t\}}, \tag{1}$$

where β is a real parameter. We suppose that for each $t \in T$ all moments are finite:

$$\left\langle \left|\Phi_{\{t\}}\right|^k \right\rangle_0 < \infty, \quad k > 0. \tag{2}$$

Let, moreover, a compact function h be defined on the space S (see §1, Chapter 1) such that

$$\langle h(x(t)) \rangle_0 < \infty \tag{2'}$$

for each $t \in T$.

Theorem 1. *Let the conditions (2) and (2′) be satisfied and let some class G of local functions exist so that*

a) the linear span of $G \cap C(\Omega)$ is everywhere dense in $C(\Omega)$;

b) $h_{\{t\}} \in G$ for each $t \in T (h_{\{t\}}(x) = h(x(t)))$.

Let the following estimates be valid for any function $F_A \in G$: for each $n = 1, 2, \ldots$ and arbitrary integers $k_1 > 0, \ldots, k_n > 0$, it is

$$\sum_{(t_1, \ldots, t_n)} \left| \left\langle F_A, \Phi'^{k_1}_{\{t_1\}}, \ldots, \Phi'^{k_n}_{\{t_n\}} \right\rangle_0 \right| \leqslant n! C(F_A) k_1! \ldots k_n! \overline{C}^{(k_1 + \cdots + k_n)}, \tag{3}$$

where the sum is over all ordered collections $(t_1, \ldots, t_n)$ of mutually distinct points, the constant $C(F_A)$ depends only on the function F_A, and $\overline{C}$ does not depend on n, F_A, and $k_1, \ldots, k_n$. Then, for sufficiently small $\beta > 0$, the system of measures $\{\mu_\Lambda, \Lambda \subset T\}$ admits a cluster expansion.

Proof. We make use of formula (21) in §1, Chapter 2, to infer that

$$\begin{aligned} \langle F_A \rangle_\Lambda &= \sum_{n=0}^{\infty} \frac{1}{n!} \langle F_A, U_\Lambda'^n \rangle_0 \\ &= \langle F_A \rangle_0 + \sum_{\{(t_1,k_1),\ldots,(t_s,k_s)\}} \prod_{i=1}^{s} \frac{(-\beta)^{k_i}}{k_i!} \left\langle F_A, \Phi_{\{t_1\}}'^{k_1}, \ldots, \Phi'^{k_s}\{t_s\} \right\rangle_0, \end{aligned} \tag{4}$$

where the sum $\sum_{\{(t_1,k_1),\ldots,(t_s,k_s)\}}$, $s > 0$, is taken over all unordered collections of pairs (t_i, k_i) with $t_i \in \Lambda$, k_i and integer, and $t_i \neq t_j$ for $i \neq j$. We set

$$b_R(F_A) = \sum_{(k_1,\ldots,k_s)} \prod \frac{(-\beta)^{k_i}}{k_i!} \left\langle F_A, \Phi_{\{t_1\}}'^{k_1}, \ldots, \Phi_{\{t_s\}}'^{k_s} \right\rangle,$$

where $R = \{t_1, \ldots, t_s\}$. It follows from (4) that

$$\langle F_A \rangle = b_\emptyset(F_A) + \sum_{R \subset \Lambda, R \neq \emptyset} b_R(F_A), \tag{5}$$

and from (3) we imply that the numbers $b_R(F_A)$ obey the conditions (18) and (19) of §1 in Chapter 1. Since for each $A \subset T$ the function $h_A(x) = \sum_{t \in A} h(x(t))$, $x \in S^A$, is a compact function on S^A and it follows from the hypothesis of the theorem and the preceding estimates that $\langle h_A \rangle_\Lambda < M_A$, where M_A depends only on A and does not depend on Λ, we may conclude from Lemma 3, §1, Chapter 1, that the family $\{\mu_\Lambda, \Lambda \subset T\}$ is locally weakly compact. The theorem is proved.

Remark. Expansion (5) can also be obtained for the potentials $\{\Phi_A, A \subset T\}$ of a general form, if we write the Hamiltonian U_Λ in the form $U_\Lambda = \sum \Phi_{\{t\}}^{(\Lambda)}$, where $\Phi_{\{t\}}^{(\Lambda)} = \sum_{A \subset \Lambda : t \in A} (1/|A|) \Phi_A$. Supposing that $\Phi_{\{t\}}^{(\Lambda)}$ obeys the estimate (2) and uniformly in Λ the estimate (3), we again get (5).

Theorem 1 may be applied only rarely since the estimate (3) can be deduced, roughly speaking, only for bounded potentials $\Phi_{\{t\}}$. But even in this case, difficulties arise with an estimation of semi-invariants of the form (3) for $k_i > 1$ (coinciding points). The following approach is free of both deficiencies.

Let for each $s > 0$ the moment

$$\langle e^{-s\Phi_{\{t\}}} \rangle_0 < \infty, \quad t \in T, \tag{6a}$$

be defined. Then for any Λ, the partition function $Z_\Lambda < \infty$, which may be represented in the cluster form (1) of §1, Chapter 3, with the quantities

$$k_A = \langle \Psi'_A \rangle_0, \quad A \subset T, \tag{6b}$$

Where Ψ'_A is the system of variables $\{\Psi_t = \exp\{-\beta\Phi_t\}, t \in A\}$. This fact follows from the formula for the expansion of moments in semi-invariants:

$$Z_\Lambda = \left\langle \prod_{t\in\Lambda} \Psi_t \right\rangle_0 = \sum_{\{A_1,\dots,A_k\}} \langle \Psi'_{A_1} \rangle_0 \cdots \langle \Psi'_{A_k} \rangle_0 \tag{6c}$$

(see (6), §1, Chapter 2).

We now consider a local function F_A of the form

$$F_A = \prod_{t\in A} f_t, \tag{7}$$

where f_t is a $\sum_{\{t\}}$-measurable function such that

$$\langle (|f_t| \Psi_t)^k \rangle_0 < \infty \tag{8a}$$

for all $t \in A, k > 0$.

Introducing the notation

$$R_t = \begin{cases} f_t \Psi_t, & t \in A, \\ \Psi_t, & t \in \Lambda \backslash A, \end{cases}$$

and expanding $\langle F_A \rangle_\Lambda$ into semi-invariants, we get

$$\begin{aligned} \left\langle \prod_{t\in A} f_t \right\rangle_\Lambda &= Z_\Lambda^{-1} \left\langle \prod_{t\in\Lambda} R_t \right\rangle_0 = Z_\Lambda^{-1} \sum_{\{A_1,\dots,A_k\}} \langle R'_{A_1} \rangle_0 \dots \langle R'_{A_k} \rangle_0 \\ &= \sum_{\Gamma}{}^{(\Lambda)} \left(\prod_{B\in T} g_B \right) f_{\tilde{\Gamma}}^{(\Lambda)}, \end{aligned} \tag{8b}$$

with the sum taken over all collections $\Gamma = \{B_1, \dots, B_k\}$, $B_i \subset \Lambda$, $B_i \cap A \neq \emptyset$, $B_i \cap B_j \neq \emptyset$, $i \neq j$,and $A \subseteq \tilde{\Gamma}$.

Here, also

$$g_B = \langle R'_B \rangle_0 \Big/ \prod_{t\in B} \langle \Psi_t \rangle_0, \tag{9}$$

and $f_A^{(\Lambda)}$ is a correlation function in the ensemble of subsets defined by the quantities (6b).

Lemma 2. *Let condition (6a) hold and suppose that the following conditions are also satisfied:*

1)
$$\langle \Psi_t \rangle_0 > k > 0 \tag{10}$$
for each $t \in T$.

2)
$$\sum_{A:t\in A,|A|=n} |\langle \Psi'_A \rangle_0| < c\lambda^n, \tag{11}$$
for all n and with parameters c and λ, $0 < \lambda < k < \infty$, obeying the cluster condition (23), §1, Chapter 3.

3) For any $\sum_{\{t\}}$-measurable function f_t obeying the condition (8a), there exists a constant $c(f_t) > 0$ such that, whenever $\{f_t, t \in A\}$ is a finite collection of such functions and $t_0 \in T$ and $n > 1$ are arbitrary, one has

$$\sum_{\substack{B:t_0\in B,\\ |B|=n}} \langle R'_B \rangle_0 \Big/ \prod_{t\in B\cap A} c(f_t) < b_A \lambda^n, \tag{12}$$

with b_A not depending on the functions f_t, the number n, and the point $t_0 \in T$.

4) There exists a compact function h on S such that for each t the function $h_t = h(x(t))$ obeys the condition (8a).

Then the system of measures $\{\mu_\Lambda, \Lambda \subset T\}$ admits a cluster expansion.

Proof. From the conditions (10) and (11) and with the help of (6b), we obtain a cluster representation of Z_Λ. From (8b) and (12), we find that the expansion (5) in §3, Chapter 3, holds with

$$k_R(F_A) = \sum_{\Gamma:\widetilde{\Gamma}=R}^{(\Lambda)} \prod_{B\in\Gamma} g_B,$$

and that the estimate (6) in §3, Chapter 3, is satisfied with

$$m(F_A) = \prod_{t\in A} c(f_t), \quad r_B = \frac{|\langle R'_B \rangle_0|}{\prod_{t\in A\cap B} c(f_t) \prod_{t\in B} \langle \Psi_t \rangle_0}.$$

Since condition (8a) is satisfied for all bounded continuous functions f_t and the linear span of finite products $\prod_{t\in A} f_t$ of such functions is everywhere dense in $c(\Omega)$ (the Stone theorem [16]), we may conclude from condition 4) that the family of measures $\{\mu_\Lambda, \Lambda \in T\}$ is weakly locally compact, and thus our lemma is implied by Lemma 2 of §3, Chapter 3.

We make use of this lemma to get cluster expansions of Gibbs modifications of a Gaussian field.

Let μ_0 be a Gaussian measure on the space R^T so that for each $t \in T$ one has

$$\langle x_t \rangle_0 = 0, \quad \langle x_t^2 \rangle = 1,$$
$$\sum_{t':t'\neq t} |\langle x_t, x_{t'} \rangle_0| < d < 1, \tag{13}$$

and let the Hamiltonian U_Λ be of the form (1), where $\Phi_t = \Phi(x(t))$ obeys the condition (6a).

Theorem 3. *Under the assumptions formulated above and for sufficiently small $\beta > 0$, the partition function Z_Λ admits a cluster representation, and the measures μ_Λ admit a cluster expansion.*

Proof. It suffices to verify the assumptions of Lemma 2. Condition 1) is obvious and condition 2) follows from Theorem 1 in §2, Chapter 2, where $\lambda = C\delta(\beta)$, $C = C(d)$ is a constant and $\delta(\beta) = \langle|\Psi_t - 1|^2\rangle_0$. It follows from (6a) that $\delta(\beta) \to 0$ as $\beta \to 0$.

Condition 3) is verified in a similar way for all functions $f \in \bigcap_{p>1} L_p(R^1, e^{-x^2/2}dx)$ (these functions obey (8a)). Here, we have to set $c(f_t) = \|f_t\|_{L_p}$, where p is any number larger than two. As a compact function from 4), one may choose $h(x) = x^2$.

§7 Perturbation of a Gaussian Field with Slow Decay of Correlations

Here, we present another method of cluster expansion of Gibbs modification of a Gaussian field in Z^ν by means of a Hamiltonian of the form (1) of §6. The conditions to be satisfied by the covariance of the unperturbed field are less restrictive for this method than it was above (see (13) and Theorem 3 in §6). However, the interaction potential $\Phi_t(x)$ has to be a sufficiently smooth (analytic) function in $x \in R^1$.

Let thus $T = Z^\nu$, $S = R^1$, and μ_0 be the distribution of the Gaussian field $\{x_t, t \in Z^\nu\}$, $x_t \in R$, with vanishing mean, $\langle x_t\rangle_0 \equiv 0$, and with the covariance matrix

$$B = \{b_{t,t'}, t, t' \in Z^\nu\},$$

with $b_{t,t'} = \langle x_t x_{t'}\rangle_0$, such that for each $t \in Z^\nu$ one has

$$b_{t,t} = 1, \quad \sup_{t\in Z^\nu} \sum_{t'\in Z^\nu} |b_{t,t'}| < \infty. \tag{1}$$

We consider the Gibbs modifications $\{\mu_\Lambda, \Lambda \subset Z^\nu\}$ of the measure μ_0 specified by means of a Hamiltonian of the form (1) of §6 with a potential

$$\Phi_{\{t\}}(x) = P(x_t), \tag{2}$$

where $P(z) = z^l + a_1 z^{l-1} + \ldots$ is a polynomial of an even order, with the coefficient of the leading power being equal to 1.

Theorem 1. *For sufficiently small β, $0 < \beta < \beta_0$ (with β_0 depending on the covariance B and on the polynomial P), the partition functions Z_Λ corresponding to Gibbs modifications μ_Λ admit a cluster representation, and the measures μ_Λ admit a cluster expansion.*

Proof. For any set $D \subset Z^\nu$ and any number s, $0 \leqslant s \leqslant 1$, we use $L(D,s)$ to denote the mapping acting in the set of infinite symmetric matrices

$$C = \{C_{t,t'}\}_{t,t' \in Z^\nu}$$

according to the formula

$$[L(D,s)C]_{t,t'} = \begin{cases} C_{t,t'}, & \text{if } t,t' \in D \text{ or } t,t' \in Z^\nu \backslash D, \\ sC_{t,t'} & \text{otherwise.} \end{cases} \tag{3}$$

It is easy to verify that $L(D,s)$ transforms a positive definite matrix again into a positive definite matrix.

Further, let $\{f_t, t \in \Lambda\}$, $\Lambda \subset Z^\nu$, be a collection of infinitely differentiable functions in $x \in R^1$, bounded together with all their derivatives. Applying the formula of integration per partes (23) of §2, Chapter2, to the means $\langle \prod_{t\in\Lambda} f_t(x(t)) \rangle_0$, we get for an arbitrarily chosen site $t_0 \in \Lambda$ the equality

$$\begin{aligned} \left\langle \prod_{t\in\Lambda} f_t(x(t)) \right\rangle_0 &= \langle f_{t_0}(x(t_0))\rangle_0 \left\langle \prod_{t\in\Lambda\backslash\{t_0\}} f_t(x(t)) \right\rangle_0 \\ &+ \int_0^1 \Bigg[\sum_{t_1\in\Lambda\backslash\{t_0\}} \langle x(t_0)x(t_1)\rangle_0 \Big\langle f'_{t_0}(x(t_0)) f'_{t_1}(x(t_1)) \\ &\times \prod_{t\in\Lambda\backslash\{t_0,t_1\}} f_t(x(t)) \Big\rangle_{\{t_0\},s} \Bigg] ds, \end{aligned} \tag{4}$$

where we use $\langle\cdot\rangle_{D,s}$ to denote the mean under the Gaussian measure with the covariance matrix $L(D,s)B$ and $f' = df/dx$. By successively applying this device, we derive the following formula:

$$\langle f_\Lambda \rangle_0 = \langle f_{t_0}\rangle_0 \langle f_{\Lambda\backslash\{t_0\}}\rangle_0 + \sum_{\substack{B\subseteq\Lambda: t_0\in B, \\ |B|\geqslant 2}} k_B^{t_0} \langle f_{\Lambda\backslash B}\rangle_0, \tag{5}$$

where we denote $f_A = \prod_{t\in A} f_t$ for each $A \subseteq \Lambda$, and the quantities $k_B^{t_0}$ are defined in the following way. Let us use

$$\tau = \{(t_0,t_1),(t_{i_2},t_2),\ldots,(t_{i_n},t_n)\} \tag{5'}$$

to denote a sequence of n pairs (t_{i_m},t_m) of lattice sites of Z^ν such that $t_i \neq t_j$ for $i \neq j$, $i,j = 1,\ldots,n$, $i_m < m$, $m = 2,\ldots,n$, and $s = (s_1,\ldots,s_n)$ to denote a sequence of n numbers $0 \leqslant s_i \leqslant 1$. For any sequence τ and any $k \leqslant n$, we use $D_k(\tau) = D_k$ to denote the set $D_k = \{t_0, \ldots, t_k\}$ and $\mu_{\tau,s}$ to denote the Gaussian measure with vanishing mean and the covariance matrix $\mathcal{B}_{\tau,s} = L(D_{n-1}, s_n) \ldots L(D_0, s_1)\mathcal{B}$ (here, we use the notation $\langle\cdot\rangle_{\mu_{\tau},s} = \langle\cdot\rangle_{\tau,s}$). In this notation

$$k_B^{t_0} = \sum_{\tau} \prod_{m-1}^{n} \langle x(t_{i_m}) x(t_m)\rangle_0 \int_0^1 \varphi(\tau, s) \left\langle \frac{\partial}{\partial x} \prod_{i=0}^{n} f_{t_i} \right\rangle_{\tau,s} ds, \tag{6}$$

where the sum is taken over all τ such that $D_0(\tau) = \{t_0\}$ and

$$D_n(\tau) \equiv D_n = B \; (|B| = n+1).$$

Here also

$$\varphi(\tau, s) = \prod_{m=2}^{n} \prod_{i=i_m+1}^{m-1} s_i \tag{7}$$

(we put $\prod_{i=i_m+1}^{m-1} s_i = 1$ for $i_m + 1 = m$) and

$$\frac{\partial}{\partial x_\tau} = \prod_{m=1}^{n} \frac{\partial^2}{\partial x(t_{i_m}) \partial x(t_m)} = \frac{\partial^{2n}}{\prod_{i=1}^{n} \partial x(t_i)^{n_{t_i}}}, \tag{8}$$

where n_i is the number of pairs in τ containing the site t. Finally,

$$\int_0^1 \ldots ds = \underbrace{\int_0^1 \cdots \int_0^1}_{n \text{ times}} \cdots \prod_{i=1}^{n} ds_i.$$

We now introduce a lexicographic order in the lattice Z^ν and from the recursive relation (5) derive

$$Z_\Lambda = k_{\{t_0\}} Z_{\Lambda \setminus \{t_0\}} + \sum_{\substack{B \subseteq \Lambda : t_0 \in B, \\ |B| \geqslant 2}} k_B Z_{\Lambda \setminus B}, \tag{9}$$

where t_0 is the smallest site in Λ, and the quantity $k_B = k_B^{t_0}$, $|B| \geqslant 2$, is defined according to formula (6) with

$$\begin{aligned} f_t(x) &= \exp\{-\beta P(x)\}, \\ k_{\{t_0\}} &= \langle \exp\{-\beta P(x(t_0))\}\rangle_0. \end{aligned} \tag{10}$$

Formula (9) is equivalent (see Lemma 1 of §1, Chapter 3) to the cluster representation (1) in §1, Chapter 3, of the partition function Z_Λ with the quantities k_B.

We pass now to the cluster estimate.

In the first place, with the help of the Jensen inequality, we get for sufficiently small $\beta > 0$ the bound

$$k_{\{t\}} > \exp\{-\beta\langle P(x(t))\rangle_0\} > K > 0. \tag{11}$$

Secondly, we observe that for any n, one has

$$\frac{\partial^n}{\partial x^n} e^{-\beta P(x)} = \beta^{n/l} \frac{\partial^n}{\partial \xi^n} e^{-P^\beta(\xi)} = \beta^{n/l} \frac{n!}{2\pi i} \int\limits_{|z|=1} \frac{\exp(-P^\beta(\xi+z))}{z^{n+1}}\, dz, \tag{12}$$

where $\xi = \beta^{1/l}x$, $P^\beta(x) = \xi^l + a_1\beta^{1/l}\xi^{l-1} + a_2\beta^{2/l}\xi^{l-2} + \dots$.

Since $\operatorname{Re} P^\beta(s) > -C$ in the strip $|\operatorname{Im} z| \leqslant 1$, we get from the representation (12) the estimate

$$\left|\frac{\partial^n}{\partial x^n} \exp\{-\beta P(x)\}\right| < \beta^{n/l} e^C n!. \tag{13}$$

From this estimate, we get

$$\left|\left\langle \frac{\partial}{\partial x_\tau} \prod_{t\in D(\tau)} e^{-\beta P(x(t))} \right\rangle_{\tau,s}\right| < (C_1\beta^{1/l})^{2n} \prod_{t\in D(\tau)} n_t!, \quad C_1 = e^C. \tag{14}$$

For each sequence τ, we use $\mathcal{T}(\tau)$ to denote the tree with the set $D(\tau)$ of vertices among which the vertex $t_0 \in D(\tau)$ is singled out, and with the set of edges $\{t_{i_m}, t_m\} \in \tau$.

Lemma 2. *Let a set $D \subset Z^\nu$ (with a site $t_0 \in D$ singled out) and a tree $\mathcal{T}$ whose set of vertices coincides with D, be fixed. Then,*

$$\sum_{\tau:\mathcal{T}(\tau)=\mathcal{T}} \int\limits_0^1 \varphi(\tau, s)\, ds = 1. \tag{15}$$

Proof. We notice that for a fixed $\mathcal{T}$ and a selected site t_0, a sequence τ such that $\mathcal{T}(\tau) = \mathcal{T}$ is uniquely determined by an enumeration $\widehat{\tau} : t_1, \dots, t_n$ of the points of the set $D\backslash\{t_0\}$; here the enumeration should be such that for each site $t_m \in D$, there exists a unique edge of the tree $\mathcal{T}$ that joins t_m with some $t_j \in D$ labelled by a smaller number: $j < m$. Each such enumeration $\widehat{\tau}$ may be viewed as a trajectory of a random walk on D that starts at t_0 and is determined by the following rule: Supposing that during the first k steps we visited sites $t_1, \dots, t_k \in D$, the next, $(k+1)$-st, site is chosen with the uniform probability among the vertices of $\mathcal{T}$ that have not been visited up to now and that are joined by an edge with at least one of the sites $t_0, \dots, t_k$. The number of such vertices is $n_{t_0} + n_{t_1} + \dots + n_{t_k} - 2k$, with n_t denoting the degree of the vertex $t \in \mathcal{T}$, and thus the probability of a trajectory $\hat{\tau}$ equals

$$\Pr(\hat{\tau}) = \frac{1}{n_{t_0}} \cdot \frac{1}{n_{t_0} + n_{t_1} - 2} \cdot \dots \cdot \frac{1}{n_{t_0} + \dots + n_{t_{n-1}} - 2(n-1)}. \tag{16}$$

On the other hand, it is easy to verify that

$$\varphi(\tau, s) = s_1^{n_{t_0}-1} s_2^{n_{t_0}+n_{t_1}-3} \dots s_n^{n_{t_0}+\dots+n_{t_{n-1}}-2(n-1)-1}, \tag{17}$$

where τ is a sequence of the form (5′) corresponding to the enumeration $\hat{\tau}$. From (16) and (17), we conclude that

$$\Pr(\hat{\tau}) = \int_0^1 \varphi(\tau, s)\, ds,$$

which implies (15).

Using (6), (14), and (15), we find the bound

$$|k_B| < \sum_{\mathcal{T}} (C_1^2 \beta^{2/l})^{|B|-1} \prod_{(t,t') \in \mathcal{T}} |b_{t,t'}| \prod_{t \in B} (n_t!), \tag{18}$$

where the summation is taken over all trees $\mathcal{T}$, with the set of vertices coinciding with B. From the estimate (1) and Lemma 10 in §4, Chapter 2, we get a cluster estimate for the quantities k_B with the cluster parameter $\overline{C}\beta^{1/l}$.

To deduce a cluster expansion of the measures μ_Λ, we consider the functions F_A of the form

$$F_A = \prod_{t \in A} g_t(x(t)), \tag{19}$$

where $g_t(x)$, $x \in R^1$, is an analytic and bounded function in the half plane $(\mathrm{Im}\, z) \leqslant 1$:

$$|g_t(z)| < L, \quad |\mathrm{Im}\, z| \leqslant 1, \quad t \in A. \tag{20}$$

Repeating the preceding arguments we obtain a cluster expansion of the mean $\langle F_A \rangle_\Lambda$ of the form (5) of §3, Chapter 3, with the bound (6) of §3, Chapter 3 with

$$m_A(F_A) = L^{|A|},$$

and

$$r_B = \sum_{\tau} (C_1 \beta^{1/l})^{\tilde{n}} \prod_{(t,t') \in \tau} |b_{t,t'}| \prod_{t \in B} n_t!,$$

where the sum $\sum_\tau$ has the same meaning as in (18) and $\tilde{n} = \sum_{t\in B\setminus A} n_t$. Since $\tilde{n} \geqslant |B - A|$, we get, again with the help of Lemma 10 in §4 of Chapter 2, the estimate (7), §3, Chapter 3.

Linear combinations of functions of the form (19) obey the condition (20) and are everywhere dense in the space $C(\Omega)$ (the Stone theorem). Besides, with the help of the preceding construction, we find that for the function $h(x) = x^2$, the mean is bounded:

$$\langle h(x(t))\rangle_\Lambda < C,$$

with the constant C not depending on either t or Λ. Thus, the family of measures $\{\mu_\Lambda, \Lambda \subset T\}$ is weakly locally compact, and by applying Lemma 2 of §3, Chapter 3, we get the assertion of the theorem.

§8 Modifications of d-Markov Gaussian Fields (Interpolation of Inverse Covariance)

The method presented here allows us to obtain a cluster expansion of the mean $\langle F_A\rangle_\Lambda$ of a function F_A without any assumption about smoothness of interactions that was needed in the preceding section. Nevertheless, we suppose that the correlations $\langle x(t)x(t')\rangle_0$ of the original Gaussian field decay exponentially: $\langle x(t)x(t')\rangle_0 \sim e^{-\alpha\rho(t,t')}$ as $\rho(t,t') \to \infty$ with an arbitrary small exponent $\alpha > 0$.

The method of interpolations presented here is based on the following simple formula.

Lemma 1. *Let $f(s_1, \ldots, s_n)$ be a smooth real function defined for $0 \leqslant s_i \leqslant 1$, $i = 1, \ldots, n$; then*

$$f(1, \ldots, 1) = f(0, \ldots, 0) + \sum_A \int_0^1 \frac{\partial}{\partial s_A} f(s_A, 0)\, ds_A, \tag{1}$$

where the summation is over all nonempty subsets $A \subseteq \{1, \ldots, n\}$, and we use the notation

$$\frac{\partial}{\partial s_A} = \prod_{i\in A} \frac{\partial}{\partial s_i}, \quad ds_A = \prod_{i\in A} ds_i, \quad f(s_A, 0) = f(s'_1, \ldots, s'_n),$$

and

$$s'_i = \begin{cases} s_i, & i \in A, \\ 0 & i \notin A. \end{cases}$$

The proof is obtained by applying the Newton–Leibnitz formula

$$\psi(1) = \psi(0) + \int_0^1 \frac{d\psi}{ds}\, ds \tag{2}$$

subsequently to each of the variables $s_1, \dots, s_n$.

Let us consider the Gaussian field $\{x(t), t \in Z^\nu\}$ with vanishing mean and the covariance matrix $B = \{b_{t,t'}, t, t' \in Z^\nu\}$, $b_{t,t'} = \langle x(t)x(t')\rangle_{\mu_0}$, where μ_0 is the measure on the space $\Omega = R^{Z^\nu}$ corresponding to this field. We shall suppose that the field is d-Markov; it means that the matrix $A = B^{-1} = \{a_{t,t'}, t, t' \in Z^\nu\}$ of the inverse covariance is of finite range, i.e.,

$$a_{t,t'} = 0 \quad \text{for} \quad \rho(t,t') > d \tag{3}$$

for some number d, $0 < d < \infty$. The metric ρ is defined by the formula (1) of §0, Chapter 1.

Further, we shall always suppose that the matrix A obeys the following conditions:

$$\sum_{t':t' \neq t} |a_{t,t'}| < \delta\, |a_{t,t}| \tag{4}$$

for some $0 < \delta < 1$,

$$a_{t,t} > \kappa_1 > 0 \tag{5}$$

and

$$a_{t,t} < \kappa_2 < \infty \tag{6}$$

uniformly in $t \in Z^\nu$

Lemma 2. *Let a positive definite matrix $A = \{a_{t,t'}, t, t' \in T\}$ be given, where T is a finite or countable set with some metric ρ, and A obeys the conditions (3)–(5). Then there exists the inverse matrix $A^{-1} = \{a^{(-1)}_{tt'}\}$ that is positively definite. Its matrix elements satisfy the conditions*

$$\left| a^{(-1)}_{t,t'} \right| \leqslant \sigma e^{-m_0 \rho(t,t')}, \tag{7}$$

with $\sigma > 0$ and $m_0 > 0$ depending only on d, δ, and κ_1.

Proof. We use ξ to denote the class of matrices $A = \{a_{t,t'}, t, t' \in T\}$ with finite norm

$$||A|| = \sup_{t \in T} \sum_{t' \in T} |a_{t,t'}|. \tag{8}$$

It is easy to verify that ξ forms a Banach algebra with respect to this norm. An arbitrary matrix A obeying the conditions of the lemma can be represented in the form

$$A = A_0 + A_1 = A_0(E + D), \tag{8'}$$

where A_0 is a diagonal matrix, the diagonal elements of A_1 vanish, and $D = A_0^{-1}A_1$.

It follows from condition (4) that

$$\|D\| \leqslant \delta. \tag{9}$$

Hence,

$$A^{-1} = (E+D)^{-1}A_0^{-1} = \sum_{n=0}^{\infty}(-D)^n A_0^{-1}. \tag{10}$$

As follows from (3), the matrix elements

$$((-D)^n A_0^{-1})_{t,t'} = 0, \tag{11}$$

whenever $\rho(t,t') > nd$. Hence,

$$\left|a_{t,t'}^{(-1)}\right| = \left|\sum_{n \geqslant (1/d)\rho(t,t')} ((-D)^n A_0^{-1})_{t,t'}\right| \leqslant \frac{\delta^{\frac{1}{d}\rho(t,t')}}{\kappa_1(1-\delta)}. \tag{12}$$

The lemma is proved.

Remark. The expansion (10) rewritten in the form

$$a_{t,t'}^{(-1)} = \sum (a_{t_1,t_1})^{-1}(-a_{t_1,t_2})(a_{t_2,t_2})^{-1} \dots (a_{t_n,t_n})^{-1},$$

where the sum is over the collections $(t_1,\dots,t_n)$ with $t_1 = t$ and $t_n = t'$, $t_i \neq t_{i+1}$, is sometimes called a representation of the covariance matrix in terms of random walks (see [83]).

Let us consider the Gaussian measure μ_Λ^0 on $R^\Lambda, \Lambda \subset Z^\nu$, defined by the density (with respect to the Lebesgue measure $(dx)^\Lambda$ on R^Λ),

$$D_\Lambda \exp\{-(A_\Lambda x^\Lambda, x^\Lambda)\}, \tag{13}$$

where A_Λ is the restriction of the matrix A onto Λ, i.e.,

$$a_{t,t'}^\Lambda = \begin{cases} a_{t,t'}, & t,t' \in \Lambda \\ 0 & \text{otherwise,} \end{cases}$$

and D_Λ is the normalization factor.

Theorem 3. *Let $\Phi(x)$ be a function of $x \in R$ bounded from below and let $\beta > 0$. Then the system of measures μ_Λ defined by the density*

$$\frac{d\mu_\Lambda}{d\mu_\Lambda^0} = Z_\Lambda^{-1} \exp\left\{-\beta \sum_{t\in\Lambda} \Phi(x(t))\right\} \tag{14}$$

admits a cluster expansion.

Proof. Let $\eta = \{\Delta_n, n \in Z^\nu\}$ be a partition of the lattice into disjoint congruent cubes of a sufficiently large side $L > d$ to be specified later. Let Λ be a union of cubes Δ: $\Lambda = \bigcup_{\Delta\in\eta_\Lambda} \Delta$, $\eta_\Lambda \subset \eta$.

Let further $\bar{s} = \{s_\Delta, \Delta \in \eta_\Lambda\}$ be a collection of numbers $0 \leqslant s_\Delta \leqslant 1$. For each $\bar{s}$, we define the matrix

$$A_\Lambda(\bar{s}) = \{a^\Lambda_{t,t'}(\bar{s}), t, t' \in \Lambda\}, \tag{15}$$

with

$$a^\Lambda_{t,t'}(\bar{s}) = a_{t,t'}[\delta_{\Delta(t),\Delta(t')} + s_{\Delta(t)}s_{\Delta(t')}(1 - \delta_{\Delta(t),\Delta(t')})], \tag{16}$$

where $\Delta(t)$ is the cube from η_Λ containing the site t. The matrix $A_\Lambda(\bar{s})$ is positively definite. This fact follows from the following representation of the matrix $A_\Lambda(\bar{s})$:

$$A_\Lambda(\bar{s}) = \left(\prod_{\Delta\in\eta_\Lambda} L(\Delta, s_\Delta)\right) A_\Lambda, \tag{17}$$

where $L(\Delta, s_\Delta)$ is the transformation in the space of symmetric matrices introduced in the preceding section (see (3), §7); the transformation $L(\Delta, s_\Delta)$ preserves the positive definiteness of matrices. Moreover, $A_\Lambda(\bar{s})$ obeys the requirements (3)–(6). We shall use $\mu^0_\Lambda(\bar{s})$ to denote the Gaussian measure with vanishing mean and the covariance matrix $(A_\Lambda(\bar{s}))^{-1}$; clearly, $\mu^0_\Lambda = \mu^0_\Lambda(\bar{1})$, where $\bar{1}$ is the collection $\bar{s}$ with $s_\Delta \equiv 1$. We use the notation $\langle\cdot\rangle_{\mu^0_\Lambda(\bar{s})} = \langle\cdot\rangle_{\bar{s}}$.

The partition function $Z_\Lambda = \langle\exp\{-U_\Lambda\}\rangle_{\mu^0_\Lambda}$, $U_\Lambda = \beta\sum_{t\in\Lambda}\Phi(x(t))$, can be represented in the form (see Lemma 1)

$$Z_\Lambda = \langle\exp\{-U_\Lambda\}\rangle_{\bar{1}} = \langle\exp\{-U_\Lambda\}\rangle_{\bar{0}} + \sum_{\gamma\subseteq\eta_\Lambda} \int \frac{\partial}{\partial\bar{s}_\gamma}\langle\exp\{-U_\Lambda\}\rangle_{\bar{s}_\gamma}\, d\bar{s}_\gamma, \tag{18}$$

where the sum is taken over all nonempty $\gamma \subseteq \eta_\Lambda$ and $\bar{s}_\gamma$ is the collection $\{s_\Delta\}$ with $s_\Delta = 0$ whenever $\Delta \notin \gamma$. We use $\bar{0}$ to denote the collection with $s_\Delta \equiv 0$.

Two cubes $\Delta, \Delta' \in \eta$ are called neighbouring if their distance does not exceed ν, $\rho(\Delta, \Delta') = \min_{t,t'} \rho(t, t') \leqslant \nu$. A collection $\gamma \subseteq \eta_\Lambda$ of cubes is called ν-connected, if for any splitting into two disjoint collections γ' and γ'' we shall find in them neighbouring cubes $\Delta' \in \gamma'$ and $\Delta'' \in \gamma''$.

Clearly,

$$\langle e^{-U_\Lambda}\rangle_{\bar{0}} = \prod_{\Delta\in\eta_\Lambda} \langle e^{-U_\Delta}\rangle_{\bar{0}}, \tag{19}$$

and

$$\int_0^1 \frac{\partial}{\partial\bar{s}_\gamma}\left\langle e^{-U_\Lambda}\right\rangle_{\bar{s}_\gamma} d\bar{s}_\gamma = \prod_{\Delta\in\eta_\Lambda\setminus\gamma} \langle e^{-U_\Delta}\rangle_{\bar{0}} \prod_{i=1}^{s} \int \frac{\partial}{\partial\bar{s}_{B_i}} \left\langle e^{-\sum_{\Delta\in b_i} U_\Delta}\right\rangle_{\bar{s}_{B_i}} d\bar{s}_{B_i}, \tag{20}$$

where $B_1, \ldots, B_s \in \eta_\Lambda$ are connected components of the collection γ, $|B_i| \geqslant 2$ (in case there exists a connected component of γ consisting of a single cube Δ, $\partial/(\partial\bar{s}_\gamma)\left\langle e^{-U_\Delta}\right\rangle_{\bar{s}_\gamma} = 0$).

Putting now for $B \subset \eta$,

$$k_B = \begin{cases} \displaystyle\int \frac{\partial}{\partial\bar{s}_B}\left\langle \exp\left\{-\sum_{\Delta\in B} U_\Delta\right\}\right\rangle_{\bar{s}_B} d\bar{s}_B, & \text{if the collection } B \subset \eta \text{ is } \nu\text{-connected and } |B| \geqslant 2, \\ \left\langle e^{-U_\Delta}\right\rangle_{\bar{0}}, & \text{if } B = \{\Delta\}, \\ 0 & \text{otherwise,} \end{cases} \tag{21}$$

we get from (18), (19), and (20) the cluster representation of Z_Λ:

$$Z_\Lambda = \sum_{\{B_1,\ldots,B_s\}} k_{B_1} \ldots k_{B_s} \prod_{\Delta\in\eta_\Lambda\setminus(\cup B_i)} k_{\{\Delta\}}, \tag{22}$$

with the sum taken over all $(\nu+1)$-admissible unordered collections $\{B_1, \ldots, B_s\}$ of γ-connected subsets $B_i \subseteq \eta_\Lambda$. We pass now to the derivation of a cluster estimate.

Lemma 4. *For sufficiently small $\beta > 0$ and a suitable choice of $L = L(\beta)$ for any $B \subset \eta$, $|B| \geqslant 2$, we have*

$$|k_B| < (x(\beta))^{|B|}, \tag{23}$$

$$\left|k_{\{\Delta\}}\right| > K > 0, \tag{24}$$

with a constant K and $x(\beta) \to 0$ as $\beta \to 0$.

Proof. Whenever $\gamma \subset \eta$ is a nonempty collection, we set $\tilde{k}_\gamma = \prod_{\Delta\in\gamma} \tilde{k}_\Delta$, $\tilde{k}_\Delta = e^{-U_\Delta} - 1$. Passing to the usual expansion of the exponent, $e \exp\{-\sum_{\Delta\in B} U_\Delta\} = 1 + \sum_{\substack{\gamma\subseteq B \\ \gamma\neq\emptyset}} \tilde{k}_\gamma$, we get

$$k_B = \sum_{\substack{\gamma\subseteq B, \\ \gamma\neq\emptyset}} \int \frac{\partial}{\partial\bar{s}_B} \langle \tilde{k}_\gamma \rangle_{\bar{s}_B} \, d\bar{s}_B. \tag{25}$$

For any collection $\varepsilon_B = \{\varepsilon_\Delta, \Delta \in \eta_\Lambda\}$ with ε_Δ sufficiently small for $\Delta \in B$ and $\varepsilon_\Delta = 0$ for $\Delta \in \eta_\Lambda \setminus B$, we have

$$\langle \tilde{k}_\gamma \rangle_{\bar{s}_B+\varepsilon_B} = \frac{\left\langle \tilde{k}_\gamma \exp\left(-\sum_{(\Delta,\Delta')} a_{\Delta,\Delta'}\delta_{\Delta,\Delta'}\right)\right\rangle_{\bar{s}_B}}{\left\langle \exp\left(-\sum_{(\Delta,\Delta')} a_{\Delta,\Delta'}\delta_{\Delta,\Delta'}\right)\right\rangle_{\bar{s}_B}}, \tag{26}$$

where

$$a_{\Delta,\Delta'} = \sum_{t\in\Delta, t'\in\Delta'} a_{t,t'} \, |x(t)x(t')| \, ,$$

$$\delta_{\Delta,\Delta'} = s_\Delta \varepsilon_{\Delta'} + s_{\Delta'}\varepsilon_\Delta + \varepsilon_\Delta \varepsilon_{\Delta'},$$

and the sum $\sum_{\Delta,\Delta'}$ is taken over all unordered pairs of neighbouring cubes $\Delta, \Delta' \in B$.

Using the formula (21) of §1, Chapter 2, we expand the mean (26) into a formal power series in the variables $\delta_{\Delta,\Delta'}$:

$$\langle \tilde{k}_\gamma \rangle_{\bar{s}_B+\varepsilon_B} = \langle \tilde{k}_\gamma \rangle_{\bar{s}_B} + \sum \frac{1}{n!} \sum_\sigma \prod_{i=1}^n \delta_{\Delta_i,\Delta_i'} \times \left\langle k_\gamma, a_{\Delta_1,\Delta_1'}, a_{\Delta_2,\Delta_2'}, \ldots, a_{\Delta_n,\Delta_n'} \right\rangle_{\bar{s}_B}, \tag{27}$$

with the sum taken over all ordered collections $\sigma = (\langle \Delta_1, \Delta_1' \rangle, \ldots, \langle \Delta_n, \Delta_n' \rangle)$ of pairs of neighbouring cubes. Rewriting this series as a power series in the variables ε_Δ and observing that the derivative $(\partial/\partial\bar{s}_B)\langle \tilde{k}_\gamma \rangle_{\bar{s}_B}$ equals the coefficient of the monominal $\prod_{\Delta\in B} \varepsilon_\Delta$, we get

$$\frac{\partial}{\partial \bar{s}_B} \langle \tilde{k}_\gamma \rangle_{\bar{s}_B} = \sum_n \sum_\sigma{}' C_{n,\sigma} \frac{1}{n!} \left\langle k_{\gamma'}, a_{\Delta_1,\Delta_1'}, \ldots, a_{\Delta_n,\Delta_n'} \right\rangle_{\bar{s}_B}, \tag{28}$$

with $|C_{n,\sigma}| \leqslant 1$ and $\sum'_\sigma$ taken over all σ such that each cube $\Delta \in B$ enters at least one and at most $(r+1)$ pairs from the collection σ (r is the number of cubes $\Delta' \in \eta$ neighbouring the given cube $\Delta \in \eta$).

Hence, $2n \geqslant |B| \geqslant 2n/(r+1)$ and the derivative (28) can be estimated:

$$\begin{aligned} \left| \frac{\partial}{\partial \bar{s}_B} \langle \tilde{k}_\gamma \rangle_{\bar{s}_B} \right| &\leqslant C_1^{|B|} \max_\sigma \left| \left\langle \tilde{k}_\gamma, a_{\Delta_1,\Delta_1'}, \ldots, a_{\Delta_n,\Delta_n'} \right\rangle_{\bar{s}_B} \right| \\ &\leqslant C_2^{|B|} (L^\nu)^{|B|(r+1)} \max_{\hat{\sigma}} \left| \langle \tilde{k}_\gamma, x(t_1)x(t_1'), \ldots, x(t_n)x(t_n') \rangle_{\bar{s}_B} \right|, \end{aligned} \tag{29}$$

with $\max_{\hat{\sigma}}$ taken over all collections $\hat{\sigma} = ((t_1, t_1'), \ldots, (t_n, t_n'))$ of pairs of sites such that $\Delta(t_i)$ and $\Delta(t_i')$ are neighbouring cubes and the collection $\sigma = (\langle \Delta(t_1), \Delta(t_1') \rangle, \ldots, \langle \Delta(t_n), \Delta(t_n') \rangle)$ enters the sum (28).

To estimate the semi-invariant, we expand $\tilde{k}_\gamma$ in Wick polynomials. But first we pass from the Gaussian system of random variables $\{x(t), t \in \tilde{\gamma}\}$ to the system $\{\hat{x}(t), t \in \tilde{\gamma}'\}$ of independent Gaussian variables with vanishing mean and the dispersion equaling unity:

$$\hat{x}(t) = \sum_{t'\in\tilde{\gamma}} C_{t,t'} x(t'), \tag{30}$$

where the matrix $C_{\tilde{\gamma}} = C = \{C_{t,t'}, t, t' \in \tilde{\gamma}\}$ should be chosen so that

$$CBC^T = E, \tag{31}$$

where $B = B_{\tilde{\gamma}} = \{\langle x_t x_{t'} \rangle_{\bar{s}_B}, t, t' \in \tilde{\gamma}\}$ is the covariance matrix. Writing now B^{-1} as in (10),

$$B^{-1} = A_0 + A_1 = A_0(E + D), \quad A_0 = A_0^{\tilde{\gamma}}, \quad \text{etc.}, \tag{32}$$

we put

$$C_{\tilde{\gamma}} = A_0^{1/2}(E + D)^{1/2}, \tag{33}$$

with $(E + D)^{1/2}$ being defined by the expansion into series in powers of $D = D^{\tilde{\gamma}}$. This expansion is based on the estimate $\|D^{\tilde{\gamma}}\| < \delta$ (see (9)), which follows from Lemma 5 below.

With the help of the representation $B = (E + D)^{-1} A_0^{-1}$ and the equation

$$A_0^{-1}(E + D^T)^{1/2} = (E + D)^{1/2} A_0^{-1},$$

following from the obvious relation $A_0^{-1} D^T A_0 = D$, we conclude that C satisfies (31).

Lemma 5. *Let the inverse covariance matrix $B_\Lambda^{-1} = A_\Lambda$ of a Gaussian field $\{x(t), t \in \Lambda\}$ in Λ satisfy the conditions (3)–(6). Then for each $\Lambda' \subset \Lambda$, the inverse covariance matrix $A_{\Lambda'} = B_{\Lambda'}^{-1}$ (where $B_{\Lambda'}$ is the restriction of B_Λ to Λ') satisfies the conditions (4) and (6) with the same constants δ and κ_2 and the condition (5) with the constant $\kappa_1' = \kappa_1(1 - \delta)$.*

Furthermore,

$$|\text{Det}\, A_{\Lambda'}| \leqslant M^{|\Lambda'|}, \tag{34}$$

where M is a constant depending on κ_1, κ_2 and δ.

Moreover, the norm (8) of the matrix $C_{\Lambda'}$, constructed by the formulas (32) and (33) from the matrix $B_{\Lambda'}$, satisfies the estimate

$$\|C_{\Lambda'}\| < K_1, \tag{35}$$

with K_1 depending only on δ and κ_2.

Proof. To prove the first assertion, it suffices to consider the case $\Lambda' = \Lambda \setminus \{t_0\}$. Straightforward computations show that for $A_{\Lambda'} = (a_{t,t'}^{\Lambda'}, t, t' \in \Lambda')$,

$$a_{t_1,t_2}^{\Lambda'} = a_{t_1,t_2}^{\Lambda} - \frac{a_{t_1,t_0}^{\Lambda} a_{t_2,t_0}^{\Lambda}}{a_{t_0,t_0}^{\Lambda}}, \quad t_1, t_2 \in \Lambda', \tag{36}$$

and, in particular,

$$a_{t,t}^{\Lambda'} = a_{t,t}^{\Lambda} - (a_{t,t_0}^{\Lambda})^2 / a_{t_0,t_0}^{\Lambda}. \tag{37}$$

For any $t_1 \in \Lambda'$, we denote

$$a_1 = \sum_{\substack{t_2: t_2 \in \Lambda', \\ t_2 \neq t_1}} \left| a^{\Lambda'}_{t_1,t_2} \right|, \quad a_2 = \sum_{\substack{t_2: t_2 \in \Lambda \\ t_2 \neq t_1}} \left| a^{\Lambda}_{t_1,t_2} \right|,$$

$$b = \left| a^{\Lambda}_{t_1,t_0} \right|, \quad d = a^{\Lambda}_{t_0,t_0}, \quad c = \sum_{t: t \in \Lambda'} |a_{t,t_0}|, \quad d' = a^{\Lambda}_{t_1,t_1}.$$

Then from (36) and (37), it follows that

$$a_1 \leqslant a_2 - b + (bc - b^2)/d \leqslant \delta \left(d' - b^2/d \right) - (1 - \delta) b^2/d \leqslant \delta a^{\Lambda'}_{t_1,t_1},$$

since $a_2 < \delta d'$, $c < \delta d$. Thus $A_{\Lambda'}$ obeys condition (4).

Condition (6) follows from the inequality $a^{\Lambda'}_{t,t'} \leqslant a^{\Lambda}_{t,t}, t \in \Lambda'$. This and also the estimate (12) imply $a^{\Lambda'}_{t,t} \geqslant a^{\{t\}}_{t,t} = b^{-1}_{t,t} > \kappa_1(1 - \delta)$, i.e., condition (5).

To prove (34), we notice that

$$\operatorname{Det} A_{\Lambda'} = \left(\frac{1}{2\pi} \right)^{|\Lambda'|} \left(\int_{R^{\Lambda'}} \exp \left(-\frac{1}{2} \sum a^{\Lambda'}_{t,t'} x(t) x(t') \right) \prod_{t \in \Lambda'} dx(t) \right)^{-2}. \quad (38)$$

From the condition (4), it is easy to deduce that

$$\frac{1}{2} \sum_{t,t' \in \Lambda'} a_{t,t'} x(t) x(t') \leqslant \frac{1}{2} \sum_{t \in \Lambda'} a_{t,t} (1 + \delta) x^2(t) < \frac{1}{2} \kappa_2 (1 + \delta) \sum_{t \in \Lambda'} x^2(t),$$

and thus,

$$\int_{R^{\Lambda'}} \exp \left(-\frac{1}{2} \sum_{t,t' \in \Lambda'} a^{\Lambda}_{t,t'} x(t) x(t') \right) \prod_{t \in \Lambda'} dx(t) > (\text{const})^{|\Lambda'|}. \quad (39)$$

The equality (38) together with (39) implies (34). Furthermore, (4) implies $\|(A_0^{\Lambda'})^{-1} A_1^{\Lambda'}\| = \|D^{\Lambda'}\| < \delta$ (see (8')) and thus $\|C^{\Lambda'}\| \leqslant \kappa_2^{1/2} (1+\delta)^{1/2}$, which means (35). The lemma is proved.

From the lemma proved above and Lemma 2, it follows that

$$|\langle \widehat{x}(t') x(t) \rangle_{\bar{s}_B}| \leqslant \sum_{\bar{t} \in \widetilde{\gamma}} \left| C_{t',\bar{t}} \right| \left| b_{\bar{t},t} \right| \leqslant K_1 e^{-m_0 \rho(t, \widetilde{\gamma})}, \quad t' \in \widetilde{\gamma}, \quad t \in \Lambda, \quad (40)$$

and

$$\max_{\substack{t \in \widetilde{\gamma}, \\ \bar{t} \in \Lambda}} \left\{ \sum_{t' \in \Lambda} |\langle \widehat{x}(t) x(t') \rangle_{\bar{s}_B}|, \sum_{t' \in \widetilde{\gamma}} |\langle \widehat{x}(t') x(\bar{t}) \rangle_{\bar{s}_B}| \right\} \leqslant \|C^{\widetilde{\gamma}}\| \|\mathcal{B}_\Lambda(\bar{s}_B)\| = K_2. \quad (41)$$

By expanding $\tilde{k}_\gamma$ into normalized Wick polynomials in variables $\{\hat{x}(t), t \in \tilde{\gamma}\}$, we get

$$\tilde{k}_\gamma = \sum_{\{n_t\}} \frac{b\{n_t\}}{\prod\limits_{t \in \tilde{\gamma}} (n_t!)^{1/2}} : \prod_{t \in \tilde{\gamma}} (\hat{x}(t))^{n_t} : . \tag{42}$$

Writing $x(t_1)x(t_2) = :x(t_1)x(t_2): + \text{const}$, we get for each $\hat{\sigma}$:

$$\langle \tilde{k}_\gamma, x(t_1)x(t_1'), \dots, x(t_n)x(t_n') \rangle_{\bar{s}_B}$$
$$= \sum_{\{n_t\}} \frac{b\{n_t\}}{\prod\limits_{\tau \in \tilde{\gamma}} (n_t!)^{1/2}} \left\langle : \prod_{t \in \gamma} (\hat{x}(t))^{n_t} :, :x(t_1)x(t_1'):, \dots, :x(t_n)x(t_n'): \right\rangle_{\bar{s}_B} . \tag{43}$$

The semi-invariant in (43) equals the sum $\sum_G I(G)$ over all connected diagrams without loops (see §2, Chapter 2). Every such diagram splits up into $N/2$ chains, where $N = \sum_{t \in \tilde{\gamma}} n_t$ is even; otherwise the semi-invariant (43) vanishes. Any chain is determined by a sequence of edges

$$(t_0, t_{i_1}), (t'_{i_1}, t_{i_2}), \dots, (t'_{i_q}, \bar{t}_0),$$
$$1 \leqslant i_s \leqslant n, \quad i_s \neq i_{s'}, \quad t_0, \bar{t}_0 \in \tilde{\gamma},$$

and by a label of one among n_{t_0} and $n_{\bar{t}_0}$ legs attached to the vertices $t_0 \in \tilde{\gamma}$ and $\bar{t}_0 \in \tilde{\gamma}$, respectively ($n_{t_0} \neq 0$ and $n_{\bar{t}_0} \neq 0$). Attributing to each chain h the contribution

$$J(h) = \langle \hat{x}(t_0) x(t_{i_1}) \rangle_{\bar{s}_B} \langle x(t'_{i_1}) x(t_{i_2}) \rangle_{\bar{s}_B} \dots \langle x(t'_{i_q}) \hat{x}(\bar{t}_0) \rangle_{\bar{s}_B},$$

we get the contribution $I(G)$ of a diagram G in the form of the product of the contributions of its chains. Introducing the length of the chain h by

$$S(h) = \rho(\tilde{\gamma}, t_{i_1}) + \rho(t'_{i_1}, t_{i_2}) + \dots + \rho(t'_{i_q}, \tilde{\gamma}),$$

it follows from (40) and (7) that

$$|J(h)| \leqslant C^q \exp\left(-\frac{m_0}{2} S(h)\right) |J(h)|^{1/2}, \tag{44}$$

with a constant C.

Lemma 6. *If $|B| > 2r|\gamma|$, where r is the number of cubes $\Delta' \in \eta$ neighbouring a given cube $\Delta \in \eta$, then*

$$\sum_h S(h) \geqslant \frac{1}{2r} |B| \cdot L \tag{45}$$

for each diagram G.

Proof. We shall consider a maximal set of cubes $R=\{\bar{\Delta}_1,\dots,\bar{\Delta}_m\}\subset B\backslash\gamma$ such that $\rho(\bar{\Delta}_i,\bar{\Delta}_j)\geqslant L$ and $\rho(\bar{\Delta}_i,\widetilde{\gamma})\geqslant L$. The number of such cubes clearly satisfies $|R|\geqslant(|B|-r|\gamma|)(1/r)$. The length of each chain passing through k cubes from R is of length of at least kL, and consequently, $\sum_h S(h)\geqslant|R|L$. Hence, the estimate (45) follows for $|B|>2r|\gamma|$. The lemma is proved.

Following (44) and (45), the contribution of any diagram, under the condition $|B|>2r|\gamma|$, does not exceed

$$\exp\left(-\frac{m_0}{2}\frac{1}{2r}|B|\cdot L\right)|I(G)|^{1/2}. \tag{46}$$

Thus, the semi-invariant in (43) does not exceed $\exp\{-\alpha|B|L\}\cdot\sum_G|I(G)|^{1/2}$ ($\alpha>0$) in case $|B|>2r|\gamma|$ and $\sum_G|I(G)|$ in case $|B|\leqslant 2r|\gamma|$. The estimates for the sums $\sum|I(G)|^{1/2}$ and $\sum|I(G)|$ are analogous to (2) of §3, Chapter 2: $\sum|I(G)|\leqslant C^N2^n\prod_{t\in\widetilde{\gamma}}(n_t)!!$ and a similar estimate for $\sum|I(G)|^{1/2}$.

Hence from (46) and (29), and observing that for $\sum n_t>n$ the semi-invariant in (43) vanishes and

$$\left|\sum_{\substack{\{n_t\}:\Sigma n_t\leqslant n\\ t\in\widetilde{\gamma}}} b_{\{n_t\}}C^N\right|\leqslant\left(\sum_{\{n_t\}}\left|b_{\{n_t\}}\right|^2\right)^{1/2}\widetilde{C}^{|B|}=\langle|\tilde{k}_\gamma|^2\rangle_{\bar{s}_B}^{1/2}\widetilde{C}^{|B|},$$

we find that

$$\left|\frac{\partial}{\partial\bar{s}_B}\langle\tilde{k}_\gamma\rangle_{\bar{s}_B}\right|=\begin{cases}(CL^\nu e^{-\alpha L})^{|B|}\langle|\tilde{k}_\gamma|^2\rangle_{\bar{s}_B}^{1/2} & \text{for}\quad |B|>2r|\gamma|,\\ (CL^\nu)^{|B|}\langle|\tilde{k}_\gamma|^2\rangle_{\bar{s}_B}^{1/2} & \text{for}\quad |B|<2r|\gamma|.\end{cases} \tag{47}$$

Lemma 7. *With a suitable choice of $L=L(\beta)>d$, the estimate*

$$\langle|\tilde{k}_\gamma|^2\rangle_{\bar{s}_B}^{1/2}\leqslant\widehat{\kappa}(\beta)^{|\gamma|}, \tag{48}$$

with $\widehat{\kappa}(\beta)\to 0$ as $\beta\to 0$, is valid for every nonempty collection $\gamma\subseteq B$.

Proof. For every cube Δ, we introduce the set

$$A_1^\Delta=\left\{x^\Delta\in R^\Delta:\sum_{t\in\Delta}\Phi(x(t))<L^\nu\beta^{-1/2}\right\}\subset R^\Delta,$$

with $x^\Delta=\{x(t),t\in\Delta\}$. Then, supposing $L^\nu\beta^{1/4}\leqslant 1$, we have

$$\left|\tilde{k}_\Delta(x^\Delta)\right|<C_1\beta^{1/4} \tag{49}$$

for $x^\Delta \in A_1^\Delta$, and

$$\left|\tilde{k}_\Delta(x^\Delta)\right| \leqslant 1, \tag{50}$$

for $x^\Delta \in R^\Delta \backslash A_1^\Delta = A_2^\Delta$. Hence,

$$\left|\langle |\tilde{k}_\gamma|^2\rangle_{\bar{s}_B}\right| \leqslant \sum_{\gamma' \subseteq \gamma} (C_1\beta^{1/4})^{2|\gamma'|} \mu_\Lambda^0(\bar{s}_B)\{x^\Delta \in A_2^\Delta, \Delta \in \gamma - \gamma'\}. \tag{51}$$

Furthermore, it is obvious that

$$\mu_\Lambda^0(\bar{s}_B)\{x^\Delta \in A_2^\Delta, \Delta \in \gamma\backslash\gamma'\} \leqslant (L^\nu)^{|\gamma-\gamma'|} \max_{\tau=\{t_1,\ldots,t_r\}} \mu_\Lambda^0(\bar{s}_B)\{\Omega_\tau\}, \tag{52}$$

where the maximum is over all collections $\tau = \{t_1, \ldots, t_r\}$ of points taken one from each cube $\Delta \in \gamma\backslash\gamma'$, and Ω_τ is the event

$$\Omega_\tau = \{\Phi(x(t_1)) > \beta^{-1/2}, \ldots, \Phi(x(t_r)) > \beta^{-1/2}\}.$$

Then,

$$\begin{aligned}\mu_A^0(\bar{s}_B)\{\Omega_\tau\} &= \frac{(\mathrm{Det}\, A_\tau)^{1/2}}{(2\pi)^{|\tau|/2}} \int\limits_{\Omega_\tau} \exp\left\{-\frac{1}{2}\sum_{t,t'\in\tau} a_{t,t'}^\tau x(t)x(t')\right\} \prod_{t\in\tau} dx(t) \\ &\leqslant \frac{(\mathrm{Det}\, A_\tau)^{1/2}}{(2\pi)^{|\tau|/2}} \left(\int\limits_{\Phi(x)>\beta^{-1/2}} e^{-(1/2)\kappa_1(1-\delta)x^2} dx\right) \leqslant \varepsilon(\beta)^{|\gamma\backslash\gamma'|}.\end{aligned} \tag{53}$$

Here we used the inequality (34) and the estimate

$$\frac{1}{2}\sum_{t,t'\in\tau} a_{t,t'}^\tau x(t)x(t') > \frac{1}{2}\sum_{t\in\tau} a_{t,t}^\tau (1-\delta)x^2(t) \geqslant \frac{1}{2}\kappa_1(1-\delta)^2 \sum_{t\in\tau} x^2(t),$$

following from Lemma 5. We wrote

$$\varepsilon(\beta) = \left(\frac{M}{2\pi}\right)^{1/2} \int\limits_{\Phi(x)>\beta^{-1/2}} e^{-\frac{1}{2}\kappa_1(1-\delta)^2x^2} dx, \tag{54}$$

with M defined in (34).

Further, putting $\widehat{\kappa}(\beta) = (C_1^2\beta^{1/2} + L^\gamma\varepsilon(\beta))^{1/2}$ and

$$L = L(\beta) = [\min\{\beta^{-1/4}, \varepsilon^{-1/(2\nu)}(\beta), (\widehat{\kappa}(\beta))^{-1/(4r)}\}]^{1/\nu},$$

we imply that for sufficiently small β the estimates $L > d$, $\widehat{\kappa}(\beta) < 1$, and also (48) are fulfilled. Lemma 7 is proved.

From Lemma 7, the expansion (25), and the estimate (47), we get the estimate (23) with $\kappa(\beta) = 2C \max\{\widehat{\kappa}(\beta)^{1/4r}, L^\nu e^{-\alpha L}\}$. The estimate (24) is obvious, since $\langle e^{-U_\Delta}\rangle_{\bar{0}} \to 1$ as $\beta \to 0$. Lemma 4 is proved.

It follows from the estimates (23), (24), and Lemma 3 in §4, Chapter 2, that for sufficiently small β the quantities k_B and $k_{\{\Delta\}}$ satisfy the conditions (2) and (3) in §1, Chapter 3, and also the cluster condition (23) in §1, Chapter 3.

Let $F_A, A \subseteq \Lambda$, be a local bounded function of the form $F_A = \prod_{t\in A} f_t$, $f_t(x) = f_t(x(t)), t \in A$. Repeating the preceding arguments, we obtain for the mean $\langle F_A\rangle_\Lambda$ and expansion of the form (5), §3, Chapter 3,

$$\langle F_A\rangle_\Lambda = k_\emptyset(F_A) \cdot f_{\eta_A}^{(\Lambda)} + \sum_{\substack{R\subseteq\eta_\Lambda \\ R\neq\emptyset}} k_R(F_A) f_{R\cup\eta_A}^{(\Lambda)}, \tag{55}$$

where

$$k_\emptyset(F_A) = \left\langle F_A e^{-\sum_{\Delta\in\eta_A} U_\Delta}\right\rangle_{\bar{0}} \left(\prod_{\Delta\in\eta_A} k_{\{\Delta\}}\right)^{-1},$$

$$k_R(F_A) = \int_0^1 \frac{\partial}{\partial \bar{s}_R} \left\langle F_A e^{-\sum_{\Delta\in R\cup\eta_A} U_\Delta}\right\rangle_{\bar{s}_R} d\bar{s}_R \left(\prod_{\Delta\in R\cup\eta_A} k_\Delta\right)^{-1}.$$

Here $f_B^{(\Lambda)}$, $B \subseteq \eta_\Lambda$, is the correlation function corresponding to the cluster representation (22), $\eta_A \subseteq \eta_\Lambda$ is the collection of cubes intersecting A, and the sum in (55) is over all collection of cubes $R \subseteq \eta_\Lambda$ any of whose ν-connected components $\gamma \subseteq R$ intersects η_Λ and $|\gamma| \geqslant 2$. With the help of the preceding constructions, it is easy to get the estimate (6) in §3, Chapter 3, for $k_R(F_A)$ with

$$r_B = \begin{cases} \left(\dfrac{\kappa(\beta)}{K}\right)^{|B|} & \text{for connected } B \subset \eta, |B| \geqslant 2, \\ 0 & \text{otherwise,} \end{cases}$$

and

$$m(F_A) = \prod_{t\in A} \sup|f_t| \cdot \prod_{\Delta\in\eta_A} l_\Delta,$$

$$l_\Delta = \max\left\{1, \langle|k_\Delta|^2\rangle_{\bar{0}}^{1/2} (|\langle k_\Delta\rangle_{\bar{0}}|)^{-1}\right\}.$$

Choosing further the compact function $h_A = \sum_{t\in A} x^2(t)$ for every finite $A \subset Z^\nu$ and with the help of the preceding arguments, we get the estimate

$$\langle h_A\rangle_\Lambda < C_A \tag{56}$$

for any $\Lambda \supseteq A$ with a constant C_A not depending on Λ. Thus, the family of measures $\{\eta_\Lambda\}$, where Λ is an arbitrary set composed entirely from cubes $\Delta \in \eta$, is weakly locally compact, and thus the assertion of Theorem 3 is valid for it. In particular, the limit

$$\lim_{\Lambda \nearrow Z} \mu_\Lambda = \mu \tag{57}$$

exists and

$$\lim_{\Lambda \nearrow Z^\nu} \langle F_A \rangle_\Lambda = \langle F_A \rangle_\mu = \sum_{R \subset \eta} k_R(F_A) f_{R \cup \eta_A} + k_\emptyset(F_A) f_{\eta_A}, \tag{58}$$

when Λ is running through sets of the above mentioned type.

We shall now consider a general case with Λ being no more a union of cubes $\Delta \in \eta$. We shall call boundary any cube $\Delta \in \eta$ such that $\Delta \cap \Lambda \neq \emptyset$, $\Delta \cap (Z^\nu \backslash \Lambda) \neq \emptyset$. A cube $\Delta \subset \Lambda$ neighbouring a boundary cube will be called a bordering cube. The remaining cubes are interior cubes for Λ. Numbering now in an arbitrary way the bordering cubes $\Delta_1, \dots, \Delta_p$, we shall associate with the bordering cube Δ_1 all sites in Λ contained in bordering cubes that neighbour Δ_1; in this way, we get the set $\tilde{\Delta}_1$. With the cube Δ_2, we associate all remaining points in Λ contained in bordering cubes neighbouring Δ_2; we obtain the set $\tilde{\Delta}_2$. Going on with this procedure, we finally obtain the bordering sets $\tilde{\Delta}_1, \dots, \tilde{\Delta}_p$. Together with the interior cubes, they specify a partition $\tilde{\eta}_\Lambda$ of the set Λ.

Repeating now the preceding considerations, we obtain the cluster representation (22) of the partition function Z_Λ and the cluster expansion (55) of the means $\langle F_A \rangle_\Lambda$ with the help of the quantities $k_B^{(\Lambda)}$, $B \subseteq \tilde{\eta}_\Lambda$, and $k_R^{(\Lambda)}(F_A)$, $R \subseteq \tilde{\eta}_\Lambda$, that depend, in general on Λ. However, these quantities satisfy (uniformly in Λ) the estimates obtained above, and moreover, $k_B^{(\Lambda)}$ and $k_R^{(\Lambda)}(F_A)$ depend on Λ only for those collections B and R that contain bordering sets $\tilde{\Delta}_i, i = 1, \dots, p$. For each $\Lambda \subset Z^\nu$, the estimate (56) is also satisfied, and consequently, according to Remark 1 in §3, Chapter 3, the limits (57) and (58) exist for every sequence $\Lambda \nearrow Z^\nu$. The theorem is proved.

Now we shall consider the following generalization of Theorem 3. Let $\Lambda \subset Z^\nu$ and $y = y^{\partial\Lambda} = \{y(t), t \in \partial_d\Lambda\}$ be a boundary configuration defined on the set $\partial_d\Lambda = \{t \in Z^\nu \backslash \Lambda, \rho(t, \Lambda) \leqslant d\}$ and $\mu^0_{\Lambda,y}$ be the Gaussian measure with the same inverse covariance matrix $A_\Lambda = (\mathcal{B}^{-1})_\Lambda$ as in Theorem 3 and with the boundary conditions y. The density $P_{\Lambda,y}(x)$ of this measure with respect to the Lebesgue measure $(dx)^{|\Lambda|}$ in R^Λ is

$$\frac{d\mu^0_{\Lambda,y}}{(dx)^{|\Lambda|}} = D^{-1}_{\Lambda,y} \exp\left\{ -\frac{1}{2} \sum_{\{t,t'\} \subset \Lambda} a^\Lambda_{t,t'} x(t) x(t') - \sum_{\substack{t \in \Lambda \\ \bar{t} \in \partial_d \Lambda}} a^\Lambda_{t,\bar{t}} x(t) y(\bar{t}) \right\} \tag{59}$$

(here $D_{\Lambda,y}$ is the normalization factor). For $y \equiv 0$, we have $\mu^0_{\Lambda,y} = \mu^0_\Lambda$.

Theorem 8. *Let $\{\mu_{\Lambda,y}\}$ be a family of Gibbs modifications of the measures $\mu^0_{\Lambda,y}$ by means of the Hamiltonian $U_\Lambda = \beta\sum_{t\in\Lambda}\Phi(x(t))$ and with a boundary configuration $y = y^{\partial\Lambda}$ obeying for every Λ the condition*

$$\max_{t\in\partial_d\Lambda}|y(t)| \leqslant m(\beta) = \frac{1}{2C}\inf\{|x| : \Phi(x) > \beta^{-1/2}\}, \tag{60}$$

with $C = \|A_\Lambda^{-1}\|\|A_\Lambda\|$ (the norm is defined by formula (8)). Then, for sufficiently small $\beta > 0$, the family $\{\mu_{\Lambda,y}\}$ admits a cluster expansion, and in particular, the weak local limit

$$\lim_{\Lambda\nearrow Z^\nu}\mu_{\Lambda,y} = \mu \tag{61}$$

exists and coincides with the limit (57).

Proof. We pass to the new Gaussian field

$$\widehat{x}(t) = x(t) - m_t(y), \quad t\in\Lambda,$$

with

$$m_t(y) = \langle x(t)\rangle_{\mu^0_{\Lambda,y}} \equiv \langle x(t)\rangle_{\Lambda,y}. \tag{62}$$

The distribution for the new field $\{\widehat{x}(t), t\in\Lambda\}$ coincides with the distribution (59) for $y\equiv 0$, i.e., with the measure μ^0_Λ. Here the measure $\mu_{\Lambda,y}$ is obtained as the Gibbs modification of μ^0_Λ by means of the Hamiltonian

$$\widehat{U}_\Lambda = \beta\sum_{t\in\Lambda}\Phi(\widehat{x}(t) + m_t(y)). \tag{63}$$

Thus, all arguments used in the proof of Theorem 3 in this case also apply to the quantity $\varepsilon(\beta)$ defined by formula (54) replaced by the quantity

$$\widehat{\varepsilon}(\beta) = \max_{t\in\Lambda}\int_{\Phi(x+m_t(y))>\beta^{-1/2}}\exp\left\{-\frac{1}{2}\kappa_1(1-\delta)^2x^2\right\}dx.$$

Since

$$m_t(y) = \sum_{\substack{t'\in\Lambda,\\ \bar{t}\in\partial_d\Lambda}} a^{(-1)}_{t,t'}a_{t',\bar{t}}\,y(\bar{t}), \tag{64}$$

we find from condition (60) that $|m_t(y)| < Cm(\beta)$. Thus, $\widehat{\varepsilon}(\beta)\to 0$ as $\beta\to 0$ (and also $m(\beta)\to\infty$ as $\beta\to 0$).

Furthermore, we may obtain a cluster representation of the partition function Z_Λ and a cluster expansion of the mean $\langle F_A\rangle_{\Lambda,y}$, with the quantities $k_B = k_B^\Lambda(y)$ and $k_R(F_A) = k_R^{(\Lambda)}(F_A; y)$ entering the expansion that depend, in general, on Λ and y uniformly in Λ for all y and that obey condition (60) as well as the estimates established in Theorem 3. Condition (56) is also fulfilled uniformly in Λ and y. Moreover, (64) and (7) imply that

$$|m_t(y)| < C_1 \exp\{-m_0 \rho(t, \partial\Lambda)\}, \quad C_1 = C_1(\beta).$$

Hence, it is easy to deduce that for any fixed $B \subset \eta$ and $R \subset \eta$, the limits

$$\lim_{\Lambda \nearrow Z^\nu} k_B^{(\Lambda)}(y) = k_B \quad \text{and} \quad \lim_{\Lambda \nearrow Z^\nu} k_R^{(\Lambda)}(F_A; y) = k_R(F_A) \tag{65}$$

exist with k_B and $k_R(F_A)$ defined in the preceding theorem. Thus, according to Remark 1 in §3, Chapter 3, the weak local limit (61) exists and, in view of (65) and (58), coincides with the limit (57). Theorem 8 is proved.

Let us now consider the Gaussian measure μ^0 on R^{Z^ν} with vanishing mean and with the covariance matrix B that is inverse to a matrix A obeying conditions (3)–(6).

Theorem 9. *Let us consider the Gibbs modifications μ_Λ of the measure μ^0 defined by the equalities*

$$\frac{d\mu_\Lambda}{d\mu^0} = Z_\Lambda^{-1} \exp\left\{-\beta \sum_{t\in\Lambda} \Phi(x(t))\right\}. \tag{66}$$

Then the measures μ_Λ admit a cluster expansion, and the limit measure

$$\mu = \lim_{\Lambda \nearrow Z^\nu} \mu_\Lambda$$

coincides with the measure (57).

Proof. In this case, we may apply all considerations used in the proofs of Theorems 3 and 8 supplemented with the following lemma. We use the notation $\bar{A}_\Lambda = (\mathcal{B}_\Lambda)^{-1} = \{\bar{a}_{t,t'}^\Lambda, t, t' \in \Lambda\}$, with $\mathcal{B}_\Lambda$ being the restriction of the covariance matrix $\mathcal{B}$ to $\Lambda \subset Z^\nu$.

Lemma 10. *The matrix elements $\bar{a}_{t,t'}^\Lambda$ of the matrix $\bar{A}_\Lambda$ are of the form*

$$\bar{a}_{t,t'}^\Lambda = a_{t,t'} + c_{t,t'}^\Lambda, \quad t, t' \in \Lambda,$$

where $a_{t,t'}$ are the matrix elements of Λ, and $c_{t,t'}^\Lambda$ are such that

$$\begin{aligned} c_{t,t'}^\Lambda = 0 \quad &\textit{if } \max\{\rho(t, Z^\nu\backslash\Lambda), \rho(t', Z^\nu\backslash\Lambda)\} > d, \textit{ and} \\ \left|c_{t,t'}^\Lambda\right| &< K e^{-m_0 \rho(t,t')}, \end{aligned} \tag{67}$$

where K and m_0 are constants not depending on Λ.

Proof. We use $A_{i,i'}$, $B_{i,i'}$, $i, i' = 1, 2$, to denote blocks of matrices A, B, respectively, corresponding to the decomposition $Z^\nu = \Lambda \cup (Z^\nu\backslash\Lambda)$; for example

$A_{1,1} = \{a_{t,t'}, t, t' \in \Lambda\}$, etc. From the equalities $A_{11}B_{11} + A_{12}B_{21} = E_{11}$ and $A_{21}B_{12} + A_{22}B_{21} = 0$, we find that $\bar{A}_\Lambda = B_{11}^{-1} = A_{11} - A_{12}A_{22}^{-1}A_{21}$. Hence, the assertion of the lemma follows with

$$c_{t,t'}^{\Lambda} = - \sum_{t_1,t_2 \in Z^\nu \setminus \Lambda} a_{t,t'} a_{t_1,t_2}^{(-1)} a_{t_2,t'}. \tag{68}$$

To see this, one uses condition (3) and the estimate (7).

In the following chapter, we shall use the generalization of Theorem 8 presented below. Let $\mu_{\Lambda,y}^0$ be the Gaussian measure (59) with a boundary configuration y on $\partial_d\Lambda$ and $\mu_{\Lambda,y}$ its Gibbs modification by means of the Hamiltonian

$$U_{\Lambda,y} = \beta\Bigg(\sum_{t\in\Lambda}\Phi^{(1)}(x(t);\beta) + \sum_{\{t_1,t_2\}\subset\Lambda}\Phi_{t_1-t_2}^{(2)}(x(t_1),x(t_2);\beta) + \sum_{\substack{t\in\Lambda,\\ \bar{t}\in\partial_d\Lambda}}\Phi_{t-\bar{t}}^{(2)}(x(t),y(\bar{t});\beta)\Bigg), \tag{69}$$

where $\Phi^{(1)}(\cdot;\beta)$ and $\Phi_\tau^{(2)}(\cdot,\cdot;\beta), \tau \in Z^\nu$, are families of single site and pair potentials, respectively, that depend on the parameter β, $0 < \beta < \beta_0$, are bounded from below, and are allowed also to take the value $+\infty$. Moreover, the pair potential is supposed to be of finite range d:

$$\Phi_\tau^{(2)} \equiv 0 \quad \text{for} \quad |\tau| > d. \tag{70}$$

We assume that for each $C_1 > 0$, the condition

$$\max\left\{\int_{\Gamma_1(\beta)} e^{-C_1x^2}\,dx, \iint_{\Gamma_2^\tau(\beta)} e^{-C_1(x^2+y^2)}\,dx\,dy\right\} \to 0, \tag{71}$$

as $\beta \to 0$, is satisfied by

$$\Gamma_1(\beta) = \{x \in R^1 : \Phi^{(1)}(x;\beta) > \beta^{-1/2}\},$$
$$\Gamma_2^\tau(\beta) = \{(x_1,x_2)\} \in R^2 : \Phi_\tau^{(2)}(x_1,x_2:\beta) > \beta^{-1/2}\}.$$

Furthermore, we use the notation

$$m(\beta) = \frac{1}{2C}\min(\rho^{(1)}(0,\Gamma_1(\beta)), \rho^{(2)}((0,0),\Gamma_2^\tau(\beta)), |\tau| < d, \tag{72}$$

where $\rho^{(i)}$ is the usual metric in R^i, $i = 1,2$, and the constant C is defined in Theorem 8.

Theorem 11. *In case β is sufficiently small, the interaction potentials $\Phi^{(1)}$ and $\Phi_T^{(2)}$ obey the condition (71), and the boundary conditions y obey the estimate*

$$\max_{t\in\partial_d(\Lambda)} |y(t)| \leqslant m(\beta), \tag{73}$$

then the measures $\mu_{\Lambda,y}$ admit a cluster expansion, and in particular, the weak local limit

$$\mu = \lim_{\Lambda\nearrow Z^\nu} \mu_{\Lambda,y}$$

exists and is the same for all configurations $y = y^{\partial\Lambda}$ satisfying condition (73).

The proof is analogous with the preceding proof.

Remark. An analogy of Theorem 9 for the case of the Hamiltonian (69) satisfying condition (70) and (71) is also valid.

CHAPTER 5

EXPANSIONS AROUND GROUND STATES (LOW-TEMPERATURE EXPANSIONS)

§1 Discrete Spin: Countable Number of Ground States

The cluster expansion technique was used in the previous chapter in the case of Gaussian random fields obtained as perturbations of either independent or Gaussian free fields. In the present chapter, we shall consider perturbations of fields taking a constant value. Namely, perturbations of so-called ground states. As always, let T be a countable set, S a spin space, and $U_\Lambda(\cdot/y^{\partial\Lambda})$, the energy in $\Lambda \subset T$ determined by an interaction potential $\{\Phi_A, A \subset T\}$ and by a fixed boundary configuration $y^{\partial\Lambda}$ outside Λ (or by the "empty" boundary conditions). *A ground state* in a finite set $\Lambda \subset T$ is any configuration $x^0 = x^0(\Lambda, y^{\partial\Lambda})$ for which the Hamiltonian $U_\Lambda(x^0/y^{\partial\Lambda})$ reaches its minimum. *A ground state in the whole set* T is a configuration $x^0 \in S^T$ such that for each configuration $\hat{x} \in S^T$, differing from x^0 only on a finite number of points $t \in T$, the difference of energies satisfies the inequality

$$\Delta U = \lim_{\Lambda \nearrow T} (U_\Lambda(\hat{x}) - U_\Lambda(x^0)) \geqslant 0 \tag{1}$$

(the limit in (1) in the case, say, of finite-range potential $\{\Phi_A\}$ always exists). There is a more general notion of ground state in T (see [90]), but the above definition will be sufficient for our purposes.

A basic heuristic rule in the theory of Gibbs random fields consists of the assertion that (under certain additional conditions) every ground state x^0 in T generates, for sufficiently large values of parameter β, a family $\{\mu_\beta\}$ of limit Gibbs states on S^T such that $\mu_\beta \to \delta_{x^0}$ as $\beta \to \infty$, where δ_{x^0} is the probability measure on S^T concentrated on the single configuration x^0.

In all the examples we shall present in this chapter, this rule will be confirmed and the limit Gibbs fields we shall construct for large β will be obtained as a perturbation of one or the other ground state.

Let us consider a Z-model on a lattice $T = Z^\nu$, i.e., a system with the spin space $S = Z$ (the set of all integers) and with the counting measure λ_0 on Z. Let the formal Hamiltonian be of the form

$$U = \sum_{|t-t'|=1} \Phi(x_t - x_{t'}), \tag{2}$$

where $\Phi(n) \geqslant 0$ is a nonnegative function defined on Z, attaining its unique minimum at zero, and such that $\Phi(0) = 0$, the sum

$$F(\beta) = \sum_{n=-\infty}^{\infty} e^{-\beta\Phi(n)} < \infty \tag{3}$$

is finite for all $\beta > \beta_0$, and

$$F(\beta) \to 1 \quad \text{as} \quad \beta \to \infty. \tag{4}$$

We notice that every configuration

$$x^m \equiv m, \quad m \in Z,$$

is a ground state for the interaction (2).

For any Λ and $m \in Z$, let us consider the Gibbs modification

$$\frac{d\mu_{\Lambda,m}}{d\lambda_0^\Lambda} = Z_{\Lambda,m}^{-1} \exp\{-\beta U(x/m)\}, \tag{5}$$

corresponding to the interaction (2) with the boundary configuration $y^{\partial\Lambda} \equiv m$, and

$$Z_{\Lambda,m} = \sum_{x \in Z^\Lambda} \exp\{-\beta U_\Lambda(x/m)\}. \tag{6}$$

We shall convince ourselves later that for sufficiently large $\beta > 0$ and any m and Λ, the stability condition $Z_{\Lambda,m} < \infty$ is fulfilled.

Theorem 1. *Whenever $\beta > 0$ is large enough and an arbitrary $m \in Z$ is fixed, the family of measures $\{\mu_{\Lambda,m}\}$ admits a cluster expansion. The limit measures $\mu_m = \lim_{\Lambda \nearrow Z^\nu} \mu_{\Lambda,m}$ differ for different m.*

Proof. For simplicity, we shall consider the case $\nu = 2$. Let first $m = 0$. For every configuration of the field $x = \{x(t), t \in \Lambda\}$ in Λ, we denote by $\widehat{x}$ the configuration in $\Lambda \cup \partial\Lambda$ that coincides with x in Λ and equals $\widehat{x}(t) = 0$ at sites $t \in \partial\Lambda$. The collection of bonds of the dual lattice $\widetilde{Z}^2$ that separate neighbouring sites $t, t' \in \Lambda \cup \partial\Lambda$ for which $\widehat{x}(t) \neq \widehat{x}(t')$ is called the *boundary* $\gamma(x)$ of the configuration x, and its connected components $\Gamma_1, \ldots, \Gamma_s$ are called the *contours* of this configuration (cf. analogous definitions in §0, Chapter 1, also in §5, Chapter 3).

Each contour Γ splits up the lattice Z^2 into a finite number of 1-connected subsets $\theta_1, \dots, \theta_p$; exactly one among those sets, θ^{ext}, is infinite and is called the *exterior*, $\operatorname{Ext}\Gamma$, of the contour Γ. The remaining ones are called the *interior components* of this contour, and their union is the *interior*, $\operatorname{Int}\Gamma$, of the contour Γ. An integer-valued function $n = \{n_t, t \in Z^\nu\}$ such that it is constant on each component θ_i, $i = 1, \dots, p$ associated with a contour Γ, vanishes on its exterior component θ^{ext} and takes different values on adjacent components θ_i and θ_j, is called a *label* of the contour Γ. A pair $\{\Gamma, n\}$, where Γ is a contour and n any its label, is called a *labelled contour*. Clearly, to each configuration $x \in S^\Lambda$ (under vanishing boundary conditions) there corresponds a family $\alpha = (\{\Gamma_1, n_1\}, \dots, \{\Gamma_s, n_s\})$ of mutually disjoint labelled contours, where $\Gamma_1, \dots, \Gamma_s$ are components of the boundary $\gamma(x)$ and labels n_i, $i = 1, 2, \dots, s$, are uniquely determined from the equalities

$$\widehat{x}(t) = \sum_{i=1}^{s} n_i(t), \quad t \in \Lambda \cup \partial\Lambda. \tag{7}$$

Conversely, a configuration $x \in S^\Lambda$ is assigned to each such family α.

For each labelled contour (Γ, n), we put

$$W(\Gamma, n) = \sum_{\substack{t,t' \in \Lambda \cup \partial\Lambda \\ |t-t'|=1}} \Phi(n(t) - n(t')). \tag{8}$$

Here the sum is actually taken over all pairs (t, t') of neighbouring sites separated by the contour Γ. For each configuration x, it follows from (2) and (8) that

$$U_{\Lambda,0}(x) = U_\Lambda\,(x/y^{\partial\Lambda} \equiv 0) = \sum_{(\Gamma,n)\in\alpha} W(\Gamma, n), \tag{9}$$

where $\alpha = (\{\Gamma_1, n_1\}, \dots, \{\Gamma_s, n_s\})$ is the family of labelled contours corresponding to the configuration $x \in S^\Lambda$. Moreover,

$$Z_{\Lambda,0} = \sum_{\alpha} \prod_{(\Gamma,n)\in\alpha} \exp\{-\beta W(\Gamma, n)\}, \tag{10}$$

with the sum taken over all families of mutually disjoint labelled contours in $\tilde{\Lambda} \subset \tilde{Z}^2$ ($\tilde{\Lambda}$ is the set of bonds of the dual lattice separating the sites from $\Lambda \cup \partial\Lambda$). We set now

$$k_\Gamma = \sum_{n} \exp\{-\beta W(\Gamma, n)\}, \tag{11}$$

for any contour Γ with the sum taken over all labels n of the contour Γ.

From (10) and (11), we get the cluster representation of the partition function

$$Z_{\Lambda,0} = \sum_{\{\Gamma_1,\dots,\Gamma_s\}} k_{\Gamma_1} \dots k_{\Gamma_s}, \tag{12}$$

with the sum taken over all families of mutually disjoint contours $\Gamma_1, \dots, \Gamma_s$ in $\tilde{\Lambda}$. From (11), the cluster estimate

$$|k_\Gamma| \leqslant \prod_{\xi\in\Gamma} \sum_{n_\xi\neq 0} \exp\{-\beta\Phi(n_\xi)\} \leqslant (F(\beta)-1)^{|\Gamma|} \tag{13}$$

follows (the product is over all bonds ξ from the contour Γ).

The probability distribution on the set $\mathfrak{A}_\Lambda$ of all families of contours $\gamma = \{\Gamma_1, \dots, \Gamma_s\}$ in $\tilde{\Lambda}$ defined by

$$P^{(\Lambda)}(\gamma) = Z_{\Lambda,0}^{-1} \prod_{\Gamma\in\gamma} k_\Gamma \tag{14}$$

is called a *contour ensemble* in $\tilde{\Lambda}$ (cf. the definition of an ensemble of sets in §1, Chapter 3).

A contour $\Gamma \in \gamma$ in a family of contours $\gamma = \{\Gamma_1, \dots, \Gamma_s\}$ is called *external* if it is not contained inside any other contour $\Gamma_i \in \gamma$. Let $\gamma^{\text{ext}} = \{\Gamma_{i_1}, \dots, \Gamma_{i_m}\} \subseteq \gamma$ be the family of all external contours from γ. We shall also introduce an external contour ensemble in $\tilde{\Lambda}$ as the probability distribution, on the set of families $\tilde{\gamma} = \{\Gamma_1, \dots, \Gamma_m\}$ consisting only of external contours, given by

$$P_{\text{ext}}^{(\Lambda)}(\tilde{\gamma}) \stackrel{\text{def}}{=} \sum_{\gamma:\gamma^{\text{ext}}=\tilde{\gamma}} P^{(\Lambda)}(\gamma) = Z_{\Lambda,0}^{-1} \prod_{\Gamma_i\in\tilde{\gamma}} Z^{\text{ext}}(\Gamma_i), \tag{15}$$

with

$$Z^{\text{ext}}(\Gamma) = k_\Gamma \sum_{\{\Gamma'_1,\dots,\Gamma'_l\}} \prod_{j=1} k_{\Gamma'_j}, \tag{16}$$

where the sum is taken over all configurations of contours $\{\Gamma'_1, \dots, \Gamma'_l\}$ inside the contour Γ.

The proof of the following theorem is analogous to the proof of Theorem 1, §1, Chapter 3 (see also [48], [49]).

Theorem 2. *Let β be sufficiently large. Then*

a) there exists a probability distribution P on the set $\mathfrak{A}$ of all finite or countable configurations of contours $\gamma = \{\Gamma_1, \dots, \Gamma_s, \dots\}$ in the infinite lattice $\tilde{Z}^2$ such that the correlation function of any finite family of mutually disjoint contours $\{\Gamma_1, \dots, \Gamma_s\}$ is given by

$$\rho(\Gamma_1, \dots, \Gamma_s) = \lim_{\Delta\nearrow Z^2} \rho^{(\Lambda)}(\Gamma_1, \dots, \Gamma_s), \tag{17}$$

where

$$\rho^{(\Lambda)}(\Gamma_1,\ldots,\Gamma_s) = \Pr^{(\Lambda)}(\{\Gamma_1,\ldots,\Gamma_s\} \subseteq \gamma) = \sum_{\gamma:\{\Gamma_1,\ldots,\Gamma_s\}\subseteq\gamma} P^{(\Lambda)}(\gamma) \quad (18)$$

is the probability (computed in the ensemble (14)) that the family $\{\Gamma_1,\ldots,\Gamma_s\}$ *is contained in a configuration* $\gamma \in \mathfrak{A}_\Lambda$ *and* $\rho(\Gamma_1,\ldots,\Gamma_s)$ *is defined in an analogous way in terms of* P*;*

b) for almost all configurations $\gamma \in \mathfrak{A}$ *in the limit ensemble, each contour* $\Gamma \in \gamma$ *either is itself external or is contained in some external contour* $\widetilde{\Gamma} \in \gamma$*. Thus the distribution* P *on* $\mathfrak{A}$ *induces a probability distribution* P_{ext} *on the set* $\mathfrak{A}^{\text{ext}}$ *of configurations of external contours in* $\widetilde{Z}^2$*. Moreover, the limits analogous to (17) exist:*

$$\rho_{\text{ext}}(\widetilde{\Gamma}_1,\ldots,\widetilde{\Gamma}_s) = \lim_{\Delta\nearrow Z^2} \rho^{(\Lambda)}_{\text{ext}}(\widetilde{\Gamma}_1,\ldots,\widetilde{\Gamma}_s). \quad (19)$$

Here $\{\widetilde{\Gamma}_1,\ldots,\widetilde{\Gamma}_s\}$ *is an arbitrary family of (mutually external) contours and*

$$\rho^{(\Lambda)}_{\text{ext}}(\widetilde{\Gamma}_1,\ldots,\widetilde{\Gamma}_s) = Pr^{(\Lambda)}_{\text{ext}}(\{\widetilde{\Gamma}_1,\ldots,\widetilde{\Gamma}_s\} \subseteq \widetilde{\gamma}) = \sum_{\widetilde{\gamma}\in\mathfrak{A}^{\text{ext}}_\Lambda:\{\widetilde{\Gamma}_1,\ldots,\widetilde{\Gamma}\}\subseteq\widetilde{\gamma}} P^{(\Lambda)}_{\text{ext}}(\widetilde{\gamma}). \quad (20)$$

To finish the proof of Theorem 1, we make an observation that for a local function of the form $F_A = \prod_{t\in A} f_t$, $A \subset Z^2$ (f_t is a $\sum_{\{t\}}$-measurable function, $t \in A$), the mean $\langle F_A\rangle_{\Lambda,0}$ equals

$$\langle F_A\rangle_{\Lambda,0} = \sum_{\{\widetilde{\Gamma}_1,\ldots,\widetilde{\Gamma}_s\}} \prod_{t\in A\setminus\cup\operatorname{Int}\widetilde{\Gamma}_i} f_t(0) \prod_{i=1}^{s} \left\langle \prod_{t\in A\cap\operatorname{Int}\widetilde{\Gamma}_i} f_t \right\rangle_{\operatorname{Int}\widetilde{\Gamma},0} \rho^{(\Lambda)}_{\text{ext}}(\widetilde{\Gamma}_1,\ldots,\widetilde{\Gamma}_s). \quad (21)$$

Here the sum is taken over all families of external contours $\{\widetilde{\Gamma}_1,\ldots,\widetilde{\Gamma}_s\}$ such that the interior $\operatorname{Int}\widetilde{\Gamma}_i$ of each of them intersects A, and $\langle\cdot\rangle_{\operatorname{Int}\widetilde{\Gamma},0}$ is the mean with respect to the Gibbs distribution $\mu_{\operatorname{Int}\widetilde{\Gamma},0}$ in the set $\operatorname{Int}\widetilde{\Gamma} \subset Z^2$ with the zero boundary configuration.

The formula (21) represents a cluster expansion for the means $\langle F_A\rangle_{\Lambda,0}$ from which, taking into account Theorem 2 and the estimate (13), the existence of the limit measure μ_0 for $m = 0$ follows. Moreover, by Theorem 2, for almost all configurations $x \in \Omega = Z^\nu$, the set

$$\mathcal{L}_0 = \{t : x(t) = 0\} \quad (22)$$

has exactly one infinite 1-connected component and all 1-connected components of the sets

$$\mathcal{L}_m = \{t; x(t) = m\}, \quad m \in Z,\ m \neq 0, \quad (23)$$

are finite.

It is easy to see that the map

$$G_m : \Omega \to \Omega, \quad x(t) \mapsto x(t) + m, \quad t \in Z^2,$$

transforms the measure $\mu_{\Lambda,0}$ into the measure $\mu_{\Lambda,m}$ and the limit measure μ_0 into the measure μ_m that is the limit of measures $\mu_{\Lambda,m}$. The fact that $\mu_{m_1} \neq \mu_{m_2}$ for $m_1 \neq m_2$ follows from the observations (22) and (23). The theorem is proved.

§2 Continuous Spin: Unique Ground State

We now consider the case with $S \subset R^1$ being a finite interval of the real axis which is symmetrical with respect to zero and $d\lambda_0 = |S|^{-1}dx$ being the normalized Lebesgue measure on S ($|S|$ denotes the length of S). Let, for every finite $\Lambda \subset Z^\nu$ and every boundary configuration $y = y^{\partial\Lambda}$ on $\partial\Lambda$, the energy $U_\Lambda(x/y)$ be of the form

$$U_\Lambda(x/y) = \sum_{\substack{\{t_1,t_2\}\subset\Lambda,\\ |t_1-t_2|=1}} \Phi(x(t_1), x(t_2)) + \sum_{\substack{t\in\Lambda,\bar{t}\in\partial\Lambda,\\ |t-\bar{t}|=1}} \Phi(x(t), y(\bar{t})), \tag{1}$$

where $\Phi(x_1, x_2)$, $x_i \in S$, $i = 1, 2$, is a smooth symmetric function defined on the square $S \times S$ such that

1. $\Phi(0,0) = 0$, $\Phi(x_1, x_2) > 0$ in remaining points;
2. Φ may be represented in the form

$$\Phi(x_1, x_2) = \alpha(x_1^2 + 2vx_1x_2 + x_2^2) + \Phi'(x_1, x_2),$$
$$\alpha > 0, \quad |v| < 1, \tag{2}$$
$$|\Phi'(x_1, x_2)| < C(|x_1|^3 + |x_2|^3), \quad C > 0. \tag{3}$$

It is easy to see that the configuration $x \equiv 0$ is a ground state for the Hamiltonian (1) which is unique in the class of all periodic ground states (i.e., such configurations $x \in S^{Z^\nu}$ that $\tau_s x = x$ for shifts $s \in G$ from some subgroup $G \subset Z^\nu$ of finite rank; here $(\tau_s x)(t) = x(t - s)$, $t, s \in Z^\nu$).

Let us for each finite $\Lambda \subset Z^\nu$ and each boundary configuration $y = y^{\partial\Lambda}$ on $\partial\Lambda$ consider the Gibbs modification $\mu_{\Lambda,y}$ of the measure $\mu^0 = \lambda_0^{Z^\nu}$ on the space S^{Z^ν} (a power of the measure λ_0):

$$\frac{d\mu_{\Lambda,y}}{d\mu^0} = Z_{\Lambda,y}^{-1} \exp\{\beta U_\Lambda(x/y)\}. \tag{4}$$

Theorem 1. *For sufficiently large $\beta > 0$, there exists a limit measure μ on the space S^{Z^ν}:*

$$\mu = \lim_{\Lambda \nearrow Z^\nu} \mu_{\Lambda,y} \tag{5}$$

(the limit does not depend on the choice of a boundary configuration $y^{\partial\Lambda}$). The limit measure μ and also the measures $\mu_{\Lambda,y}$ under the condition that the boundary configuration y is small:

$$\max_{t\in\partial\Lambda} |y(t)| < \beta^{-1/2} \ln \beta, \tag{6}$$

admit a cluster expansion.

Proof. We consider first the case with a boundary configuration y satisfying condition (6).

We go over to the field with values in $\widehat{S} = \beta^{1/2} S$ by substituting

$$\xi(t) = \beta^{1/2} x(t), \quad t \in \Lambda, \tag{7}$$

with the boundary configuration

$$\eta(t) = \beta^{1/2} y(t), \quad t \in \partial\Lambda, \tag{8}$$

satisfying the condition

$$\max_{t\in\partial\Lambda} |\eta(t)| < \ln \beta. \tag{9}$$

We use $\widehat{\mu}_{\Lambda,\eta}$ to denote the measure on $\widehat{S}^\Lambda$ obtained from $\mu_{\Lambda,y}$ after the substitution (7), (8).

We now take the Gaussian measure $\widehat{\mu}^0$ on R^{Z^ν} with mean 0 and the inverse $A = \{a_{t,t'} t, t' \in Z^\nu\}$ of the covariance matrix determined by

$$a_{t,t'} = \begin{cases} 2\nu\alpha, & t = t', \\ 2v\alpha, & |t - t'| = 1, \\ 0, & |t - t'| > 1. \end{cases} \tag{10}$$

Let $\widehat{\mu}^0_{\Lambda,\eta}$ be the conditional distribution of this Gaussian field in Λ with a given boundary configuration η on $\partial\Lambda$. Then

$$\frac{d\widehat{\mu}_{\Lambda,\eta}}{d\widehat{\mu}^0_{\Lambda,\eta}} = \widehat{Z}^{-1}_{\Lambda,\eta} \exp\{-\beta^{-1/2} \widehat{U}_\Lambda\} \prod_{t\in\Lambda} \chi_{\widehat{S}}(\xi(t)), \tag{11}$$

where

$$\widehat{U}_\Lambda = \sum_{\substack{t,t'\in\Lambda, \\ |t-t'|=1}} \widetilde{\Phi}(\xi(t), \xi(t'); \beta) + \sum_{\substack{t\in\Lambda, \\ t\in\partial\Lambda, \\ |t-t'|=1}} \Phi(\xi(t), \eta(t); \beta), \tag{12}$$

and

$$\widetilde{\Phi}(\xi_1, \xi_2; \beta) = \beta^{3/2} \Phi' \left(\frac{\xi_1}{\sqrt{\beta}}, \frac{\xi_2}{\sqrt{\beta}} \right),$$

$$\chi_{\widehat{S}}(\xi) = \begin{cases} 1, & \xi \in \widehat{S}, \\ 0, & \xi \notin \widehat{S}, \end{cases} \tag{13}$$

$$\widehat{Z}_{\Lambda,\eta} = \left\langle \exp\{-\beta^{-1/2} \widehat{U}_\Lambda\} \prod_{t\in\Lambda} \chi_{\widehat{S}}(\xi(t)) \right\rangle_{\widehat{\mu}^0_{\Lambda,\eta}}. \tag{14}$$

From (3) we get

$$|\widetilde{\Phi}(\xi_1, \xi_2; \beta)| < C(|\xi_1|^3 + |\xi_2|^3). \tag{15}$$

According to (11), the measure $\widehat{\mu}_{\Lambda,\eta}$ is a Gibbs modification of the measure $\widehat{\mu}^0_{\Lambda,\eta}$ by means of the Hamiltonian from Theorem 11, §8, Chapter 4, with the small parameter $\beta^{-1/2}$ instead of β and with

$$\Phi^{(1)}(\xi, \beta) = \begin{cases} 0, & \xi \in \widehat{S}, \\ \infty, & \xi \notin \widehat{S}, \end{cases}$$

and $\Phi^{(2)}(\cdot, \cdot; \beta)$ coinciding with $\widetilde{\Phi}(\cdot, \cdot; \beta)$.

It is easy to verify that for $\beta \to \infty$ the conditions (70, §8, Chapter 4) and (71, §8, Chapter 4) are fulfilled; and for sufficiently large β, it is

$$\ln \beta < m(\beta^{-1/2}),$$

where m is determined from (72, §8, Chapter 4).

Thus according to Theorem 11, §8, Chapter 3, the measures $\widehat{\mu}_{\Lambda,\eta}$ admit a cluster expansion (constructed in §8, Chapter 3) and there exists the limit

$$\widehat{\mu} = \lim_{\Lambda \nearrow Z^\nu} \widehat{\mu}_{\Lambda,\eta},$$

which does not depend on the boundary configuration η and also admits a cluster expansion. It is obvious that also the original family of measures $\{\mu_{\Lambda,y}\}$ with y satisfying condition (6) admits a cluster expansion and that the limit (5) exists.

We shall prove now that the limit (5) exists for arbitrary boundary conditions $y^{\partial\Lambda}$. By the same token, the uniqueness of the Gibbs distribution will be proved (cf. Corollary of Proposition 1, §2, Chapter 1).

Let $\Lambda \subset Z^\nu$ be a (hyper)cube, $\bar{\Lambda} = \Lambda \cup \partial\Lambda$; for any configuration $x \in S^\Lambda$, we use $\bar{x} \in S^{\bar{\Lambda}}$ to denote the configuration in $\bar{\Lambda}$ coinciding with x in Λ and with $y^{\partial\Lambda}$ in $\partial\Lambda$. For every configuration $x \in S^\Lambda$, we write

$$D(x) = D = \{t \in \bar{\Lambda} : |\bar{x}(t)| > \beta^{-1/2} \ln \beta\} \subseteq \bar{\Lambda},$$

and let $D_1, \ldots, D_k$ be 1-connected components of D. A component $D_i \subseteq D$ will be called bordering if it intersects $\partial\Lambda$; let $D^{\text{bor}}(x)$ be the union of all bordering components of D.

We use $N = N(\Lambda)$ to denote the length of the sides of the cube Λ and $K(\Lambda)$ to denote the cube, the center of which coincides with the center of Λ and the sides of which are of length $[N/2] + 1$. Let $\mathcal{E}_\Lambda \subset S^\Lambda$ be the set of those configurations $x \in S^\Lambda$ for which $D^{\text{bor}}(x) \cap K(\Lambda) \neq \emptyset$.

Lemma 2. *For any cube Λ and any boundary condition $y = y^{\partial\Lambda}$, we have*

$$\mu_{\Lambda,y}(\mathcal{E}_\Lambda) < Be^{-\hat{c}N}, \tag{16}$$

where $B > 0$ and $\hat{c} > 0$ are constants which do not depend on Λ and $y^{\partial\Lambda}$.

We shall present a proof of this lemma later but now, relying on it, we shall finish the proof of the theorem.

Let F_A be a bounded local function, and let Λ be a cube such that $A \subset K(\Lambda)$. Then

$$\langle F_A \rangle_{\mu_{\Lambda,y}} = \int_{\mathcal{E}_\Lambda} F_A d\mu_{\Lambda,y} + \int_{S^\Lambda \backslash \mathcal{E}_\Lambda} F_A d\mu_{\Lambda,y}. \tag{17}$$

It is obvious that for each $x \in S^\Lambda \backslash \mathcal{E}_\Lambda$ the set $\Lambda \backslash D^{\mathrm{bor}}(x)$ contains the cube $K(\Lambda)$, and let $R(x)$ be the 1-connected component of $\Lambda \backslash D^{\mathrm{bor}}(x)$ containing $K(\Lambda)$. Thus

$$\int_{S^\Lambda \backslash \mathcal{E}_\Lambda} F_A d\mu_{\Lambda,y} = \mu(S_\Lambda \backslash \mathcal{E}_\Lambda) \sum_R \langle F_A / R(x) = R \rangle_{\mu_{\Lambda,y}} \mu_{\Lambda,y}(\{x : R(x) = R\}), \tag{18}$$

where the sum is over all 1-connected sets $R \subset \bar{\Lambda}$ containing the cube $K(\Lambda)$ and $\langle F_A | R(x) = R \rangle_{\mu_{\Lambda,y}}$ is the conditional mean of F_A under the condition $R(x) = R$.

Since $A \subset K(\Lambda)$, and taking into account the 1-Markov property of the distribution $\mu_{\Lambda,y}$ (see §2, Chapter 1), we get

$$\langle F_A / R(x) = R \rangle_{\mu_{\Lambda,y}} = \int \langle F_A \rangle_{\mu_{\hat{R},\hat{x}}} d\bar{\mu}_{\Lambda,y}(\hat{x} / R(x) = R), \tag{19}$$

where $\widehat{R} = R \backslash (\partial_i R \cap \Lambda)$ and $\widehat{x}$ is a configuration on $\partial \widehat{R}$ coinciding at the sites $t \in \partial \widehat{R} \cap \partial \Lambda$ with the boundary configuration y and taking at the remaining sites $t \in \partial \widehat{R}$ arbitrary values satisfying the condition

$$|\widehat{x}(t)| < \beta^{-1/2} \ln \beta, \quad t \in \partial \widehat{R} \cap \Lambda. \tag{20}$$

We use the notation $\widehat{\mu}_{\Lambda,y}(\cdot / R(x) = R)$ for the measure on the space of configurations $\widehat{x}$ on $\partial \widehat{R}$ induced by the conditional distribution $\mu_{\Lambda,y}(\cdot / R(x) = R)$ under the condition $R(x) = R$. The integral in (19) is taken over the set of all configurations $\widehat{x}$ (taking into account condition (20)). For any configuration $x \in S^\Lambda \backslash \mathcal{E}_\Lambda$ with Λ large, the set $\widehat{R}(x)$ is also large and thus the mean $\langle F_A \rangle_{\mu_{\hat{R},\hat{x}}}$, by virtue of what was proved above, is near the limit mean $\langle F_A \rangle_\mu$ uniformly with respect to all $\widehat{x}$ satisfying condition (20). From this, from (17)–(19), and from Lemma 2, the convergence

$$\langle F_A \rangle_{\mu_{\Lambda,y}} \to \langle F_A \rangle_\mu, \quad \Lambda \nearrow Z^\nu,$$

follows. The theorem is proved.

Proof of Lemma 2. For simplicity of notation, we take $S = [-1,1]$. Let $x \in \mathcal{E}_\Lambda$, $D_1 \subseteq D^{\text{bor}}(x)$ be a 1-connected bordering component of $D(x)$ intersecting $K(\Lambda)$, and $\widehat{D} \subset D_1 \cap (\Lambda \backslash K(\Lambda))$ be a 1-connected component of the set $D_1 \cap (\Lambda \backslash K(\Lambda))$ such that

$$\widehat{D} \cap \partial K(\Lambda) = \emptyset \quad \text{and} \quad \widehat{D} \cap \partial(\Lambda) \neq \emptyset.$$

We notice now that for all sufficiently small $\varepsilon > 0$, the inequalities

$$\begin{aligned} &\Phi(x_1, \varepsilon x_2) \leqslant \Phi(x_1, x_2), \quad |x_1| < \varepsilon, \quad \varepsilon < |x_2| \leqslant 1, \\ &\Phi(\varepsilon x, \varepsilon x_2) \leqslant \Phi(x_1, x_2) - \widetilde{C}\varepsilon^2, \quad \varepsilon < |x_1| < 1, \quad \varepsilon < |x_2| < 1, \end{aligned} \tag{21}$$

are fulfilled with some constant $\widetilde{C} > 0$.

Further, we put

$$W = \max_{x, x' \subset [-1,1]} \Phi(x, x')/2\bar{\bar{c}}\varepsilon^2,$$

where $\bar{\bar{c}} < \widetilde{C}$; we shall choose a number $0 < \kappa < 1/(3v)$ in such a way that κN is an integer.

Lemma 3. *Let N be sufficiently large. Then for each configuration $x \in \mathcal{E}_\Lambda$, there exists a 1-connected set $B \subseteq \widehat{D}$ such that*

$$|B| \geqslant \kappa N, \quad |\widehat{D} \cap (\partial B \cap \Lambda)| < |B|/W. \tag{22}$$

Proof. Let $\gamma = \{t_1, \ldots, t_n\} \subset \widehat{D}$ be a 1-connected sequence of points from $\widehat{D}$ such that $t_1 \in \partial\Lambda, \ldots, t_n \in \partial K$. Let us consider a point $\bar{t} \in \gamma$, the distance of which from both $\partial\Lambda$ and K is larger than $1/4N$, and let $T_m^{\bar{t}}$ be the cube $\{t : \rho(t, \bar{t}) \leqslant m\}$, $m = 1, 2, \ldots$. By $B_m \subseteq T_m^{\bar{t}} \cap \widehat{D}$, we shall denote the connected component of the set $T_m^{\bar{t}} \cap \widehat{D}$ containing the point $\bar{t}$. One may choose m such that $\kappa N \leqslant m \leqslant \nu\kappa K$ and

$$|\widehat{D} \cap (T_{m+1} \backslash T_m)| < (\kappa N/W)^n, \tag{23}$$

where $n = [m/(\kappa N)]$. Indeed, for $m = \nu\kappa N < N/3$ and N large enough, it is

$$\begin{aligned} |\widehat{D} \cap (T_{m+1} \backslash T_m)| &< \{[2(m+1)]^\nu - (2m)^\nu\} \\ &< 2\nu(2m+2)^{\nu-1} < 2^\nu \nu(N/3+1)^{\nu-1} < (\kappa N/W)^\nu \end{aligned}$$

(in fact for any fixed constant N).

Let $m_0 \geqslant \kappa N$ be the smallest of integers for which (23) is fulfilled and put $n_0 = [m_0/(\kappa N)]$. We shall show that B_{m_0} satisfies the inequalities (22). Since B_{m_0} contains at least m_0 points from the chain γ, we have $|B_{m_0}| \geqslant m_0 \geqslant \kappa N$. The second inequality in (22) for $n_0 = 1$ follows directly from (23) and the first inequality in (22). Let now $n_0 \geqslant 2$ and $n_0 \kappa N \leqslant m_0 \leqslant (n_0+1)\kappa N$. It means that for all m such that $\kappa N \leqslant m < m_0$, it is

$$|\widehat{D} \cap (T_{m+1} \cap T_m)| \geqslant (\kappa N/W)^n.$$

Hence,

$$\begin{aligned} |B_{m_0}| = |\widehat{D} \cap T_{m_0}| &> \sum_{m=\kappa N}^{m_0-1} \left|\widehat{D} \cap (T_{m+1} \backslash T_m)\right| \geqslant (\kappa N/W)^{n_0-1} \kappa N \\ &= W(\kappa N/W)^{n_0}. \end{aligned}$$

Thus (23) implies (22). Lemma 3 is proved.

For each $x \in \mathcal{E}_\Lambda$, we shall now fix a set $B = B(x)$ indicated in the previous lemma. Denoting by $V_B \subseteq \mathcal{E}_\Lambda$ the set of those configurations $x \in \mathcal{E}_\Lambda$ for which $B(x) = B$, where $B \subset \Lambda$ is a 1-connected set, $|B| > \kappa N$, we shall estimate the probability $p(V_B)$. To this end, we define a map G_ε^B, $\varepsilon = \beta^{-1/2} \ln \beta$, in the space of configurations by

$$(G_\varepsilon^B x)(t) = \begin{cases} \varepsilon x(t), & t \in B, \\ x(t), & t \notin B. \end{cases}$$

Notice that for any $x \in V_B$, it is

$$\begin{aligned} &- [U_\Lambda(x/y) - U_\Lambda(G_\varepsilon^B x/y)] \\ &= \sum_{t,t' \in B} [\Phi(\varepsilon x(t), \varepsilon x(t')) - \Phi(x(t), x(t'))] \\ &+ \sum_{t \in B, t' \in \partial B \cap \widehat{D}} [\Phi(\varepsilon x(t); x(t')) - \Phi(x(t), x(t'))] \\ &+ \sum_{t \in B, t' \in \partial B, t' \bar{\in} \widehat{D}} [\Phi(\varepsilon(x(t), x(t')) - \Phi(x(t), x(t'))] \\ &+ \text{ analogous terms containing differences} \\ &[\Phi(\varepsilon x(t), y(\bar{t})) - \Phi(x(t), y(\bar{t}))], \quad t \in B, \quad \bar{t} \in \partial\Lambda. \end{aligned} \tag{24}$$

By (21), the first sum may be bounded by $(-\widetilde{C}\varepsilon^2 N_2(B))$, where $N_2(B)$ is the number of pairs of nearest neighbours in B. By (22), the second sum does not exceed

$$2 \max |\Phi(x, x')| \frac{|B|}{W} < |B| \cdot c\varepsilon^2,$$

and the third sum is, again due to (21), negative (analogous estimates hold for the remaining terms in (24)).

Since one has $N_2(B) \geqslant |B| - 1$ for any 1-connected set B, we find from the preceding estimates that

$$-U_\Lambda(x) \leqslant -U_\Lambda(G_\varepsilon^B x) - \hat{c}\varepsilon^2 B,$$

where $\hat{c} > 0$ is some positive constant.

Hence

$$P(V_B) = Z_\Lambda^{-1} \int_{V_B} e^{-\beta U_\Lambda(x/y)} \prod_{t\in\Lambda} dx(t)$$
$$< e^{-\beta\hat{c}\varepsilon^2|B|} Z_\Lambda^{-1} \int e^{-\beta U_\Lambda(G_\varepsilon^B x/y)} \prod dx(t) < \varepsilon^{-|B|} e^{-\beta\hat{c}\varepsilon^2|B|}.$$

The number of 1-connected sets $B \subset \Lambda$ of the cardinality l does not exceed $N^\nu \tilde{c}^l$, where $\tilde{c}$ is an absolute constant; thus we get

$$p(\mathcal{E}_\Lambda) < N^\nu \sum_{l>\kappa N} \varepsilon^{-|l|} e^{-\hat{c}\varepsilon^2\beta l} \tilde{c}^l.$$

The estimate (16) then follows easily.

§3 Continuous Spin: Two Ground States

Let, as above, $S \subset R^1$ be an interval symmetrical with respect to zero, $\mu_0 = \lambda_0^{Z^\nu}$, $\lambda_0 = dx/|S|$, and the Hamiltonians U_Λ, for finite $\Lambda \subset Z^\nu$ and for a boundary condition y, be of the form (1) in §2. We shall suppose here that $\Phi(x_1, x_2)$ in (1) in §2 is a symmetric even function with exactly two minima at the points (x_0, x_0) and $(-x_0, -x_0)$, $x_0 > 0$. Moreover, let $\Phi(x_0, x_0) = 0$, and suppose that in a small neighbourhood $\widehat{U}_\varepsilon^+ = U_\varepsilon^+ \times U_\varepsilon^+$ of the point (x_0, x_0), where $U_\varepsilon^+ = \{x : |x - x_0| < \varepsilon\}$, the function $\Phi(x_1, x_2)$ may be expressed in the form

$$\Phi(x_1, x_2) = \alpha((x_1 - x_0)^2 + (x_2 - x_0)^2 + 2v(x_1 - x_0)(x_2 - x_0)) + \Phi'(x_1, x_2),$$
$$\alpha > 0, \quad |v| < 1, \tag{1}$$
$$|\Phi'(x_2, x_2)| < C(|x_1, -x_0|^3 + |x_2 - x_0|^3), \quad c > 0.$$

In view of the fact that Φ is even, a similar expression holds in a neighbourhood $\widehat{U}_\varepsilon^-$ of the point $(-x_0, -x_0)$. We shall suppose that $\widehat{U}_\varepsilon^+ \widehat{U}_\varepsilon^- = \emptyset$ and that $\Phi(x_1, x_2) > \delta > 0$ outside those neighbourhoods.

It is easy to see that the configurations $y^\pm \in S^{Z^\nu}$ defined by

$$y^+ \equiv x_0, \quad y^- \equiv -x_0$$

are ground states for the Hamiltonians U_Λ.

Whenever $\Lambda \subset Z^\nu$, we use $\mu_{\Lambda,\pm}$ to denote the Gibbs modification corresponding to the boundary configurations $y^\pm$, respectively,

$$\frac{d\mu_{\Lambda,\pm}}{d\mu_0} = (Z_{\Lambda,\pm})^{-1} \exp\{-\beta U_\Lambda(x/y^\pm)\}. \tag{2}$$

Notice that the space S^Λ as well as the measure μ_0 are invariant with respect to the mapping $G: x^\Lambda \mapsto -x^\Lambda$; hence,

$$Z_{\Lambda,+} = Z_{\Lambda,-}, \quad \mu_{\Lambda,+}(A) = \mu_{\Lambda,-}(GA), \quad U_\Lambda(x^\Lambda/y^+) = U_\Lambda(-x^\Lambda/y^-). \quad (3)$$

Theorem 1. *For large enough $\beta > 0$, the partition functions $Z_{\Lambda,\pm}$ admit a cluster representation and the measures $\mu_{\Lambda,\pm}$ and $\mu_\pm = \lim_{\Lambda \nearrow Z^\nu} \mu_{\Lambda,\pm}$ a cluster expansion. Moreover, the measures μ_+ and μ_- differ.*

Proof. By the symmetry (3), it is enough to consider any of the measures (2), say, $\mu_{\Lambda,+}$.

Let $x \in S^\Lambda$ be a configuration in Λ and $\hat{x}$ the configuration in $\overline{\Lambda} = \Lambda \cup \partial\Lambda$ equal to x in Λ and to y^+ in $\partial\Lambda$.

We shall put $\varepsilon = C\beta^{-1/2}\ln\beta$, with $C > 0$ to be chosen later, and call a site $t \in \Lambda$ a *plus-regular* site of the configuration x if $|x(t) - x_0| < \varepsilon$ and $|x(t') - x_0| < \varepsilon$ for all nearest neighbours $t' \in \overline{\Lambda}$ of t. In a similar way, *minus-regular* sites are defined. The set of remaining sites is called the *boundary* of the configuration x and is denoted by $\gamma(x)$. Its 1-connected components $\Gamma_1, \ldots, \Gamma_s$ are called *contours.*

Whenever Γ is a contour, we use $\theta_1(\Gamma), \ldots, \theta_p(\Gamma)$ to denote the 1-connected components of the set $\Lambda \backslash \Gamma$. Obviously, the values attained by the configuration x on each of the sets $\partial\Gamma \cap \theta_j(\Gamma)$ belong either all to the neighbourhood U_ε^+ of the point x_0 or all to the neighbourhood U_ε^- of the point $-x_0$. Accordingly, we shall assign to each component $\theta_j(\Gamma)$ the sign $\sigma_j = \pm 1$. The family of signs $\sigma(\Gamma) = \{\sigma_j, j = 1, \ldots, p\}$ is called a *label* of the contour Γ and the pair (Γ, σ) a *labelled* contour. The labels $\sigma(\Gamma_1)$ and $\sigma(\Gamma_2)$ of two contours $\Gamma_1, \Gamma_2 \subset \gamma(x)$ are clearly matching in the sense that if $\Gamma_1 \subset \theta_i(\Gamma_2)$ and $\Gamma_2 \subset \theta_j(\Gamma_1)$, then

$$\sigma_j(\Gamma_1) = \sigma_i(\Gamma_2). \quad (4)$$

Thus, to every configuration x, there corresponds a family $\{(\Gamma_1, \sigma_1), \ldots, (\Gamma_s, \sigma_s)\}$ of labelled contours with matching labels.

Let $D_1, \ldots, D_l$ be 1-connected components of the complement of the boundary of a configuration x, i.e., components of $\Lambda \backslash \gamma(x)$.

We say that a component D is bordering a contour $\Gamma \subset \gamma(x)$ if $\partial D \cap \Gamma \neq \emptyset$. We put $\sigma(D) = \sigma_j(\Gamma)$ if $D \subseteq \theta_j(\Gamma)$ and observe that all values attained by the configuration x in D belong either to U_ε^+ or to U_ε^- according to the sign $\sigma(D)$.

The partition function $Z_{\Lambda,+}$ may be expressed in the form

$$Z_{\Lambda,+} = \sum_{\{(\Gamma_i,\sigma_i)\}} \int \prod_i \exp\left\{-U_{\Gamma_i}(x^{\Gamma_i}/y^+)\right\} \prod Z_D(x^\gamma) d\mu_0(x^\gamma), \quad (5)$$

where the sum is over all 1-admissible families of labelled contours (with matching labels) and the integration is over the set of all configurations x^γ in

$\gamma = \cup\Gamma_i$, consistent with a given label; the partition function

$$Z_D(x^\gamma) = \int\limits_{S^D} \exp\{-\beta U_D(x^D/x^\gamma \cup y^+)\}d\mu_0(x^D),$$
$$S^D = \{x^D : x(t) \in U_\varepsilon^{\sigma(D)}, t \in D\}, \tag{6}$$

depends only on the values of configuration x^γ in the set $\partial D \subset \gamma$.

Breaking up the Hamiltonian U_D in (6) into the quadratic part $U_D^{(q)}$ and a new Hamiltonian U'_D defined by the potential Φ' in the expression (1) and performing the substitution $\xi(t) = (x(t) - x_0)\beta^{1/2}$, analogous to the substitutions (7) in §2 and (8) in §2, we rewrite the integral (6) in the form

$$Z_D(\xi^{\partial D}) = \beta^{-|D|/2}$$
$$\times \left\langle \prod_{t\in D} \chi_\varepsilon(\xi(t)) \exp\left\{-\beta U'_D\left(\frac{\xi^D}{\sqrt{\beta}} \Big/ \frac{\xi^{\partial D}}{\sqrt{\beta}}\right)\right\}\right\rangle_{0,D,\xi^{\partial D}} Z_D^0(\xi^{\partial D}), \tag{7}$$

where $\langle\cdot\rangle_{0,D,\xi^{\partial D}}$ is the mean under the conditional Gaussian measure $\mu_{D,\xi^{\partial D}}$ on R^D, with the inverse covariance matrix (10) in §2 and the boundary configuration $\xi^{\partial D}$; $Z_D^0(\xi^{\partial D})$ is the normalization factor of this Gaussian measure

$$Z_D^0(\xi^{\partial D}) = \int\limits_{R^D} \exp\{-U_D^{(q)}(\xi^D/\xi^{\partial D})\}d\xi^D, \tag{7'}$$

where $U_D^{(q)}$ is the quadratic part of the Hamiltonian U_D; χ_ε is the characterstic function of the interval $[-\beta^{1/2}\varepsilon, \beta^{1/2}\varepsilon]$. Moreover, in view of the definition of D, one has

$$|\xi^{\partial D}(t)| < \varepsilon\beta^{1/2}, \quad t \in \partial D. \tag{7''}$$

For the mean $\langle\cdot\rangle_{0,D,\xi^{\partial D}}$ in (7), we shall derive a cluster expansion similar to that obtained in the proof of Theorem 3 in §8, Chapter 4 (cf. (22), §8, Chapter 4).

Let $B_D = \{b_{t,t'}^D, t, t' \in D\} = A_D^{-1}$ be the covariance matrix of the Gaussian field $\mu_{D,\xi^{\partial D}}^0$, where, according to ((7), §8, Chapter 4),

$$|b_{t,t'}^D| < Ce^{-m_0|t-t'|}, \quad t, t' \in D, \tag{8}$$

and $C > 0$, $m_0 > 0$ are constants. We shall choose a number $d > 0$ such that $Ce^{-m_0 d} < 1/2$ and consider for each component D its interior d-boundary layer

$$\partial_i^d D = \{t \in D; \rho(t, Z^\nu \backslash D) \leqslant d\}.$$

Let us split up the lattice Z^ν into cubes $\eta = \{\Delta\}$ as shown in §8, Chapter 4, and represent the Hamiltonian in the form

$$U'_D = U'_{\partial_i^d D} + \sum_{\Delta,\Delta'} U'_{\Delta,\Delta'} + \sum_{\Delta} U'_\Delta, \tag{9}$$

where $U'_{\partial_i^d D}$ denotes the energy of a configuration inside the d-boundary layer due to self-interaction and interaction with a configuration outside the layer; $U'_{\Delta,\Delta'}$ is the interaction energy of the configurations in neighbouring cubes $\Delta, \Delta' \subset D \backslash \partial_i^d D$, and U'_Δ is the energy inside the cube $\Delta \subset D \backslash \partial_i^d D$.

Further, using again the interpolation formula ((1), §8, Chapter 4), and repeating with some modifications the derivation of the formula ((22), §8, Chapter 4), we obtain the equality

$$Z_D(\xi^{\partial D}) = Z_D^0(\xi^{\partial D}) \sum \prod_i \bar{k}_{\widehat{B}_i}(\xi^{\partial D}) \prod_j k_{B_j}, \tag{10}$$

where the sum is over families $\{B_1, \ldots, B_p\}$ of connected sets B_i consisting of interior cubes $\Delta \in \eta_{D \backslash \partial_i^d D}$ (i.e., such that they are placed outside the boundary layer $\partial_i^d D$ and are not bordering the cubes intersecting the layer; cf. the final part of the proof of Theorem 3 in §8, Chapter 4) and over families $\{\widehat{B}_1, \ldots, \widehat{B}_s\}$ of mutually disjoint sets $\widehat{B}_j$ of cubes such that:

(1) there exists a family $\alpha = \{A_1, \ldots, A_l\}$ of connected sets of cubes, each of which contains at least one bordering cube (i.e., a cube bordering a cube intersecting $\partial_i^d D$), and such that each $\widehat{B}_j$ is a union of some $A_i \in \alpha$;

(2) $\widehat{B}_j$ contains with each $A_i \in \alpha$ all other connected sets of cubes $A_k \in \alpha$ adjoining every contour Γ which, in its turn, adjoins A_i;

(3) there is no smaller subset of cubes in $\widehat{B}_j$ fulfilling the conditions (1) and (2).

The sets $\widehat{B}_j$ will be called *bordering*. The quantities $\bar{k}_B$ are defined similarly as in ((21), §8, Chapter 4); they do not depend on the boundary configuration $\xi^{\partial D}$. The quantities $\bar{k}_{\widehat{B}}(\xi^{\partial D})$ are also introduced in analogy with ((21), §8, Chapter 4), with the only change that the factor $\exp\{-\beta U'_{\text{bor}}(\widehat{B})\}$ is added to the exponent $\exp\{-\beta(\sum_{\Delta,\Delta'} U'_{\Delta,\Delta'} + \sum_\Delta U'_\Delta)\}$, where $U'_{\text{bor}}(\widehat{B})$ is part of the Hamiltonian $U'_{\partial_i^d D}$ associated with the contours $\Gamma \subset \gamma$ adjoining the set $\widehat{B}$.

The bounds on the quantities $\bar{k}_B$ are obtained in exactly the same way as they were derived in §8, Chapter 4. When deriving the bounds on $\bar{k}_{\widehat{B}}$, one passes to the Gaussian field $\hat{\xi}(t) = \xi(t) - \langle \xi(t) \rangle$ in $\widehat{B}$ with vanishing mean. Moreover, at the sites $t \in \widehat{B}$, the distance of which from ∂D is smaller than d, one has $|\langle \xi(t) \rangle| < \varepsilon \beta^{1/2}/2$ (according to (7) and the choice of d); and thus the condition (71) in §8, Chapter 4, is satisfied and the estimate on $\bar{k}_{\widehat{B}}(\xi^{\partial D})$ turns out to be the same as in §8, Chapter 4. Thus, we get

$$\prod_j Z_{D_j}(\xi^{\partial D_j}) = \prod Z^0_{D_j}(\xi^{\partial D_j}) \sum_{\substack{\{B_1,\ldots,B_s\} \\ \{\widehat{B}_1,\ldots,\widehat{B}_p\}}} \prod_i \bar{k}_{\widehat{B}_i} \prod_j \bar{k}_{B_j}. \tag{11}$$

It is easy to verify that

$$Z^0_D(\xi^{\partial D}) = Z^0_D(0)\exp\left\{\sum_{t,t'\in\partial D} C_{t,t'}\xi_t\xi_{t'}\right\}, \tag{12}$$

where the matrix $\{C_{t,t'}\}$ is of a form similar to (68) in §8, Chapter 4), and $|C_{t,t'}| < Ke^{-m_0|t-t'|}$ with some constant K.

To get a cluster representation $Z^0_D(0)$ we use again Lemma 1, §8, Chapter 4, and write down

$$\ln Z^0_D(0) = \sum_{\Delta\subset D} \ln Z^0_\Delta(0) + \sum_{\hat\gamma}\int_0^1 \frac{\partial}{\partial \bar s_{\hat\gamma}} \ln Z^0_{\hat\gamma,\bar s_{\hat\gamma}}(0)\, d\bar s_{\hat\gamma}, \tag{13}$$

where $\bar s_{\hat\gamma}$ are introduced in the same way as in §8, Chapter 4. We notice that the sum $\sum_{\hat\gamma}$ is taken only over connected families $\hat\gamma \subset \eta_D$ of cubes (since for the remaining (nonconnected) families the integrand vanishes).

The integrand may be split up into a sum of semi-invariants with respect to the Gaussian measure $\mu^0_D(\bar s_{\hat\gamma})$ (see §8, Chapter 4). Hence, with the help of estimates from §8, Chapter 4, we get the expression for the logarithm:

$$\ln Z^0_D(0) = CN(D) + \sum_{\hat\gamma}^{\text{bor}} a(\hat\gamma), \tag{14}$$

where $N(D)$ is the number of interior cubes in D, $\sum_{\hat\gamma}^{\text{bor}}$ is taken over all connected families of cubes $\hat\gamma$ containing bordering cubes, and

$$C = \ln Z^0_\Delta(0) + \sum_{\hat\gamma}^{(l)} \frac{1}{|\hat\gamma|}\int \frac{\partial}{\partial \bar s_{\hat\gamma}} \ln Z^0_{\hat\gamma,\bar s_{\hat\gamma}}(0)\, d\bar s_{\hat\gamma}, \tag{15}$$

with the sum over all connected families $\hat\gamma \subset \eta$ containing (an arbitrary) given site t; the "residual" term $a(\hat\gamma)$ admits an obvious estimate

$$|a(\hat\gamma)| \leqslant \left[BL^\nu e^{-(1/2)m_0 L}\right]^{|\hat\gamma|}, \tag{16}$$

where L is the length of side of cube Δ and B is an absolute constant.

We choose further a constant L' and call two contours Γ and Γ' neighbouring if $\rho(\Gamma,\Gamma') < L'$. With the help of this notion, we define the L'-connectedness of a set of contours. It is clear that any collection γ of labelled contours splits up into L'-connected components $g_1,\ldots,g_s$.

Substituting (14) into (12), we get

$$\prod_j Z^0_{D_j}(\xi^{\partial D}) = \widetilde C^{N(D)} \prod_i e^{V(\xi^{g_i})} \prod_{i,k} e^{V(\xi^{g_i},\xi^{g_k})} \prod_{\{g\}\subseteq\{g_1,\ldots,g_s\}} e^{V(\{g\})}, \tag{17}$$

with the product $\prod_{\{g\}}$ taken over all finite subsets $\{g\}$ of the collection of components $\{g_1, \ldots, g_s\}$ with $|\{g\}| > 2$.

Further,

$$V(\xi^g) = \sum_{t,t' \in \tilde{g}} C_{t,t'} \xi_t \xi_{t'} + \sum_{\hat{\gamma}}^{(g)} a(\widehat{\gamma}), \tag{18}$$

with the sum $\sum_{\hat{\gamma}}^{(g)}$ taken over all collections of cubes, $\widehat{\gamma}$, which contain only cubes adjoining contours from g,

$$V(\xi^g, \xi^{g'}) = \sum_{t \in \tilde{g}, t' \in \tilde{g}'} C_{t,t'} \xi_t \xi_{t'} + \sum_{\hat{\gamma}}^{(g,g')} a(\widehat{\gamma}), \tag{19}$$

where we use $\sum_{\hat{\gamma}}^{(g,g')}$ to denote the sum over all connected components $\widehat{\gamma}$ containing cubes adjoining contours from g as well as cubes adjoining contours from g'. Finally,

$$V(\{g\}) = \sum_{\hat{\gamma}}^{\{g\}} a(\widehat{\gamma}), \tag{20}$$

with the sum over all connected collections $\widehat{\gamma}$ that, for each g from the family $\{g\}$, contain at least one cube adjoining some contour from g.

Let us introduce the quantities

$$k(\xi^g, \xi^{g'}) = e^{V(\xi^g, \xi^{g'})} - 1, \quad k(\{g\}) = e^{V(\{g\})} - 1. \tag{21}$$

They obey, as is not difficult to verify, the following estimates:

$$\begin{aligned} |k(\xi^g, \xi^{g'})| &< k(\ln \beta)^2 (C\beta^\alpha)^{\sum_{\Gamma \in g \cup g'} |\Gamma|} (\lambda')^{\rho(\tilde{g}, \tilde{g}')}, \\ |k(\{g\})| &< k \cdot \overline{C}^{\sum_{\Gamma \in \cup_g} |\Gamma|} (\lambda')^{d\{g\}}. \end{aligned} \tag{22}$$

Here the constant $\alpha = \alpha(L') > 0$ can be made sufficiently small by an appropriate choice of L', and also $\lambda' = BL^\nu e^{-(m_0/2)L}$ is small if L is sufficiently large; we use $d_{\{g\}}$ to denote the length of a minimal tree constructed on the elements of the collection $\{g\}$ where we measure the length of edges by the distance $\rho(\tilde{g}, \tilde{g}')$, $g, g' \in \{g\}$. We notice that $\rho(\tilde{g}, \tilde{g}') \geqslant L'$. Further, by expanding the exponents, we get

$$\begin{aligned} \prod_{i,k} e^{V(\xi^{g_i}, \xi^{g_k})} \prod_{(\{g\})} e^{V(\{g\})} &= 1 + \sum_{\substack{\{g_1, g_1'\}, \ldots, \{g_s, g_s'\}, \\ \{g_1\}, \ldots, \{g_p\}}} \prod_{j=1}^{s} k(\xi^{g_j}, \xi^{g'_{j'}}) \prod_{m=1}^{p} k(\{g\}_m) \\ &= \sum_{\{\eta_1, \ldots, \eta_l\}} \overline{\kappa}_{\eta_1} \ldots \overline{\kappa}_{\eta_l}, \end{aligned}$$

where η_k is a connected collection consisting of groups of components of contours, $\eta_k = \{\{g\}_1, \ldots, \{g\}_{m'_k}\}$, and $\overline{\kappa}_{\eta_k}$ is the product of the corresponding quantities $k(\{g\})$ or $k(\xi^{g_i}, \xi^{g_i'})$ (some quantities $\overline{\kappa}_\eta$ thus depend on configuration ξ^γ). Thus,

$$\prod_j Z^0_{D_j}(\xi^{\partial D}) = \widetilde{C}^{N(D)} \left(\prod e^{V(\xi^g)}\right) \sum_{\{\eta_1,\dots,\eta_l\}} \overline{\kappa}_{\eta_1} \dots \overline{\kappa}_{\eta_l}. \tag{23}$$

Substituting now, for a given labelled boundary $\overline{\gamma} = \{(\Gamma_1, \sigma_1), \dots, (\Gamma_s, \sigma_s)\}$, the expansion (23) into (11), we fix any term of the expression obtained when multiplying out the right-hand side of (11). This term is labelled by the collections $\{\eta_1, \dots, \eta_l\}$, $\{\widehat{B}_1, \dots, \widehat{B}_p\}$, and $\{B_1, \dots, B_s\}$. Without changing the collection $\{B_1, \dots, B_s\}$, we shall unite the elements of the collections $\{\eta_1, \dots, \eta_l\}$ and $\{\widehat{B}_1, \dots, \widehat{B}_p\}$ into connected components $\zeta_1, \dots, \zeta_k$, where each ζ_i together with any connected collection $\eta = \{\{g\}_i\}$ contains all bordering collections of cubes $\widehat{B}_i$, which include at least one cube adjoining some contour in η.

After that, averaging in (5) over boundary configurations x^γ, we get the final cluster representation of partition functions

$$Z_{\Lambda,+} = \beta^{-|\Lambda|/2} \widetilde{C}^{|\Lambda|} \sum_{\substack{\{\zeta_1,\dots,\zeta_m\},\\ \{B_1,\dots,B_n\}}} k_{\zeta_1} \cdot \ldots \cdot k_{\zeta_m} \cdot \bar{k}_{B_1} \cdot \ldots \cdot \bar{k}_{B_n}, \tag{24}$$

with mutually disjoint sets $\tilde{\zeta}_1, \dots, \tilde{\zeta}_m$, $\tilde{B}_1, \dots, \tilde{B}_n$ such that different B_i have no neighbouring cubes and no cubes adjoining those which intersect $\tilde{\zeta}_j$.

We notice that this representation is more general than that of Remark 1 in §1, Chapter 3: The admissible mutual distance of clusters depends on their label. The quantities $\bar{k}_B$ in (24) were introduced earlier and k_ζ equals

$$k_\zeta = \int \prod_{\eta \in \xi} \overline{\kappa}_\eta \prod_g \exp\left\{-V(\xi^g) - \beta U_{\bar{g}}\left(\frac{\xi^g}{\sqrt{\beta}}\right)\right\} \prod_{\widehat{B} \in \zeta} k_{\widehat{B}}(\xi^\gamma) d\lambda_0^{\gamma \cap \tilde{\zeta}}\left(\xi^{\gamma \cap \tilde{\zeta}}\right), \tag{25}$$

with the integration performed over the set of those configurations in $\gamma \cap \tilde{\zeta}$ that are admitted by the condition that $\gamma \cap \tilde{\zeta}$ is included in the boundary with a given label; the product $\prod_g$ is taken over all g entering the elements of collections $\eta \in \zeta$.

From (25) and the preceding estimates, we get the bound on k_ζ:

$$|k_\zeta| < (C_\beta)^{-|\gamma \cap \tilde{\zeta}| h'} \lambda^{\sum_{\widehat{B} \in \zeta} |\widehat{B}|} \lambda'^{\sum_{\eta \in \zeta} d_\eta}, \tag{26}$$

with $0 < h' < 1$, $\lambda > 0$, $\lambda' > 0$ sufficiently small. In the derivation of (26), one uses the fact that the matrix $\hat{a}_{t_1,t_2} = a_{t_1,t_2} - C_{t_1,t_2}$ is positive-definite and, consequently, for each g one has

$$\exp\left\{-\beta U_{\bar{g}}\left(\frac{\xi^g}{\sqrt{\beta}}\right)+\sum_{t,t'\in\bar{g}} C_{t,t'}\xi(t)\xi(t')\right\}$$
$$\leqslant \exp\left\{-h\ln\beta\sum_{\Gamma\in g}|\Gamma|\right\}=\beta^{-h\sum_{\Gamma\in g}|\Gamma|},$$

with $h>0$ being a constant, depending on the choice of the constant C in the definition of ε, which may be made sufficiently large. Hence, choosing L, L', and β sufficiently large, we may achieve that the cluster estimate and cluster condition are fulfilled.

If F_A is a local function of the form $F_A=\prod_{t\in A}F_{\{t\}}$, we may obtain a cluster expansion of the mean $\langle F_A\rangle$ by repeating the preceding constructions and considerations leading to formula (21) in §1.

Weak local compactness of the set $\{\mu_{\Lambda,+}\}$ of the measures is implied by the fact that the space S^{Z^ν} is compact (see [16]). The theorem is proved.

CHAPTER 6

DECAY OF CORRELATIONS

§1 Hierarchy of the Properties of Decay of Correlations

We shall consider a translation invariant random field $\{x(t), t \in Z^\nu\}$ in the lattice Z^ν with values in a space S determined by a probability distribution μ on the space $\Omega = S^{Z^\nu}$. Translation invariance of the field means that

$$\mu(\tau_t A) = \mu(A)$$

for each set $A = \mathfrak{B}(\Omega)$ and each $t \in Z^\nu$. Here $(\tau_t x)(s) = x(s-t)$ is the shift of the configuration $x \in S^{Z^\nu}$ by the vector $t \in Z^\nu$. For any bounded local function F_A we use the notation $F_{A+t} = \tau_t^* F_A$ for a shift of this function

$$(\tau_t^* F_A)(x) = F_A(\tau_t^{-1} x).$$

We shall indicate here several properties evaluating the rate of decrease of dependence between values of the field $\{x(t), t \in Z^\nu\}$ at mutually distant sites that are expressed in terms of decay of semi-invariants (or mean values). These properties are listed in the order of their strength.

The first three properties are well known and are presented here for completeness. More essential for Gibbs fields are the latter four properties of decay of correlations.

I. Ergodicity

Whenever F_{A_1} and F_{A_2} are bounded local functions, let

$$\frac{1}{|\Lambda|} \sum_{t \in \lambda} \langle F_{A_1} F_{A_2+t} \rangle_\mu \to \langle F_{A_1} \rangle_\mu \langle F_{A_2} \rangle_\mu \tag{1}$$

with $\Lambda \nearrow Z^\nu$ running over expanding sequence of cubes. The condition (1) is called the *ergodicity* property of the field $\{x(t), t \in Z^\nu\}$. With the help of the equality $\langle F_{A_t} \rangle = \langle F_A \rangle$, it may be rewritten in the form

$$\frac{1}{|\Lambda|} \sum_{t \in \Lambda} \langle F_{A_1}, F_{A_2+t} \rangle_\mu \to 0, \quad \Lambda \nearrow Z^\nu. \tag{2}$$

We notice that (1) is equivalent to the usual definition of ergodicity: the measure of any set $G \in \mathfrak{B}(\Omega)$ that is invariant with respect to translations, $\tau_t G = G$ for all $t \in Z^\nu$, is either zero or one. A proof of this equivalence may be found, e.g., in [32].

II. Mixing

We say that a field satisfies the property of *mixing* if

$$\langle F_{A_1} F_{A_2+t} \rangle_\mu \to \langle F_{A_1} \rangle_\mu \langle F_{A_2} \rangle_\mu, \quad |t| \to \infty, \tag{3}$$

for any bounded local functions F_{A_1} and F_{A_2}. In the same way as in the preceding case, the condition (3) may be rewritten in the form

$$\langle F_{A_1}, F_{A_2+t} \rangle_\mu \to 0, \quad |t| \to \infty. \tag{4}$$

III. Mixing of Higher Orders

Let for any n and any collection $F_{A_1}, \ldots, F_{A_n}$ of bounded local functions

$$\langle F_{A_1+t_1} F_{A_2+t_2} \cdots F_{A_n+t_n} \rangle_\mu \to \langle F_{A_1} \rangle_\mu \langle F_{A_2} \rangle_\mu \cdots \langle F_{A_n} \rangle_\mu \tag{5}$$

if $\min_{i,j} |t_i - t_j| \to \infty$.

Using the properties D) and E) of semi-invariants (cf. §1.II), it is easy to show that (5) is equivalent to

$$\langle F_{A_1+t_1}, F_{A_2+t_2} \cdots F_{A_n+t_n} \rangle_\mu \to 0 \tag{6}$$

for any fixed collection of functions $F_{A_1}, \ldots, F_{A_n}$ and $\min_{i,j} |t_i - t_j| \to \infty$.

The following properties are related to more explicit estimates of the rate of decay of semi-invariants and do not suppose, in general, the translation invariance of the field $\{x(t), t \in Z^\nu\}$.

IV. Strong Decay of Correlations

It consists in the assumption that for any $n \geqslant 2$, any bounded functions $f_1, \ldots, f_n$ defined on the space S, any integers $k_1 \geqslant 1, \ldots, k_n \geqslant 1$, and any $t \in Z^\nu$, one has

$$\sum_{(t_2, \ldots, t_n)} \left| \langle f_1^{\prime k_1}(x(t_1)), \ldots, f_n^{\prime k_n}(x(t_n)) \rangle_\mu \right| < C \tag{7}$$

with $t_1 = t$. Here the sum is over all ordered collections $(t_2, \ldots, t_n)$ of distinct points $t_i \in Z^\nu$ and the constant $C = C(k_1, \ldots, k_n, f_1, \ldots, f_n)$ does not depend on t.

V. Strong Cluster Estimate

A typical estimate on semi-invariants $\langle f_1'^{k_1}, \ldots, f_n'^{k_n} \rangle_\mu$ implying (7) has the form

$$\left| \langle f_1'^{k_1}(x(t_1)), \ldots, f_n'^{k_n}(x(t_n)) \rangle_\mu \right| < C(f_1, \ldots, f_n; k_1, \ldots, k_n) \sum_{\mathcal{T}} \prod_{(i,j) \in \mathcal{T}} \varphi(t_i, t_j), \tag{8}$$

where the sum is taken over all trees $\mathcal{T}$ on the set of vertices $(1, \ldots, n)$ and the product $\prod_{(i,j)\in\mathcal{T}}$ is over all edges of the tree $\mathcal{T}$; the function $\varphi(t_1, t_2) > 0$ is symmetrical and satisfies the condition

$$\sup_{t_1 \in Z^\nu} \sum_{t_2 \in Z^\nu \setminus \{t_1\}} \varphi(t_1, t_2) < \infty. \tag{9}$$

The estimates (7) and (8), and also the following estimate (11), are called the *best* (maximal) ones if the constants $C(f_1, \ldots, f_n, k_1, \ldots, k_n)$ involved are of the form (with some constant $C_0 > 0$):

$$C(f_1, \ldots, f_n; k_1, \ldots, k_n) = (n-1)! C_0^{k_1+k_2+\cdots+k_n} \prod_{i=1}^n k_i! \prod_{i=1}^n \left(\sup_{x \in S} |f_i(x)| \right)^{k_i} \tag{10}$$

in the case (7) and

$$C(f_1, \ldots, f_n; k_1, \ldots, k_n) = C_0^{k_1+\cdots+k_n} \prod_{i=1}^n k_i! \prod_{i=1}^n \left(\sup_{x \in S} |f_i(x)| \right)^{k_i} \tag{10'}$$

in the cases (8) and (11).

VI. Strong Exponential Estimate of Semi-Invariants

Let there exist a connected family $\mathfrak{A}$ of subsets of Z^ν satisfying the condition 1) and 2) of §4.II and let, for any collection of bounded functions $f_1, \ldots, f_n$ on S, any collection of integers $k_1 \geqslant 1, \ldots, k_n \geqslant 1$, and collection $T = \{t_1, \ldots, t_n\}$ of distinct points, the estimate

$$\left| \left\langle f_1'^{k_1}(x(t_1)), \ldots f_n'^{k_n}(x((t_n)) \right\rangle_\mu \right| < C(f_1, \ldots, f_n; k_1, \ldots, k_n) \lambda^{\hat{d}_T(\mathfrak{A})} \tag{11}$$

be fulfilled with $0 < \lambda < 1$, $C(f_1, \ldots, f_n; k_1, \ldots, k_n)$ a constant not depending on T, and $\hat{d}_T(\mathfrak{A})$ the quantity defined by the formula (1.1.4.II). We notice that if the assumptions of Lemma 7.4.II are fulfilled, we have

$$c\hat{d}_T \leqslant \hat{d}_T(\mathfrak{A}) \leqslant c' \hat{d}_T, \tag{12}$$

where $c \leqslant c'$ are constants and $\hat{d}_T$ denotes the minimal length of the tree $\mathcal{T}$ with the set of vertices T; the length of edges from $\mathcal{T}$ is measured in the metric (14'.4.II).

VII. Uniformly Strong Exponential Estimate of Semi-Invariants

Let again $\mathfrak{A}$ be a connected family of subsets of Z^ν as above and let the estimate

$$\left|\left\langle F'^{k_1}_{A_1},\ldots,F'^{k_n}_{A_n}\right\rangle_\mu\right| < \prod_{i=1}^n 3k_i! \prod_{i=1}^n c_0^{k_1+\hat{d}_{A_i}(\mathfrak{A})}\lambda^{d(\mathfrak{A},\{A_1,\ldots,A_n\})}\prod_{i=1}^n(\sup|F_{A_i}|)^{k_i} \tag{13}$$

be satisfied for some family G of bounded local function F_A. Here $F_{A_1},\ldots,F_{A_n}$ is an arbitrary collection of distinct function from G, $k_1 \geqslant 1,\ldots,k_n \geqslant 1$ is an arbitrary sequence of integers, c_0 is a constant, $0<\lambda<1$, and the number $d(\mathfrak{A},\{A_1,\ldots,A_n\})$ is defined by the formula (12.4.II). If the assumptions of Lemma 7.4.II are satisfied, we have

$$c'\hat{d}_{\{A_1,\ldots,A_n\}} \leqslant d(\mathfrak{A};\{A_1,\ldots,A_n\}) \leqslant c\hat{d}_{\{A_1,\ldots,A_n\}},$$

where $c' \leqslant c$ are constants and $d_{\{A_1,\ldots,A_n\}}$ is the minimal length of the tree T constructed on the set of vertices $(1,\ldots,n)$ with $\rho(A_i,A_j)$, where ρ is the metric (14′.4.II) on Z^ν, taken for the length of the edge (i,j).

It is not difficult to show that each of the listed properties implies the preceding ones. Actually, the first property (ergodicity) is truly weaker than the second one (mixing), the second is weaker than the third (mixing of an order n, $n>2$), and the third is weaker than the forth (strong decay of correlations). Concerning relations between the remaining properties, it is unknown if they constitute a strictly reinforcing sequence; it is just possible to assume that the last property is actually stronger than all the remaining ones. An example of a field satisfying the estimate (10) but failing to satisfy (7) is the field determining a pure phase for the ferromagnetic Ising model with $h=0$ and at low temperatures $T=B^{-1}$ (see below, §5).

Remark 1. In all preceding formulations we supposed for simplicity that the functions F_A (resp. f_i) are bounded. The estimates introduced here make sense also for unbounded functions F_A (or f_i) possessing moments of a sufficiently high order: $F_A \in L_p(\Omega,\mu)$ for some $p>1$. In this case the factors $\prod_{i=1}^n(\sup|F_{A_i}|)^{k_i}$ (or $\prod_{i=1}^n(\sup|f_i|)^{k_i}$) on the right hand side of (13) (or (10) and (10′)) are replaced by some quantities depending on moments of the functions F_{A_i} (or f_i), e.g., by the quantity

$$\max\prod_{i=1}^s\left\langle\left|F_{A_1}^{p_1^{(i)}}\right|^q\cdots\left|F_{A_n}^{p_n^{(i)}}\right|^q\right\rangle^{1/q}$$

with $q\geqslant 1$ and the maximum taken over all partitions of the collection of integers $(k_1,\ldots,k_n)$ into sums of collections

$$(k_1,\ldots,k_n)=(p_1^{(1)},\ldots,p_n^{(1)})+\cdots+(p_1^{(s)},\ldots,p_n^{(s)}),$$

where $p_j^i \geqslant 0$ are integers and $s = 1, 2, \ldots$. (An analogous estimate is possible in the case (10) and (10′)).

Remark 2. In all the definitions of this section one may conceive, instead of lattice Z^ν, and arbitrary countable set T with a metric $\rho(\cdot,\cdot)$ satisfying the condition (14.4.II): all the balls of any fixed diameter are of bounded cardinality. Moreover, in the case of properties I, II, and III, one has to suppose that a group G, preserving the metric, is acting transitively on T so that the little group of any point $t \in T$ is finite and the measure μ is G-invariant.

§2 An Analytic Method of Estimation of Semi-Invariants of Bounded Quasi-Local Functionals

We consider a Gibbs field in Z^ν constructed in §1 or 2.IV in the form of a limit modification μ of an independent field with distribution μ_0 by means of a Hamiltonian $U_\Lambda = \beta \sum \Phi_A$. Here β is supposed to be small and $\{\Phi_A\}$ is a potential bounded from below satisfying the conditions (2.1.IV) and (2′.1.IV) and, possibly, also the conditions (1.2.IV) and (2.2.IV).

Let, further, G be a family of bounded local functions F_A satisfying the condition

$$\sum_{\substack{F_A \in G : t \in A, \\ |A| = n}} \sup_x |F_A(x)| < B\lambda^n \tag{1}$$

for each $t \in Z^\nu$ and $n \geqslant 1$. Here $B > 0$ is a constant not depending on t and n, and λ, $0 < \lambda < 1$, is the same parameter as that in the estimate (2.1.IV) for the potential Φ_A (we stress that, if the family G contains several (or even infinitely many) quantities F_A corresponding to a single $A \subset Z^\nu$, then all of them enter the left-hand side of the sum (1)).

Lemma 2. *Let μ be the measure on $\Omega = S^{Z^\nu}$ defined in §1.IV (or §2.IV) with the help of the Hamiltonian $U_\Lambda = \beta \sum_{A \subseteq \Lambda} \Phi_A$ with β sufficiently small, and let G be a family of bounded local functions satisfying the condition (1). Then for each finite collection $\{F_{A_i}, i = 1, \ldots, s\} \subseteq G$ of mutually distinct functions $F_{A_i} \in G$, and for an arbitrary sequence $k_1 \geqslant 1, \ldots, k_s \geqslant 1$ of integers, the estimate*

$$\left| \left\langle F_{A_1}^{\prime k_1}, \ldots, F_{A_s}^{\prime k_s} \right\rangle_\mu \right| < K C_0^{k_1 + \cdots + k_s} k_1! \ldots k_s! \tag{2}$$

is fulfilled with some constants $K > 0$ and $C_0 > 0$ (not depending on the collection $\{F_{A_i}\}$ and on the numbers $k_1, \ldots, k_s$).

Proof. Let us, for each $\Lambda \subset Z^\nu$, consider the characteristic functional of the family $\{F_{A_i}, i = 1, \dots, s\}$ computed with the measure μ_Λ:

$$\begin{aligned}\Psi_\Lambda(\{B_{A_i}\}) &= \left\langle \exp\left\{\sum_{i=1}^{s} \beta_{A_i} F_{A_i}\right\}\right\rangle_\Lambda \\ &= \frac{1}{Z_\Lambda}\left\langle \exp\left\{-\beta \sum_{A\subset\Lambda} \Phi_A + \sum_{i=1}^{s} \beta_{A_i} F_{A_i}\right\rangle_{\mu_0} = \frac{\widetilde{Z}_\Lambda}{Z_\Lambda},\end{aligned} \tag{3}$$

where

$$\widetilde{Z}_\Lambda = \left\langle \exp\left\{-\beta \sum_{A\subset\Lambda} \Phi_A + \sum_{i=1}^{s} \beta_{A_i} F_{A_i}\right\rangle \mu_0.$$

We shall introduce a new potential

$$\widetilde{\Phi}_A = \Phi_A - \beta^{-1} \sum_{i:A_i=A} \beta_{A_i} F_{A_i} \tag{4}$$

with the sum taken over those i for which $A_i = A$. Then, if

$$\max_i |\beta_{A_i}| < \delta \tag{5}$$

with a fixed small δ, the potential $\{\beta\widetilde{\Phi}A_i\}$ with sufficiently large β satisfies all the conditions (2.1.IV), (2'.1.IV) (or (1.2.IV), (2'.2.IV)), and the partition functions $\widetilde{Z}_\Lambda$ and Z_Λ admit a cluster expansion with quantities k_B satisfying the cluster condition. Hence, according to Theorem 4.4.III, the function $\ln \Psi_\Lambda = \ln \widetilde{Z}_\Lambda - \ln Z_\Lambda$ is an analytic function of complex parameters $\{\beta_{A_i}\}$ from the domain (5).

Moreover, for any A_i on has

$$\frac{\partial \ln \Psi_\Lambda}{\partial \beta_{A_i}} = \langle F_{A_i}\rangle_{\tilde{\mu}_\Lambda} \tag{6}$$

with the Gibbs modification $\tilde{\mu}_\Lambda$ of μ_0 by means of the energy function $\widetilde{U}_\Lambda = \beta \sum_{A\subseteq\Lambda} \widetilde{\Phi}_A$. Taking into account the cluster expansion (6.1.IV) for the means $\langle F_{A_i}\rangle_{\tilde{\mu}_\Lambda}$, the estimate (1) implies

$$\sup_{A_i, |\beta_{A_i}|<\delta} \left| \frac{\partial \ln \Psi_\Lambda}{\partial \beta_{A_i}} \right| < K \tag{7}$$

with an absolute constant K which does not depend on the collection $\{F_{A_1}, \dots, F_{A_s}\}$ and on $\lambda \subset Z^\nu$. Hence, with the help the Cauchy inequality (see §1.II.(H)), we find that the semi-invariant

$$\left\langle F'^{k_1}_{A_1}, \dots, F'^{k_s}_{A_s} \right\rangle_{\mu_\Lambda} = \left.\frac{\partial^{k_1+\dots+k_s} \ln \Psi_\Lambda}{\partial^{k_1}\beta_{A_1} \dots \partial^{k_s}\beta_{A_s}}\right|_{\beta_{A_1}=\dots=\beta_{A_s}=0}$$

obeys the estimate (2) for all $\Lambda \subset Z^{\nu}$.

Passing to the limit $\Lambda \nearrow Z^{\nu}$ we finally get the assertion of the lemma.

Lemma 2. *Let the potential $\{\Phi_A\}$ determining a measure μ be bounded (the condition 2.1.IV), satisfy the assumption of the preceding lemma and let β be sufficiently small. Let G be a family of bounded local quantities as in the preceding lemma. Then one has the estimate*

$$\left|\left\langle F'^{k_1}_{A_1}, \ldots, F'^{k_s}_{A_s} \right\rangle_\mu \right| < K(\overline{C}\beta)^{d(\mathfrak{A},\{A_1,\ldots,A_s\})} C_0^{k_1+\cdots+k_s} \prod_{i=1}^{s} k_1!. \tag{8}$$

Here $k_1 \geqslant 1, \ldots, k_s \geqslant 1$, the constants K and C are the same as in the estimate (2), $\overline{C}$ is an absolute constant, $\mathfrak{A}$ is a collection of supports of the potential $\{\Phi_A\}$, and the number $d(\mathfrak{A}; \{A_1, \ldots A_s\})$ is defined by the formula (12.4.II).

Proof. Repeating the reasoning from the preceding proof we show that the semi-invariant

$$\varphi(\beta) \equiv \left\langle F'^{k_1}_{A_1}, \ldots, F'^{k_s}_{A_s} \right\rangle_\mu$$

is, for $|\beta| < \beta_0$, an analytic function of β and satisfies the estimate (2):

$$|\varphi(\beta)| < K C_0^{k_1+\cdots+k_s} \prod k_i!. \tag{9}$$

Thus, with the help of the Cauchy inequalities, we get for the coefficients of the power series $\varphi(\beta) = \sum_n b_n \beta^n$ the estimate

$$|b_n| < \bar{c}^n K C_0^{k_1+\cdots+k_s} \prod k_i! \tag{10}$$

with $\bar{c} = \beta_0^{-1}$. On the other hand, from the formula (21.1.II) (the condition K), we find that

$$\begin{aligned} b_n &= \frac{1}{n!} \left\langle F'^{k_1}_{A_1}, \ldots, F'^{k_s}_{A_s}, \left(\sum_{\tilde{A} \subseteq \Lambda} \Phi_{\tilde{A}} \right)^{\prime n} \right\rangle_{\mu_0} \\ &= \frac{1}{n!} \sum_{\substack{(\tilde{A}_1, \ldots \tilde{A}_n): \\ \tilde{A}_i \subseteq \Lambda, i=1,\ldots,n}} \left\langle F'^{k_1}_{A_1}, \ldots, F'^{k_s}_{A_s}, \Phi_{\tilde{A}_1}, \ldots, \Phi_{\tilde{A}_n} \right\rangle_{\mu_0}. \end{aligned}$$

Since the semi-invariant $\left\langle F'^{k_1}_{A_1}, \ldots, F'^{k_s}_{A_s}, \Phi_{\tilde{A}_1}, \ldots, \Phi_{\tilde{A}_n} \right\rangle_{\mu_0}$ vanishes for a non-connected collection $(A_1, \ldots, A_s, \tilde{A}_1, \ldots, \tilde{A}_n)$ (the condition C, §1.II), we get

$b_n = 0$ for $n < d(\mathfrak{A} : \{A_1, \ldots, A_s\})$. From it and the estimate (10) we conclude that

$$|\varphi(\beta)| < KC_0^{k_1+\cdots+k_s} \prod k_i! \sum_{n \geqslant d(\mathfrak{A}:\{A_1,\ldots,A_s\})} (\overline{C}\beta)^n$$
$$< \overline{K}C_0^{k_1+\cdots+k_s} \prod k_i!(\overline{C}\beta)^{d(\mathfrak{A},\{A_1,\ldots,A_s\})}$$

for $|\beta| < \beta_0$. The lemma is proved.

We shall now suppose that the potential $\{\Phi_A\}$ is bounded and of finite range: $\Phi \equiv 0$ for $\operatorname{diam} A > d$, and that the family $\mathfrak{A}$ of its supports satisfies the condition 2′) of Lemma 7.4.II (i.e., the number $\hat{d}_{\{t_1,t_2\}}(\mathfrak{A})$ is uniformly bounded for all neighbouring pairs $t_1, t_2 \in Z^\nu$). Let, further, a family $G = \{F_A^{(t)}, t \in Z^\nu, |A| < \infty\}$ of mutually distinct bounded local variables ($F_A^{(t)}$ is measurable with respect to the σ-algebra $\sum_A$) be given so that the condition

$$\sup_x \left|F_A^{(t)}(x)\right| < B\bar{\lambda}^{-\hat{d}_{A\cup\{t\}}(\mathfrak{A})} \tag{11}$$

is satisfied. Here $B > 0$ is a constant, $\bar{\lambda}$, $0 < \bar{\lambda} < \bar{\lambda}_0$, is sufficiently small, and $\hat{d}_D(\mathfrak{A})$, $D \subset Z^\nu$, is defined by the formula (11.4.II). As follows from (11) and Lemma 5.4.II, the family G satisfies the condition (1) with $\lambda = C\bar{\lambda}$ and $C = C(\mathfrak{A})$ being an absolute constant.

We shall examine a family $\{W_t, t \in Z^\nu\}$ of quasi-local variables

$$W_t = \sum_{A \subset Z^\nu} F_A^{(t)}, \quad t \in Z^\nu. \tag{12}$$

Again, due to (11) and the estimate of Lemma 5.4.II, the series (12) absolutely converges, and for each $t \in Z^\nu$ one has

$$\sup_x |W_t(x)| < \overline{K}$$

with some constant $\overline{K}$.

Theorem 3. *Let μ be the measure obtained with the help of a potential $\{\Phi_A\}$ for a sufficiently small β and let the family of variables $\{W_t, t \in Z^\nu\}$ be defined by the formula (12). Then, for any collection $(k_1 \geqslant 1, \ldots, k_s \geqslant 1)$ and any $T = \{t_1, \ldots, t_s\}$, the semi-invariant $\left\langle W_{t_1}'^{k_1}, \ldots, W_{t_s}'^{k_s}\right\rangle_\mu$ obeys the estimate*

$$\left|\left\langle W_{t_1}'^{k_1}, \ldots, W_{t_s}'^{k_s}\right\rangle_\mu\right| < \widehat{K}\tilde{\lambda}^{\hat{d}_T(\mathfrak{A})} C_0^{k_1+\cdots+k_s} \prod_{i-1}^{s} k_i! \tag{13}$$

with constants $\widehat{K} > 0$, $C_0 > 0$, and $\tilde{\lambda} = \tilde{\lambda}(\bar{\lambda}, \beta) < 1$.

Proof. We put

$$\widetilde{F}_A^{(t)} = (\bar{\lambda}^{-1/2})^{\hat{d}_{A\cup\{t\}}(\mathfrak{A})} F_A^{(t)}. \tag{14}$$

With the help of, once more, the estimates (11) and the one from Lemma 5.4.II, we may show that the family of variables $\{\widetilde{F}_A^{(t)}, t \in Z^\nu, A \subset Z^\nu\}$ satisfies the condition (1) with $\lambda = C\bar{\lambda}^{1/2}$. Further, since the potential $\{\Phi_A\}$ is of finite range, for sufficiently small β the condition (2.1.IV) and (2'.1.IV) are satisfied with the same λ. Thus, whenever $t_1, \ldots, t_s$ is a collection of distinct points, $p_1, \ldots, p_s$ is a collection of integers, and $\{A_{1j}, j = 1, \ldots, p_1\}, \ldots, \{A_{sj}, j = 1, \ldots, p_s\}$ are s collection of mutually distinct sets, and also $\{k_{ij}\}$, $j = 1, \ldots, p_i$, $i = 1, \ldots, s$, is a collection of integers, we get, by Lemma 2, the estimate

$$\begin{aligned}&\left\langle \widetilde{F}_{A_{11}}^{(t_1),k_{11}}, \ldots, \widetilde{F}_{A_{1p_1}}^{(t_1),k_{1p_1}}, \widetilde{F}_{A_{21}}^{(t_2),k_{21}}, \ldots, \widetilde{F}_{A_{2p_2}}^{(t_1),k_{2p_2}}, \ldots, F_{A_{sp_s}}^{(t_s),k_{sp_s}} \right\rangle \Bigg| \\ &\quad \leqslant K \prod_{ij} k_{ij}! C_0^{\sum_{ij} k_{ij}} (\overline{C}\beta)^{d(\mathfrak{A},\{A_{ij}\})}.\end{aligned} \tag{15}$$

Inserting further the expansion (12) and the representation (14) in the semi-invariant $\left\langle W_{t_1}^{\prime k_1}, \ldots, W_{t_s}^{\prime k_s} \right\rangle$ and using the condition of multilinearity (B, §1.II) and the estimate (15), we get

$$\begin{aligned}&\left\langle W_{t_1}^{\prime k_1}, \ldots, W_{t_s}^{\prime k_s} \right\rangle \\ &< \overline{K} \sum_{\substack{(p_1,\ldots,p_s):\\ 1\leqslant p_i \leqslant k_i}} \sum_{\substack{\{A_{ij}\}:\\ i=1,\ldots,s,\\ j=1,\ldots,p_i}} \sum_{\{k_{ij}\}}' \prod_{ij} k_{ij}! C_0^{\sum_{ij} k_{ij}} (\overline{C}_\beta)^{d(\mathfrak{A};\{A_{ij}\})} \prod_{ij} \bar{\lambda}^{k_{ij}/2\hat{d}_{A_{ij}\cup\{t_i\}}}\end{aligned} \tag{16}$$

where the sum $\sum'_{\{k_{ij}\}}$ is over all collections $\{k_{ij}\}$ such that $\sum_j k_{ij} = k_i$, $i = 1, \ldots, s$. Further

$$(\overline{C}\beta)^{d(\mathfrak{A};\{A_{ij}\})} \prod_{ij} \bar{\lambda}^{k_{ij}/2\hat{d}_{A_{ij}\cup\{t_i\}}} < \tilde{\lambda}^{\hat{d}_T(\mathfrak{A})} \prod_{ij} \bar{\lambda}^{k_{ij}/4\hat{d}_{A_{ij}\cup\{t_i\}}}. \tag{17}$$

with $\tilde{\lambda} = \max\{\overline{C}\beta, \bar{\lambda}^{1/4}\} < 1$. Besides,

$$\prod_j k_{ij}! \leqslant k_i!, \quad i = 1, \ldots, s.$$

Inserting these inequalities into (16), summing over $\{k_{ij}\}$ with fixed collections $\{P_1, \ldots, P_s\}$ and $\{A_{ij}\}$, then over collections of sets $\{A_{ij}\}$ (bearing in mind the estimates from §4.II and supposing that $\bar{\lambda}$ is sufficiently small), and finally over collection of numbers $(p_1, \ldots, p_s)$, we get the estimate (13) and bring thus the proof of the theorem to an end.

§3 A Combinatorial Method of Estimation of Semi-Invariants in the Case of Exponentially-Regular Cluster Expansion

Let T be a countable set, μ be a measure on S^T, and suppose that, for some class G of local functions $F_{\bar{A}}$, in general unbounded but summable with respect to the measure μ, the means $\langle F_A\rangle_\mu$ admit an exponentially-regular cluster expansion (see §3.III):

$$\langle F_A\rangle_\mu = \sum_{R\subset T} b_R(F_A), \tag{1}$$

$$|b_R(F_A)| \leqslant C(F_A)\lambda^{\hat{d}_R\{\mathfrak{A}\}}. \tag{1'}$$

Theorem 1. *Let the family G be such that the product $F_1 \cdot F_2 \in G$ if F_1 and $F_2 \in G$ and let the parameter λ in the estimate (1′) of the quantities $b_R(F_A)$ be sufficiently small. Then, for each collection $\mathcal{F} = \{F_{A_1}, \ldots, F_{A_N}\}$ of (not necessarily distinct) functions from G, the following estimate of the semi-invariant $\langle F_{A_1}, \ldots, F_{A_N}\rangle_\mu$ is valid*

$$|\langle F_{A_1}, \ldots, F_{A_n}\rangle_\mu| < \hat{C}_{\mathcal{F}} C_1^{\sum_{i=1}^N \hat{d}_{A_i}(\mathfrak{A})}(\overline{C}\lambda)^{d(\mathfrak{A};\{A_1,\ldots,A_N\})}\prod_{i=1}^N n_1. \tag{2}$$

Here $\mathfrak{A}$ is a class of sets entering the definition of exponentially-regular cluster expansion (§3.III), $\hat{d}_A(\mathfrak{A})$ and $d(\mathfrak{A}, \{A_1, \ldots, A_N\})$ are the numbers defined by the formulas (11.4.II) and (12.4.II), respectively, C_1 and $\overline{C}$ are absolute constants, λ is the parameter from (1′), $n_i = n_i(A_1, \ldots, A_N)$, $i = 1, \ldots, N$, are the numbers of the sets A_j, $1 \leqslant j \leqslant N$, such that $A_i = A_j$, and

$$\hat{C}_{\mathcal{F}} = \max_{\delta} \prod_{D\in\delta} C\left(\prod_{i\in D} F_{A_i}\right), \tag{3}$$

with the maximum taken over all partition $\delta = \{D_1, \ldots, D_k\}$ of the set $\mathcal{N} = \{1, \ldots, N\}$ and $C(F_A)$, $F_A \in G$, the constants from (1′).

Remark. In the case of bounded variables F_{A_i}, $i = 1, \ldots, N$, with the estimate (1′) in the form

$$|b_R(F_A)| < \sup|F_A|\,\lambda^{\hat{d}_R(\mathfrak{A})}, \tag{3'}$$

the estimate (2) implies the uniformly strong exponential estimate (1.13) for the semi-invariant $\langle F_{A_1}, \ldots, F_{A_N}\rangle$. If some F_{A_i} are unbounded and the factor $\sup|F_{A_i}|$ in (3′) is replaced by the quantity $\langle |F_{A_i}|^q\rangle_\mu^{1/q}$, $q \geqslant 1$, the estimate (2) implies a changed estimate (1.13) with a factor of the form (1.14).

Proof of the theorem. Let $D \subseteq \mathcal{N} = \{1,\dots,N\}$ be a subset of indices $\{1,\dots,N\}$; we introduce the notation

$$b_R(D) \equiv b_R\left(\prod_{i\in D} F_{A_i}\right), \tag{4}$$

$$f(D) \equiv \left\langle \prod_{i\in D} F_{A_i} \right\rangle_\mu = \sum_R b_R(D). \tag{5}$$

Due to the condition 1) in the definition of exponentially-regular expansion we have $b_R(D) \neq 0$ only if R is an $\mathfrak{A}$-composable set (see §4.II) and the collection $\{R_1,\dots,R'_p, A(D)\}$ is connected. Here $A(D) = \bigcup_{i\in D} A_i$ and we use $R'_1,\dots,R'_p$ to denote the components with respect to the $\mathfrak{A}$-connectedness of the set R. Connected components $\Gamma_\alpha = \{\{R'_i, i \in K_\alpha\}, \{A_j, j \in D_\alpha\}\}$, $K_\alpha \subseteq [1,\dots,p]$, $\alpha = 1,\dots,m$ of the collection $\Gamma = \{\{R'_i, i = 1,\dots,p\}, \{A_j, j \in D\}\}$ yield the partitions $\{D_\alpha, \alpha = 1,\dots,m\}$ and $\{R_\alpha, \alpha = 1,\dots,m\}$ of the sets D and R, respectively; here

$$R_\alpha = \bigcup_{i\in K_\alpha} R'_i, \quad \alpha = 1,\dots,m. \tag{6}$$

These partitions will be called canonical partition of a pair of sets (D, R), $D \subseteq \mathcal{N}$, $R \subset T$ with $b_R(D) \neq 0$. Moreover, $A(D_\alpha) \neq \emptyset$, $\alpha = 1,\dots,m$, and the condition 2) in the definition of exponentially-regular expansion implies

$$b_R(D) = \prod_{\alpha=1}^{m} b_{R_\alpha}(D_\alpha). \tag{7}$$

A mapping $\xi = \xi^D : i \mapsto \xi(i) = R_i(\xi)$, $i \in D$, from the set D into a family of $\mathfrak{A}$-composable subsets of T such that for each $i \in D$ the set $R_i(\xi)$ is $\mathfrak{A}$-connected with respect to A_i will be called a section over $D \subseteq \mathcal{N}$. For each $D' \subseteq D$ we put

$$\xi^D(D') = \bigcup_{i\in D'} \xi^D(i).$$

Whenever ξ is a section over $D \subseteq \mathcal{N}$, we define the graph $G_\xi = G_\xi(D)$ with D playing the role of the set of vertices and with the edges connecting those pairs $(i,j) \subseteq D$, $(i \neq j)$, for which the set $R_i(\xi) \cup R_j(\xi)$ is $\mathfrak{A}$=connected with respect to $\{A_i, A_j\}$. We use G^α_ξ, $\alpha = 1,\dots,m$, to denote the connected components of the graph G_ξ and $D_\alpha = D_\alpha(\xi) \subseteq D$ the set of vertices of G^α_ξ, and $R_\alpha = R_\alpha(\xi) = \xi(D_\alpha)$, $R(\xi) = \xi(D)$. It is easy to see that the partitions $\{D_1,\dots,D_m\}$ and $\{R_1,\dots,R_m\}$ are canonical partitions of the pair of sets $\{D, R(\xi)\}$.

Let now $\xi = \xi^n$ be a section over $\mathcal{N}$. We define a virtual field $\{f_\xi(D), D \subseteq \mathcal{N}\}$ (see §1.II) on subsets of $\mathcal{N}$ by the following rule: if a subgraph G_ξ^D of the graph $G_\xi^{\mathcal{N}}$ constructed on the set of vertices D is disconnected, we put

$$f_\xi(D) = \prod_{\alpha=1}^{m} f_\xi(D_\alpha), \tag{8}$$

with $D_\alpha \subset D$ denoting the sets of vertices of connected components of the graph G_ξ^D. If G_ξ^D is a connected graph, we put

$$f_\xi(D) = \frac{1}{m_\xi(D)} b_{\xi(D)}(D), \tag{9}$$

where $m_\xi(D)$ is the number of distinct sections $\tilde{\xi} = \tilde{\xi}^D$ over the set D such that the graph $G_{\tilde{\xi}}^D(D)$ is connected and $\tilde{\xi}(D) = \xi(D)$. It is easy to see that the virtual field $\{f_\xi(D), D \subseteq \mathcal{N}\}$ does not depend on the graph $G_\xi^{\mathcal{N}}$ for all sections ξ over $\mathcal{N}$ (see §1.II,I). For a virtual field $\{f_\xi(D), D \subseteq \mathcal{N}\}$ we define its semi-invariants $\{g_\xi(D), D \subseteq \mathcal{N}\}$ by the formula (16.1.II).

Lemma 2. *The equality*

$$\langle F_{A_1}, \ldots, F_{A_N} \rangle_\mu = \sum_\xi g_\xi(\mathcal{N}) \tag{10}$$

holds true with the sum $\sum_\xi$ *taken over all sections over the set* $\mathcal{N}$.

Proof. According to the formula (9.1.II) we have

$$\langle F_{A_1}, \ldots, F_{A_N} \rangle_\mu = \sum_{\delta=\{D_1,\ldots,D_k\}} f(D_1)\ldots f(D_k)(-1)^{k-1}(k-1)! \tag{11}$$

with the sum taken over all partitions $\delta = \{D_1, \ldots, D_k\}$ of the set $\mathcal{N}$. On the other hand

$$g_\xi(\mathcal{N}) = \sum_{\delta=\{D_1,\ldots,D_k\}} f_\xi(D_1)\ldots f_\xi(D_k)(-1)^{k-1}(k-1)!. \tag{12}$$

Thus, (10) follows from the equality

$$f(D_1)\ldots f(D_k) = \sum_\xi f_\xi(D_1)\ldots f_\xi(D_k) \tag{13}$$

for each partition δ of the set $\mathcal{N}$.

To prove (13) we notice that for a given partition $\delta = \{D_1, \ldots, D_k\}$ of the set $\mathcal{N}$, each section ξ over $\mathcal{N}$ is uniquely determined by a sequence $(\xi^{D_1}, \ldots, \xi^{D_k})$ of sections over the sets D_i such that

$$\xi\big|_{D_i} = \xi^{D_i}, \quad i = 1, \ldots, k,$$

where $\xi\big|_{D_i}$ is the restriction of ξ over $\mathcal{N}$ to the set $D_i \subseteq \mathcal{N}$ and the order in $\{\xi^{D_1}, \ldots, \xi^{D_k}\}$ is arbitrary. Since also

$$f_\xi(D_i) = f_{\xi^{D_i}}(D_i), \quad i = 1, \ldots, k,$$

the right-hand side of the equality (13) equals

$$\prod_{i=1}^{k} \sum_{\xi^{D_i}} f_{\xi^{D_i}}(D_i)$$

with the sum $\sum_{\xi^{D_i}}$ taken over all sections over the set D_i, $i = 1, \ldots, k$. Thus the equality (13) is implied by the following one:

$$f(D) = \sum_{\xi} f_\xi(D) \tag{14}$$

whenever $D \subseteq \mathcal{N}$ and with the sum taken over all sections over the set D. Further, from (8) and (9) we get

$$\sum_{\xi} f_\xi(D) = \sum_{\xi} \frac{1}{\prod m_{\xi|_{D_\alpha}}(D_\alpha)} \prod_{\alpha=1}^{m} b_{R_\alpha}(D_\alpha), \tag{15}$$

where $D_\alpha = D_\alpha(\xi) \subseteq D$, $\alpha = 1, \ldots, m$, are the sets of vertices of the connected components G_ξ^α of the graph G_ξ, $R_\alpha = R_\alpha(\xi) = \xi(D_\alpha)$, and $\xi\big|_{D_\alpha}$ is the restriction of the section ξ to the set $D_\alpha \subseteq D$.

As already said, the collections

$$\{D_1(\xi), \ldots, D_m(\xi)\}, \quad \{R_1(\xi), \ldots, R_m(\xi)\} \tag{16}$$

form a canonical partition of the pair of sets (D, R), $R = R(\xi)$, and the number of distinct sections $\tilde{\xi}$ for which the collection $\{D_\alpha(\tilde{\xi})\}$ and $\{R_\alpha(\tilde{\xi})\}$ coincide with the collections (16) is $\prod_\alpha m_{\xi|_{D_\alpha}}(D_\alpha)$. Hence, from (15) and (7), $\sum_\xi f_\xi(D) = \sum_R b_R(D) = f(D)$. The lemma is proved.

Since for any section ξ over $\mathcal{N}$ the virtual field $\{f_\xi(D), D \subseteq \mathcal{N}\}$ does not depend on the graph G_ξ, we have

$$g_\xi(D) = 0 \tag{17}$$

if the subgraph $G_\xi^D \subseteq G_\xi$ spanned on the vertices $D \subseteq \mathcal{N}$ is disconnected (see Lemma 2.1.II). Hence, using the Möbius inversion formula in the same way as in §7.II, we may write

$$g_\xi(\mathcal{N}) = \sum_\delta \mu_\xi(\delta, \mathbb{1}) f_\xi(D_1) \dots f_\xi(D_k) \tag{18}$$

with the sum taken over all partitions $\delta = \{D_1, \dots, D_k\} \in \mathfrak{A}_{G_\xi}$ of the set $\mathcal{N}$ such that all subgraphs $G_\xi^{D_i}$ of the graph G_ξ are connected; we use $\mathfrak{A}_{G_\xi}$ to denote the lattice of such partitions and μ_ξ to denote its Möbius function. Further, using (17) with $D = \mathcal{N}$, Lemma 2, Lemma 2.7.II, and the formula (14.7.II), we get

$$|\langle F_{A_1}, \dots, F_{A_N} \rangle_\mu| \leqslant \sum_\xi{}' \sum_{\delta = \{D_1, \dots, D_k\} \in \mathfrak{A}_{G_\xi}} \prod_{i=1}^k v_i(\xi, \delta)\, |f_\xi(D_i)| \tag{19}$$

with the sum taken over all sections $\xi = \xi^{\mathcal{N}}$ over $\mathcal{N}$ for which the graph G_ξ is connected; $v_i(\xi, \delta)$, $1 \leqslant i \leqslant k$, equals the number of edges (i, j) attached to the vertex i in the connected graph $G^\delta = G_\xi^\delta$ with the set of vertices $(1, \dots, k)$ and with the edges (i, j) for which the set $\xi(D_i) \cup \xi(D_j)$ is $\mathfrak{A}$-connected with respect to $\{A(D_i), A(D_j)\}$. Fixing a partition δ and sets $R_i = \xi(D_i)$, $i = 1, \dots, k$, and summing in (19) over all sections $\tilde{\xi}$ for which $G_{\tilde{\xi}}$ is a connected graph, $\delta \in \mathfrak{A}_{G_{\tilde{\xi}}}$, $\tilde{\xi}(D_i) = R_i$, we estimate, with the help of (9), the right-hand side of (19) by

$$\sum_{\{(R_1, D_1), \dots, (R_k, D_k)\}} \prod |b_{R_i}(D_i)|\, v_i(G^\delta(R_1, \dots, R_n)), \tag{20}$$

where $\delta = \{D_1, \dots, D_k\}$ is a partition of the set $\mathcal{N}$, $\{R_1, \dots, R_k\}$ is a collection of $\mathfrak{A}$-composable subsets, and $G^\delta(R_1, \dots, R_k)$ is the graph with vertices $(1, \dots, k)$ and edges (i, j) for which the set $R_i \cup R_j$ is $\mathfrak{A}$-connected with respect to $\{A(d_i), A(D_j)\}$. The sum in (20) is over all collections of pairs $\{(R_1, D_1), \dots, (R_k, D_k)\}$ such that:

1) the graph $G^\delta(R_1, \dots, R_k)$ is connected:

2) the set R_i is $\mathfrak{A}$-connected with respect to the collection $\{A_j, j \in D_i\}$ for each $i = 1, \dots, k$.

According to the estimate (1′) and the last condition we get

$$\prod_{i=1}^k b_{R_i}(D_i) \leqslant \widehat{C}_F \prod_{i=1}^k \lambda^{d_{R_i}(\mathfrak{A}; \{A_j, j \in D_i\})}. \tag{21}$$

The conditions 1) and 2) imply

$$\sum_{i=1}^k d_{R_i}(\mathfrak{A}; \{A_j, j \in D_i\}) \leqslant d(\mathfrak{A} : \{A_1, \dots, A_N\}). \tag{21'}$$

We consider now some ordering in the set T and, whenever $D \subseteq \mathcal{N}$, we use $t(D) \in A(D)$ to denote the point that is minimal in $A(D)$ with respect to this order in T. Whenever $\delta = \{D_1, \ldots, D_k\}$ is a partition, we use $s(\delta)$ to denote the set of all points $t(D_i)$ for all elements D_i of the partition δ and let $u_t^\delta = u_t$ be the number of those elements for which $t = t(D_i)$, $1 \leqslant i \leqslant k$. We shall introduce the notation

$$E(\delta) = \prod_{t \in S(\delta)} u_t^{u_t}. \tag{22}$$

Notice now that $\mathfrak{A}$-connectedness of the set $R_i \cup R_j$ with respect to $\{A(D_i), A(D_j)\}$ means that

$$Q_i \cap Q_j \neq \emptyset, \tag{23}$$

where $Q = A(D_i) \cup \widetilde{R}_i$ and $\widetilde{R}_i = \bigcup_{B \in \mathfrak{A}, B \cap R_i \neq \emptyset} B$ is the union of all sets in $\mathfrak{A}$ that intersect R_i. The estimate (14.5.II) together with (23) implies

$$\prod_{i=1}^{k} v_i(G^\delta(R_1, \ldots, R_k)) \leqslant E(\delta) C^{\sum_{i=1}^{k} \hat{d}_{Q_i}(\mathfrak{A})} \tag{24}$$

with an absolute C and the number $\hat{d}_Q(\mathfrak{A})$ defined in (11.4.II). Further,

$$\begin{aligned} \hat{d}_{Q_i}(\mathfrak{A}) &\leqslant \hat{d}_{A(D_i)}(\mathfrak{A}) + \hat{d}_{\tilde{R}_i}(\mathfrak{A}) \\ &\leqslant \sum_{j \in D_i} \hat{d}_{A_j}(\mathfrak{A}) + d(\mathfrak{A}, \{A_j, j \in D_i\}) + \overline{K} D_{R_i}(\mathfrak{A}, \{A_j, j \in D_i\}) \end{aligned} \tag{25}$$

with a constant $\overline{K} = \overline{K}(\mathfrak{A})$.

When estimating $\hat{d}_{\tilde{R}_i}(\mathfrak{A})$ we used the condition 2) form §3.III, the equality (10.4.II), and also the conditions 1) and 2) from §4.II which imply that for each $\mathfrak{A}$-composable set R one has

$$\hat{d}_{\tilde{R}}(\mathfrak{A}) \leqslant K|R| \leqslant KM \sum_m d_{R'_m}(\mathfrak{A}), \tag{26}$$

where $\{R'_m, m = 1, \ldots, p\}$ are the $\mathfrak{A}$-connected components of R, K is the maximal number of sets $b \in \mathfrak{A}$ that contain a fixed point $t \in T$, and M is the maximal cardinality of the set $B \in \mathfrak{A}$.

Thus, according to (25), (24), (22), (21), (21′), (20), and (19), we get

$$\begin{aligned} &|\langle F_{A_1}, \ldots, F_{A_N} \rangle_\mu| \\ &\leqslant \widehat{C}_{\mathcal{F}} \prod_{j=1}^{N} C^{\hat{d}_{A_j}(\mathfrak{A})} (C_1 \lambda)^{D(\mathfrak{A}; \{A_1, \ldots, A_N\})} \\ &\times \sum_{\{(R_1, D_1), \ldots, (R_k, D_k)\}} E(\delta) \prod_{i=1}^{k} (C_1)^{-1/2 d_{R_i}(\mathfrak{A}; \{A_j, j \in D_i\})} C^{d(\mathfrak{A}; \{A_j, j \in D_i\})} \end{aligned} \tag{27}$$

with a sufficiently large constant $C_1 > 1$ and the summation taken over all collections of pairs $\{(R_1, D_1), \ldots, (R_k, D_k)\}$, satisfying the conditions 1) and 2) with $\delta = \{D_1, \ldots, D_k\}$. Fixing a partition δ we sum in the right-hand side of (27) over all ordered collections of sets $(R_1, \ldots, R_k)$ satisfying only the condition 2). Then

$$\sum_{(R_1,\ldots,R_k)} \prod_{i=1}^{k} (C_1\lambda)^{d_{R_i}(\mathfrak{A};\{A_j, j\in D_i\})} \leqslant \prod_{i=1}^{k} \sum_{R_i} (C_1\lambda)^{d_{R_i}(\mathfrak{A},\{A_j, j\in D_i\})}. \tag{28}$$

Using now Lemma 6.4.II and the estimate

$$|A| < M\hat{d}_A(\mathfrak{A}) \tag{29}$$

analogous to (26), we bound, for $C_1 \gg C$, the right-hand side of (27) by

$$\hat{C}_{\mathcal{F}} \prod_{j=1}^{N} C_0^{\hat{d}_{A_j}(\mathfrak{A})} (C_1\lambda)^{d(\mathfrak{A};\{A_1,\ldots,A_N\})} \sum_{\delta} E(\delta) \tag{30}$$

with constants C_0 and C_1.

Now, we shall pass to the estimate of the sum $\sum_\delta E(\delta)$. Let $S_0 = \{t_1, \ldots, t_l\} = S(\delta = \underline{0})$ be the set of points corresponding to the partition $\underline{0}$ of the sets $\mathcal{N}$ into one-point subsets, where the points t_i are enumerated according to the order in T. Notice that each point $t_i \in S_{\underline{0}}$ is the minimal point for a family of subsets $A_j, j \in \mathcal{N}$; we shall use D_i^0 to denote the corresponding set of indices $j \in \mathcal{N}$, and $\delta_0 = \{D_1^0, \ldots, D_l^0\}$ to denote the partition of $\mathcal{N}$ generated by these sets. Each partition $\delta = \{D_1, \ldots, D_k\}$ induces the partitions δ_i of elements D_i^0: $\delta_i = \{D_{i1}^0, \ldots, D_{is_i}^0\}$, where

$$D_{ij}^0 = D_i^0 \cap D_j, \quad i = 1, \ldots, l, \quad j = 1, \ldots, k,$$

under the condition $D_{ij}^0 \neq \emptyset$. Moreover, $S(\delta) \subseteq S_0$ and a point $t \in S(\delta)$ coincides with some $t_i \in S_0$ if the partition δ_i contains at least one element of the form $D_{ij}^0 = D_i^0 \cap D_j$ such that $t = t(D_j)$ and the element D_j does not intersect any of the elements D_m^0 with $m < i$. We shall use $\delta_i' \subseteq \delta_i$ to denote for each $i = 1, \ldots, l$, the collection of all such elements $D_{ij}^0 \in \delta_i$ (for $i = 1$ one has $\delta_1 = \delta_1'$). Clearly, $u_t^\delta = |\delta_i'|$ where $|\delta_i'|$ is the cardinality of the set δ_i', and $t = t_i \in S_0$. Hence the following estimate follows

$$\sum_{\delta} E(\delta) = \sum_{\delta} \prod_{t\in S(\delta)} u_t^{u_t} \leqslant \sum_{\{\delta_1,(\delta-2,\delta_2'),\ldots,(\delta_l,\delta_l')\}} \prod_{i=1}^{l} |\delta_i'|^{|\delta_i'|} = \sum_{\delta_1} \prod_{i=2}^{l} \sum_{(\delta_i,\delta_i'} |\delta_i'|^{|\delta_i'|}, \tag{31}$$

where the sum $\sum_{\delta'}$ is taken over all partitions of the set D_1^0 and the sum $\sum_{(\delta_i,\delta_i')}$, for $i \geqslant 2$, is over all pairs $\{\delta_i, \delta_i'\}$ with δ_i an arbitrary partition of the set D_i^0 and $\delta_i' \subseteq \delta_i$ an arbitrary subset of elements of the partitions δ_i.

Lemma 3. *For each set $D \subseteq \mathcal{N}$ one has*

$$\sum_{(\delta,\delta')} |\delta'|^{|\delta'|} \leqslant c^{|D|}|D|^{|D|}, \tag{32}$$

where c is an absolute constant, $|D|$ is the cardinality of the set D and the sum $\sum_{(\delta,\delta')}$ is defined above.

The proof of the lemma will be given below. But first we notice that the estimate (32) implies

$$\sum_{\delta} E(\delta) < C^N \prod_{i=1}^{l} |D_i^0|^{|D_i^0|}.$$

Further,

$$\prod_{i=1}^{l} |D_i^0|^{|D_i^0|} \leqslant \prod_{j=1}^{N} v_j,$$

where v_j is the number of sets A_m in the collection $\mathcal{A} = \{A_1, \ldots, A_N\}$ that intersect the set $A_j \in \mathcal{A}$ (see §5.II). Hence by Theorem 2.5.II and the inequality (15.4.II) we have

$$\sum_{\delta} E(\delta) \leqslant \left(\prod_{i=1}^{N} n_i\right) \widetilde{C}^{\sum_{i=1}^{N} \hat{d}_{A_i}(\mathfrak{A})} \tag{33}$$

with an absolute constant $\widetilde{C}$. Then (30) and (33) imply (2). The theorem is proved.

Proof of Lemma 3. We shall use $S_{m,k}$ to denote the number of partitions δ of a set of cardinality m consisting of k elements. Denoting $|D| = j$, we easily find that

$$\sum_{(\delta,\delta')} |\delta'|^{|\delta'|} = \sum_{\substack{m_1 m_2: \\ m_1+m_2=j}} \frac{j!}{m_1! m_2!} \sum_{k_1 \leqslant m_1} S_{m_1,k_1} \sum_{k_2 \leqslant m_2} S_{m_2 k_2} \cdot k_2^{k_2}. \tag{34}$$

For each partition δ of a set of cardinality m we define the numbers $\{p(n), n = 1, 2, \ldots\}$ that equal the number of elements of the partition δ of cardinality n. One has

$$\sum_n np(a) = m, \quad \sum_n P(n) = k = |\delta|.$$

It is easy to count the number of partitions δ with a fixed collection $\{p(n), n = 1, 2, \ldots\}$. It equals $\dfrac{m!}{\prod_n p(n)!(n!)^{p(n)}}$ and thus

$$S_{m,k} = \sum_{\substack{\{p(n)\}: \\ \Sigma np(n)=m, \\ \Sigma p(n)=k}} \frac{m!}{\prod_n p(n)!(n!)^{p(n)}}. \tag{35}$$

We notice that, by (35), the generating function is of the form

$$\sum_{m,k} \frac{S_{m,k}}{m!} z^m w^k = e^{we^z}.$$

From here it is easy to show that $S_{m,k} \leqslant m! e^k / k!$. Inserting this estimate into (34) we get (32). Lemma 3 is proved.

§4 Slow (Power) Decay of Correlations

Let $\{\mu_\Lambda, \Lambda \subset Z^\nu\}$ be the Gibbs modification

$$\frac{d\mu_\Lambda}{d\mu_0} = Z_\Lambda^{-1} \exp\left\{-\beta \sum_{A \subseteq \Lambda} \Phi_A\right\}, \tag{1}$$

where μ_0 is the probability distribution on S^{Z^ν} of the values of an independent field $\{x(t), t \in Z^\nu\}$ and $\{\Phi_A, A = \{t, t'\}, t \neq t'\}$ is a pair potential: $\Phi_A \equiv 0$ for $|A| \neq 2$. Let

$$\sup_x |\Phi_A(x),| = j_A \equiv J(t,t') < \infty \tag{2}$$

for each $A = \{t, t'\}$.

We shall consider here two cases: for each $t \in Z^\nu$ suppose

A)
$$\sum_{A: t \in A} J_A = D_1 < \infty, \tag{3}$$

B)
$$\sum_{A: t \in A} J_A^{1/2} = D_2 < \infty. \tag{4}$$

The condition B) clearly implies the condition A).

As follows from Theorem 1.1.IV, a limit Gibbs modification μ on the space S^{Z^ν} exists for both cases A) and B) providing that β is sufficiently small.

Let $f = \{f_1, \ldots, f_n\}$ be a collection of bounded real functions defined on the space S, $\kappa = (k_1 \geqslant 1, \ldots, k_n \geqslant 1)$ be a collection of integers, and $(t_1, \ldots, t_n)$ be an ordered collection of distinct lattices sites. We shall consider the semi-invariant

$$\left\langle f_1'^{k_1}(x(t_1)), \ldots, f_n'^{k_n}(x(t_n))\right\rangle_\mu . \tag{5}$$

In both cases (3) and (4) above we shall establish a strong decay of correlations, i.e., an estimate of the form (1.7) for semi-invariants (5) with the best constant having the form (1.10). In the case (4) we shall actually get an even stronger cluster estimate similar to the cluster estimate (1.8) (with the best constant (1.10′)).

Theorem 1. *Let the condition (4) be satisfied and let β be sufficiently small. Then the estimate*

$$\left\langle f_1'^{k_1}(x(t_1)), \dots, f_n'^{k_n}(x(t_n)) \right\rangle_\mu$$
$$\leqslant \prod_{i=1}^{n} C_0^{k_i} k_1! (\sup |f_i|)^{k_i} \sum_{\mathcal{T}} \prod_{(t,t') \in \mathcal{T}} (C\beta J^{1/2}(t,t')), \tag{6}$$

holds true with some absolute constants $C_0 > 0$, $C > 0$, and with the sum taken over all trees $\mathcal{T}$ constructed on lattice sites from Z^ν so that the set of their vertices contains the sites $t_1, \dots, t_n$.

Corollary. *The estimate (6) implies the property of strong decay of correlation (1.7) with a maximal constant of the form (1.10).*

Indeed, fix a tree $\mathcal{T}$ and carry out the summation over all ordered collections $(t_2, \dots, t_n)$, $t_i \in Z^\nu$, $i = 2, \dots, n$, of sites from the set $A(\mathcal{T})$ of vertices of the tree $\mathcal{T}$. It leads to the factor $(n-1)! C_{|A(\mathcal{T})|-1}^{n-1} \leqslant (n-1)! 2^{|A(\mathcal{T})|-1}$ on the right-hand side of (6). Summing over all trees $\mathcal{T}$ such that $t \in A(\mathcal{T})$, we get, with the help of the conclusion 1) from Lemma 10.4.II, the estimates (1.7) and (1.10) with a small factor $(\overline{C}\beta)^{n-1}$ appearing in the right-hand side of (1.10).

Remark. If $J(t,t')$ satisfies the estimate

$$(t,t') < c / |t - t'|^\kappa$$

with some constants $\kappa > 4\nu$ and c, we may deduce from (6) the strong cluster estimate (1.8) with the function

$$\varphi(t,t') = \bar{c} / |t - t'|^{\kappa/4} \qquad (\bar{c} \text{ is a constant})$$

with the best constant (1.10) (see [93]).

Proof of Theorem 1. We shall introduce the family of functions

$$\tilde{f}_i = \frac{f_i}{\sup\limits_{x \in S} |f_i(x)|}, \quad \left|\tilde{f}(x)\right| \leqslant 1. \tag{6'}$$

As follows from the proof of Lemma 1.2, the semi-invariant

$$I + \left\langle \prod_{i=1}^{n} \tilde{f}_{/k_i}^{t_i}(x(t_i)) \right\rangle_\mu$$

is an analytic function of the parameter β in a circle $|\beta| < \beta_0$ with $\beta_0 > 0$ not depending on sites $t_1, \dots, t_n$ and numbers $k_1, \dots, k_n$. The formula (21.1.II) yields

$$I = \sum_{\{(A_1,r_1),\dots,(A_p,r_p)\}} \prod_{j=1}^{p} \frac{\beta^{r_j}}{r_j!} \left\langle \prod_{i=1}^{n} \tilde{f}_i'^{k_i}, \Phi_{A_1}'^{k_i}, \dots, \Phi_{A_p}'^{r_p} \right\rangle_{\mu_0} \tag{7}$$

with the sum taken over all unordered collection of pairs $\{(A_1, r_1), \ldots, (A_p, r_p)\}$, where for each $j = 1, \ldots, p$, $A_j = (t_j, t_j')$ denotes a pair of sites and $r_j \geqslant 1$ is an integer, such that the sets A_j are mutually distinct, the collection $\Gamma = \{A_1, \ldots, A_p\}$ is connected, and $t_i \in \bigcup_j A_j$, $i = 1, \ldots, n$ (otherwise $\left\langle \prod_{i=1}^n \tilde{f}_i^{\prime k_1}, \prod_{j=1}^p \Phi_{A_j}^{\prime r_j} \right\rangle_{\mu_0} = 0$; see the condition C, §1.II).

To evaluate the semi-invariants

$$\left\langle \prod_{i=1}^{n}{}' \tilde{f}_i^{\prime k_1}, \prod_{j=1}^{p} \Phi_{A_j}^{\prime r_j} \right\rangle_{\mu_0},$$

we consider with each finite $\Lambda \subset Z^\nu$ the following function of complex variables $\lambda = \{\lambda_i, i = 1, \ldots, n\}$, and $\mu = \{\mu_A, A = \{t, t'\} \subset \Lambda, t \neq t'\}$,

$$F_\Lambda(\lambda, \mu) = \ln \left(\left\langle \exp \left\{ \sum_{i=1}^{n} \lambda_i \tilde{f}_i(x(t_i)) + \sum_{A \subset \Lambda} \mu_A \widetilde{\Phi}_A \right\} \right\rangle_{\mu_0} \right)$$

with the notation

$$\widetilde{\Phi}_A = \Phi_A J_A^{-1/2}, \quad A = \{t, t'\} \subset Z^\nu.$$

Theorem 1.1.IV and Lemma 2.4.III imply that $F_\Lambda(\lambda, \mu)$ is analytic in the domain

$$\max_{i,A}\{|\lambda_i|, |\mu_A|\} < \lambda_0,$$

where $\lambda_0 > 0$ does not depend on Λ and on the set $\{t_1, \ldots, t_n\}$. Moreover, in this domain

$$\max_{i,A} \left\{ \left| \frac{\partial F_\Lambda}{\partial \lambda_i} \right|, \left| \frac{\partial F_\Lambda}{\partial \mu_\Lambda} \right| \right\} < M,$$

where the constant M does not depend on Λ and $\{t_1, \ldots, t_n\}$ (see again §4.III).

From here and from the Cauchy inequalities we get

$$\left| \left\langle \prod_{i=1}^{n}{}' \tilde{f}_i^{\prime k_i}, \prod_{j=1}^{p}{}' \Phi_{A_j}^{\prime r_j} \right\rangle_{mu_0} \right| = \left| \frac{\partial^{|\kappa|+|\rho|}}{\partial^\kappa \lambda \partial^\rho \mu} F_L(\lambda, \mu) \Big|_{\substack{\lambda=0 \\ \mu=0}} \right| \leqslant M(\lambda_0^{-1})^{|\kappa|+|\rho|} \kappa! \rho!,$$

where $\kappa = (k_1, \ldots, k_n)$, $\rho = (r_1, \ldots, r_n)$, $|\kappa| = k_1 + \cdots + k_n$, $\kappa! = \prod k_i!$ (and analogously for $|\rho|$ and $\rho!$), and

$$\frac{\partial^{|\kappa|+|\rho|}}{\partial^\kappa \lambda \partial^\rho \mu} = \frac{\partial^{|\kappa|+|\rho|}}{\prod_{i=1}^{n} \partial^{k_i} \lambda_i \prod_{j=1}^{p} \partial^{r_j} \mu_{A_j}}.$$

This estimate and (7) imply, substituting f_i for $\tilde{f}_i$, the estimate

$$\left|\left\langle \prod_{i=1}^{n}{}' f_i'^{k_i}(x(t_i))\right\rangle_\mu\right| \leqslant M\prod_{i=1}^{n} k_i!(\lambda_0^{-1})^{|\kappa|}\prod_{i=1}^{n}(\sup|f_i|)^{k_i}\sum_{\substack{\Gamma=\{A_1,\ldots,A_p\}\\ t_i\in\Gamma, i=1,\ldots,n}}\prod_{j=1}^{p}(C\beta J_{1/2}^{A_j})$$

with the sum taken over collections $\Gamma = \{A_1, \ldots, A_p\}$ described above. Associating with each such a collection the connected graph with the set of vertices $\tilde{\Gamma}$ and with the edges $\{t_j, t_j'\} = A_j$, and passing from the summation over connected graphs to the summation over trees (Lemma 8.4.II), we get (6). The theorem is proved.

Theorem 2. *If the condition (3) is fulfilled and β is sufficiently small, the following estimate holds true: whenever $t = t_1 \in Z^\nu$ and n is an arbitrary integer,*

$$\sum_{(t_2,\ldots,t_n)}\left|\left\langle \prod_{i=1}^{n}{}' f_i'^{k_i}(x(t_i))\right\rangle_\mu\right| < M(jn-1)!\prod_{n}^{i=1} k_i!(\sup|f_i|)^{k_i}C^{k_1+\cdots+k_n} \tag{8}$$

with some constant M and C.

Proof. We pass again to the functions (6′) and consider the logarithm of their characteristic function with respect to the distribution μ_Λ corresponding to a finite set $\Lambda \subset Z^\nu$:

$$F_\Lambda(\lambda) = \ln\left\langle \exp\left\{\sum_{i=1}^{n}\lambda_i\tilde{f}_i(x(t_i))\right\}\right\rangle_\Lambda . \tag{9}$$

Here $t_i \in \Lambda$, $i = 1, \ldots, n$, and, as above, we use the notation $\lambda = (\lambda_1, \ldots, \lambda_n)$. Clearly,

$$F_\Lambda(\lambda) = \ln Z_\Lambda(\lambda) - \ln Z_\Lambda, \tag{10}$$

where Z_Λ is the partition function of the distribution μ_λ, and

$$Z_\Lambda(\lambda) = \left\langle \exp\left\{\sum \lambda_i\tilde{f}_i(x(t_i)) - \beta\sum_{A\subset\Lambda}\Phi_A\right\}\right\rangle_{\mu_0} . \tag{11}$$

Repeating the arguments from §1.IV, we obtain, for sufficiently small β and

$$|\lambda_i| < \lambda_0 \tag{12}$$

with some constant λ_0, a cluster representation (1.1.III) of the partition function $Z_\Lambda(\lambda)$ in terms of quantities $k_B(\lambda)$ of the form (13):

$$k_B(\lambda) = \begin{cases} 1 & \text{if} \quad B = \{t\}, t \neq t_i \quad \text{for all} \quad i = j, \dots, n \\ \left\langle e^{\lambda_i \tilde{f}_i(x(t_i))} \right\rangle_{\mu_0}, & \text{if} \quad B = \{t_i\}, \\ \sum\limits_{\widetilde{\Gamma} = B} k_\Gamma(\lambda), & |B| \geqslant 2, \end{cases} \tag{13}$$

where the sum is taken over all connected collections $\Gamma = \{A_1, \dots, A_p\}$ of two-point sets $A_j = \{t_j, t'_j\}$, and

$$k_\Gamma(\lambda) = \left\langle \prod_{j=1}^{p} (e^{-\beta \Phi_{A_j}} - 1) \prod_{i: t_i \in \widetilde{\Gamma}} e^{\lambda_i \tilde{f}(x(t_i))} \right\rangle_{\mu_0} . \tag{14}$$

The quantities $k_B(\lambda)$ satisfy, in the domain (12), the estimates (1.2.III) and the cluster condition (23.1.III). Thus $\ln Z_\Lambda(\lambda)$ admits an expansion (see (9.4.III) and (7.4.III))

$$\ln Z_\Lambda(\lambda) = \sum_\eta^{(\Lambda)} d_\eta g_\eta(\lambda) - \sum_{i=1}^{n} \ln \left\langle e^{\lambda_i \tilde{f}_i} \right\rangle_{\mu_0}, \tag{15}$$

where $\eta = \{B_1, \dots, B_s\}$ are connected collections of sets, $B_i \subseteq \Lambda$, $|B_i| \geqslant 2$,

$$g_\eta^{(\lambda)} = \prod_{B_i \in \eta} g_{B_i}(\lambda), \quad g_B(\lambda) = \frac{k_B(\lambda)}{\prod\limits_{t \in B} k_{\{t\}}(\lambda)},$$

and d_η are coefficients.

The equalities (15) and (10) imply

$$\left\langle \prod_{i=1}^{n}{}' \tilde{f}_i^{\prime k_i} \right\rangle_\Lambda = \left. \frac{\partial^{|\kappa|} f_\Lambda(\lambda)}{\partial^\kappa \lambda} \right|_{\lambda=0} = \sum_\eta^{(\Lambda)} d_\eta \left. \frac{\partial^{|\kappa|} g_\eta(\lambda)}{\partial^\kappa \lambda} \right|_{\lambda=0}, \tag{16}$$

for $n \geqslant 2$. The sum in (16) is taken over all collections $\eta = \{B_1, \dots, B_s\}$ such that $t_i \in \bigcup_{j=1}^{S} B_j$, $i, \dots, n$.

Then

$$\frac{\partial^{|\kappa|} g_\eta(\lambda)}{\partial^\kappa \lambda} = \sum_{(\kappa_1, \dots, \kappa_s)} \frac{\kappa!}{\kappa_1! \dots \kappa_s!} \prod_{j=1}^{s} \frac{\partial^{|\kappa_j|} g_{B_j}(\lambda)}{\partial^{\kappa_j} \lambda}, \tag{17}$$

where $\kappa_j = (k_1^{(j)}, \dots, k_\eta^{(j)})$, $j = 1, \dots, s$, $k_i^{(j)} \geqslant 0$, $\kappa! = \prod_{i=1}^{n} k_i!$, $|\kappa| = \sum_{i=1}^{n} k_i$, and the sum in (17) is taken over all ordered collections $(\kappa_1, \dots, \kappa_s)$ such that $\kappa = \kappa_1 + \dots + \kappa_s$. Here,

$$k_i^{(j)} \neq 0 \tag{18}$$

only if $t_i \in B_j$.

Further, the quantity $g_B(\lambda)$ is an analytic function of $\lambda = (\lambda_1, \dots, \lambda_n)$ in the domain (12), and, according to (13) and (14), satisfies the estimate

$$|g_B(\lambda)| \leqslant \sum_{\substack{\Gamma=\{A_1,\dots,A_p\},\\ \widetilde{\Gamma}=B}} \prod_{l=1}^{p} (C\beta J_{A_l}), \tag{19}$$

with the sum taken over all connected collections $\Gamma = \{A_1, \dots, A_p\}$ of two-point sets, $\widetilde{\Gamma} = B$. Since $|\Gamma| \geqslant |B|/2$, the estimate (19) may be replaced by $|g_B(\lambda)| < B^{|B|/4}\tilde{g}_B$ with the notation $\tilde{g}_B = \sum_{\{A_1,\dots,A_p\}} \prod_{l=1}^{p} (C\beta^{1/2} J_{A_l})$. Hence,

$$\left| \left. \frac{\partial^{|\kappa_j|} g_{B_j}(\lambda)}{\partial^{\kappa_j} \lambda} \right|_{\lambda=0} \right| < C_0^{|\kappa_j|} \kappa_j! \beta^{|B_j|/4} \tilde{g}_{B_j}.$$

Inserting the last estimate into (17) we get

$$\left| \left. \frac{\partial^{|\kappa|} g_{B_j}(\lambda)}{\partial^{\kappa_j} \lambda} \right|_{\lambda=0} \right| < C_0^{|\kappa|} \tilde{g}_\eta \cdot \kappa! \beta^{\sum_{B\in\eta} |B|/4} N_{\eta,\kappa}, \tag{20}$$

where we use $N_{\eta,\kappa}$ to denote the number of ordered collections $(\kappa_1, \dots, \kappa_s)$ that satisfy the condition (18) and $\kappa_1 + \cdots + \kappa_s = \kappa$. Let further, the collection η contain s_1 sets B containing the site t_1, s_2 sets containing the site t_2, etc. It is easy to estimate

$$N_{\eta,\kappa \leqslant 2}^{|\kappa|} 2^{s_1+\cdots+s_n}. \tag{21}$$

We shall use n_j to denote the number of sites from the collection $\{t_1, \dots, t_n\}$ that fall into $B_j \in \eta$. Then

$$s_1 + \cdots + s_n = \sum_{j=1}^{s} n_j \leqslant \sum_j |B_j|. \tag{22}$$

For sufficiently small β we get, by (22), (21), (20), and (16), the estimate

$$\left| \left\langle \prod_{i=1}^{n}{}' \tilde{f}_i^{\prime k_i} \right\rangle_\mu \right| \leqslant (2C_0)^{k_1+\cdots+k_n} \prod k_i! \hat{d}_{(\Lambda)}(\mathfrak{A})\eta d_\eta \tilde{g}_\eta \beta^{\sum_{B\in\eta} |B|/8} \tag{23}$$

(the summation here is the same as in (16)). Since for every fixed η there is at most $(n-1)!2^{\sum_{B\in\eta} |B|}$ of ordered collections $(t_1, \dots, t_n)$ such that $t_1 = t$ and $t_i \in \bigcup_{B\in\eta} B$, $i = 1, \dots, n$, we get, by (23), the estimate

$$\sum_{(t_2,\dots,t_n)} \left| \left\langle \prod_{i=1}^{n}{}' \tilde{f}_i^{\prime k_i} \right\rangle_\mu \right| \leqslant \widetilde{C}_0^{k_1+\cdots+k_n}(n-1) \prod k_i! \sum_{\eta, t\in\tilde{\eta}}^{(\Lambda)} d_\eta \tilde{g}_\eta \tag{24}$$

with the sum taken over all collections η such that $t \in \tilde{\eta}$. Since the quantities $\{\tilde{g}_B\}$ satisfy the cluster conditions for β small enough, we have

$$\left|\sum_{\eta, t \in \tilde{\eta}}^{(\Lambda)} d_\eta \tilde{g}_\eta\right| < M$$

uniformly in Λ and t (see §4.III). Hence, passing in (24) to the limit $\Lambda \nearrow Z^\nu$ and replacing the functions $\tilde{f}_i$ by f_i, we get (8). The theorem is proved.

Remark. We may present another proof of Theorem 2 suitable also for unbounded functions f_i on S with finite moments:

$$\langle |f_i(x(t))|^p \rangle_{\mu_0} < \infty, \quad p \geqslant 1.$$

Let us represent the semi-invariant $\left\langle f_1'^{k_1}, \ldots, f_n'^{k_n} \right\rangle_\mu$ in the form $\langle \bar{f}_1(x(\bar{t}_1)), \ldots, \bar{f}_N(x(\bar{t}_N))\rangle_\mu$ where $N = k_1 + \cdots + k_N$ and the site t_1 is met k_1 times in the collection $\{\bar{t}_1, \ldots, \bar{t}_N\}$, t_2 is met k_2 times there, etc., and $\bar{f}_j = f_i$ whenever $\bar{t}_j = t_i$. We shall expand this semi-invariant into moments

$$\left\langle \prod_{j=1}^{N}{}' \bar{f}_j \right\rangle = \sum_{\{T_1, \ldots, T_k\}} \langle \bar{f}_{T_i} \rangle (-1)^{k-1}(k-1)!, \tag{25}$$

where $\{T_1, \ldots, T_k\}$ is an arbitrary partition of the set $\mathcal{N} = \{1, \ldots, N\}$ and $\bar{f}_T = \prod_{j \in T} f_j$, $T \subseteq \mathcal{N}$. Each moment $\langle \bar{f}_T \rangle$ can be, with the help of cluster expansions from §1.IV and §3.III, represented in the form

$$\langle \bar{f}_T \rangle_\mu = \sum_\gamma b_\gamma(\bar{f}_T), \quad T \subseteq \mathcal{N} \tag{26}$$

where the sum is taken over all the class $\gamma = (\Gamma_1, \ldots, \Gamma_s)$ of connected collections $\Gamma_i = \{A_1^{(i)}, \ldots, A_p^{(i)}\}$ of two-point sets $A_l^{(i)}$ (with possible repetitions) such that the collections of sets $\{\widehat{T}, \widetilde{\Gamma}_\Lambda, \ldots, \widetilde{\Gamma}_s\}$ is connected. Here, $\widehat{T} \subseteq \{t_1, \ldots, t_n\}$ is the set of all those sites t_i that coincide with some $\bar{t}_j$, $j \in T$. Moreover, the quantities $b_\gamma(\bar{f}_T)$ obey the estimate

$$|b_\gamma(\bar{f}_T)| < \langle |\bar{f}_T|^2 \rangle_0^{1/2} \prod_{\substack{A \in \cup \Gamma \\ \Gamma \in \gamma}} (X \beta J_A) \tag{27}$$

(C is an absolute constant).

Inserting the expansion (26) for each moment $\langle \bar{f}_T \rangle_\mu$ into the right-hand side of the formula (25) we may conclude, with the help of the arguments from the preceding section using formal semi-invariants, that one may cancel all terms of the form

$$b_{\gamma_1}(\bar{f}_{T_1}) \cdot \ldots \cdot b_{\gamma_k}(\bar{f}_{T_k}),$$

for which the collection

$$\Gamma = \bigcup_{\Gamma' \in \bigcup \gamma_i} \Gamma', \tag{28}$$

obtained as the union of all collections Γ' from $\bigcup \gamma_i$ is unconnected.

Further, with the help of a combinatorial estimate on the number of sequences $(\gamma_1, \ldots, \gamma_k)$ such that the collection Γ in (28) is fixed and summing over all Γ, we get, using the estimate (27), the final result (8). However, one has to keep in mind that the factor $\prod_{i=1}^{n}(\sup |f_i(x)|)^{k_i}$ has to be replaced (cf. (14.1)) by the factor $\sup_{\{t_1,\ldots,t_k\}} \max \prod_{i=1}^{s} \left\langle \prod_{i=1}^{n} |f_i|^{2k_i^{(j)}} \right\rangle^{1/2}$, where $(k_1^{(j)}, \ldots, K_n^{(j)}) = \kappa_j$ is a collection of integers; the maximum is taken over all collections $(\kappa_1, \ldots, \kappa_s)$, $s = 1, 2, \ldots$, such that $\kappa_1 + \cdots + \kappa_s = \kappa = (k_1, \ldots, k_n)$ and the supremum is over all collections of lattice sites $t_1, \ldots, t_n \in Z^\nu$.

§5 Low-Temperature Region

We shall consider now the limit Gibbs field $\{\sigma_t, t \in Z^\nu\}$, $\sigma_t = \pm 1$, for the ferromagnetic Ising model on a square lattice Z^2 with $h = 0$ and for large values $\beta > 0$, corresponding to the pure (+)-phase. The probability distribution μ^+ on the space $\{-1, 1\}^{Z^\nu}$ of the values of this field is obtained as the limit, as $\Lambda \nearrow Z^2$, of the Gibbs measures μ_Λ^+ under (+)-boundary conditions (see §0.I and §5.III). Let us recall that in §0.1 we introduced a contour as a connected closed polygon composed of bonds of the dual lattice $\widetilde{Z}^2$. Every such contour Γ splits up the lattice into 1-connected components — one external component and one or several internal ones. We say that a contour Γ is encircling a lattice site $t \in Z^2$ if this site belongs to any of its internal components.

Theorem 1. *If $\beta > 0$ is sufficiently large, n is arbitrary, $\{t_1, \ldots, t_n\}$ is a collection of mutually distinct lattice sites, and $k_1 \geqslant 1, \ldots, k_n \geqslant 1$ are integers, then the estimate*

$$\left| \langle \sigma_{t_1}'^{k_1}, \ldots, \sigma_{t_n}'^{k_n} \rangle_{\mu^+} \right| < C(n, k_1, \ldots, k_n)(C_0 e^{-\beta})^{r(t_1,\ldots,t_n)}, \tag{1}$$

holds true with

$$C(n, k_1, \ldots, k - n) = n^{k_1 + \cdots + k_n} \prod k_i! (C_1)^{2^{k_1 + \cdots + k_n}}, \tag{1'}$$

where C_0 and C_1 are absolute constants and

$$r(t_1,\ldots,t_n)=\min\{\hat{d}_{\{t_1,\ldots,t_n\}},\min\Gamma|\}.$$

Here, $\hat{d}_{\{t_1,\ldots,t_n\}}$ is the minimal length of trees constructed on the set of sites $\{t_1,\ldots,t_n\}$ (see §4.II) and $\min|\Gamma|$ is the minimal length of contours encircling all sites $t_1,\ldots,t_n$.

From the estimate (1) we get the following conclusions.

Corollary 1. *For a fixed collection $k_1,\ldots,k_n$ and a fixed site $t=t_1$ one has*

$$\sum_{(t_2,\ldots,t_n)}\left|\langle\sigma'^{k_1}t_1,\ldots,\sigma'^{k_n}_{t-n}\rangle_{\mu+}\right|<\widetilde{C}=\widetilde{C}(n,k_1,\ldots,k_n). \tag{2}$$

Corollary 2. *The estimate*

$$\left|\langle\sigma'^{k_1}t_1,\ldots,\sigma'^{k_n}_{t_n}\rangle_{\mu+}\right|<\widetilde{C}_1(\widetilde{C}_2e^{-2\beta/n})^{\hat{d}_{\{t_1,\ldots,t_n\}}}, \tag{3}$$

with absolute constants $\widetilde{C}_2$ and $\widetilde{C}_1=\widetilde{C}_1(n,k_1,\ldots,k_n)$ holds true.

Before proving Theorem 1 with both its Corollaries we shall notice that it is not possible to get the estimates (1) and (2) with a best constant of the form (1.10) or (1.10′). More precisely, we have the following statement.

Theorem 2. *There is no constant $\kappa>0$ such that the series*

$$\sum_{n\geqslant1}\frac{1}{(n-1)!}\sum_{(t_2,\ldots,t_n)}\sum_{(k_1,\ldots,k_n)}\left|\langle\sigma'^{k_1}t_1,\ldots,\sigma'^{k_n}_{t_n}\rangle_{\mu+}\right|\prod_{i=1}^{n}\frac{\kappa^{k_i}}{k_i!} \tag{4}$$

would converge for some (and, consequently, for all) lattice sites $t_1\in Z^\nu$.

Proof. Let the contrary hold: the series (4) converges for some $\kappa>0$.

Let $\mu_{\beta,h}$, $h\neq0$, be the limit measure for the Ising model. Due to the uniqueness of such a measure for $h\neq0$ (see §0.1), one has $\mu_{\beta,h}=\tilde{\mu}_h=\lim_{\Lambda\nearrow Z^\nu}\tilde{\mu}_{\Lambda,h}$, where $\tilde{\mu}_{\Lambda,h}$ are Gibbs modifications of measures μ_Λ^+:

$$\frac{d\tilde{\mu}_{\Lambda,h}}{d\mu_\Lambda^+}=\widetilde{Z}_\Lambda^{-1}\exp\left\{h\sum_{t\in\Lambda}\sigma_t\right\}. \tag{5}$$

We expand the mean values $\langle\sigma_t\rangle_{\tilde{\mu}_{\Lambda,h}}$ in terms of semi-invariants of the measure μ_Λ^+:

$$\langle\sigma_t\rangle_{\tilde{\mu}_{\Lambda,h}}=\sum_{n\geqslant1}\frac{1}{(n-1)!}\sum_{(t_2,\ldots,t_n)}\sum_{(k_1,\ldots,k_n)}\prod_{i=1}^{n}\frac{h^{k_i}}{k_i!}\left\langle\sigma'^{k_1+1}_{t_1},\sigma'^{k_2}_{t_2},\ldots,\sigma'^{k_n}_{t_n}\right\rangle_{\mu_\Lambda^+},$$
$$t_1=t. \tag{6}$$

From the assumption about the convergence of the series (4), it is easy to conclude, passing to the limit $\Lambda \nearrow Z^2$, that the mean values $\langle\sigma_t\rangle_{\tilde{\mu}_h} = \langle\sigma_t\rangle_{\beta,h}$ can be expanded in power series in h, $|h| < \kappa$, that is similar to the series (6) (containing, however, semi-invariants of the measure μ^+) and, consequently, it is an analytic function in h for $|h| < \kappa$. On the other hand, from the proof of existence of the two limit measures μ^+ and μ^- for $h = 0$ (see the end of §0.I) it follows that the mean

$$m(h) = \langle\sigma_t\rangle_{\beta,h}$$

has a discontinuity in h for $h = 0$. This contradiction proves Theorem 2.

Proof of Theorem 1. We shall consider the ensemble of contours on the dual lattice $\widetilde{Z}_2$, i.e., a probability distribution on the set of finite or countable configurations $\alpha = \{\Gamma_i, i = 1, 2, \ldots\}$ of contours in $\widetilde{Z}^2$, that was mentioned in §0.I (see also §5.III and §1.V). We define a random field $\{\chi_{\bar{t}}\}$ on bonds $\bar{t}$ of the lattice $\widetilde{Z}^2$ with values in the set $\{0, 1\}$ by putting

$$\chi_{\bar{t}}(\alpha) = \begin{cases} 1 & \text{if } \alpha \text{ contains a contour } \Gamma \text{ passing through the bond } \bar{t}, \\ 0 & \text{otherwise.} \end{cases}$$

Using the cluster representation of the partition function Z_Λ of the contour ensemble (see §5.III), one may prove, with the help of the methods of Chapter III, that the cluster expansion for the correlation function f_A and for the mean $\langle F_A\rangle$ of any bounded local function of the field $\{\chi_t\}$ satisfies the assumptions of Lemma 4.3.III whenever β is sufficiently large. These mean values thus admit an exponentially-regular cluster expansion (see §3.III, Definition 1) with parameter $\lambda = Ce^{-\beta}$, where C is an absolute constant, and with the set of all bonds of the lattice $\widetilde{Z}^2$ taken for the family $\mathfrak{A}$ appearing in the definition of an exponentially-regular expansion.

Now we shall introduce the functions χ_Γ defined on configurations of contours:

$$\chi_\Gamma(\alpha) = \begin{cases} 1 & \text{if } \Gamma \in \alpha, \\ 0 & \text{otherwise.} \end{cases}$$

For each finite collection $\gamma = \{\Gamma_1, \ldots, \Gamma_s\}$ of contours (that may be, in general, repeated in the collection γ several times) we put

$$\chi_\gamma(\alpha) = \prod_{\Gamma\in\gamma} \chi_\Gamma(\alpha).$$

It is easy to see that χ_Γ, and also χ_γ, are local functions of the field $\{\chi_{\bar{t}}\}$. Applying again arguments similar to those from Chapter III (see also [49]), it is easy to show that for any collection of contours $\gamma = \{\Gamma_1, \ldots, \Gamma_s\}$ the estimate

$$|\langle \chi_\gamma \rangle| \leqslant \prod_{\Gamma \in \hat{\gamma}} (Ce^{-\beta})^{|\Gamma|} \tag{7}$$

holds true with C an absolute constant and the product $\prod_{\Gamma \in \hat{\gamma}}$ taken over the set $\hat{\gamma} \subseteq \gamma$ of all mutually distinct contours from γ. Thus, applying Theorem 1.3 to the collection of functions $G = \{\chi_\gamma\}$ and taking into account the inequality (7), we get the following estimate of the semi-invariants $\langle \chi_{\gamma_1}, \ldots, \chi_{\gamma_N} \rangle$, where $\gamma_1, \ldots, \gamma_N$ is an arbitrary sequence of collections of contours: $\gamma_i = \{\Gamma_1^{(i)}, \ldots, \Gamma_{s_i}^{(i)}\}$.

Theorem 3. *The estimate*

$$\langle \chi_{\gamma_1, \ldots, \gamma_{\gamma_N}} \rangle < \prod_{\Gamma}{}' (Ce^{-\beta})^{\Gamma} \prod_{i=1}^{n} \widehat{C}^{\hat{d}_{\tilde{\gamma}_i}} (Ce^{-\beta})^{d(\gamma_1, \ldots, \gamma_N)} \prod_{i=1}^{N} n_i, \tag{8}$$

holds true with some constants $C > 1$ and $\widehat{C} > 1$ and the product $\prod'_\Gamma$ taken over the family $\hat{\gamma}$ of all mutually distinct contours Γ from the collection $\gamma = \gamma_1 \cup \cdots \cup \gamma_N$, $\tilde{\gamma} = \cup \widetilde{\Gamma}_j$ is the set of vertices of contours entering γ_i $\hat{d}_{\tilde{\gamma}_i} = d_{\tilde{\gamma}_i}(\mathfrak{A})$ and $d(\gamma_1, \ldots, \gamma_N) = d(\mathfrak{A}, \gamma_1, \ldots, \gamma_N)$ are the numbers defined in §4.II, $\mathfrak{A}$ is the family of all bonds of the lattice $\widetilde{Z}^2$, and finally n_i, $i = 1, \ldots, N$, is the multiplicity of the collection γ_i in the sequence $\{\gamma_1, \ldots, \gamma_N\}$.

In a particular case, to be needed later, when all the collections γ_i from $\{\gamma_1, \ldots, \gamma_N\}$ consist of mutually distinct contours that successively encircle each other, the quantity $\hat{d}_{\tilde{\gamma}}$ may be bounded by $\sum_{\Gamma \in \gamma} |\Gamma|$ and thus the factor $\prod_i \widehat{C}^{\hat{d}_{\tilde{\gamma}_i}}$ may be omitted in the estimate (8) (replacing the constant C by $C\widehat{C}$).

We shall pass to the derivation of the estimates (1) and (1′). For each lattice site $t \in Z^2$ we shall single out, from a configuration α of contours, the set of those that encircle the site t. Clearly, these contours successively encircle each other and for almost all α their number is finite. In addition, for almost all (with respect to the measure μ^+) configurations of spins $\{\sigma_t, t \in Z^2\}$, one has

$$\sigma_t = \prod_{\Gamma} (-1)^{\chi_\Gamma} = \prod_{\Gamma} (1 - 2\chi_\Gamma), \tag{9}$$

where the product is taken over all contours Γ in $\widetilde{Z}^2$ encircling the site t. Rewriting (9) in the form

$$\sigma_t = 1 - 2\sum_{\Gamma} \chi_\Gamma + 4 \sum_{\{\Gamma_1, \Gamma_2\}, \Gamma_1 \neq \Gamma_2} \chi_{\Gamma_1} \chi_{\Gamma_2} + \cdots + (-2)^n \sum_{\substack{\{\Gamma_1, \ldots, \Gamma_n\}, \\ \Gamma_i \neq \Gamma_j, i \neq j}} \chi_{\Gamma_1} \cdots \chi_{\Gamma_n} + \cdots \tag{10}$$

and inserting (10) into (1), we conclude that the semi-invariant $\left\langle\sigma_{t_1}^{\prime k_1},\ldots,\sigma_{t_n}^{\prime k_n}\right\rangle$ equals

$$\sum_{\substack{\{\gamma_1^{(1)},\ldots,\gamma_{k_1}^{(1)}\},\\ \{\gamma_1^{(n)},\ldots,\gamma_{k_n}^{(n)}\},}} \prod_{i,j}(-2)^{|\gamma_j^{(i)}|}\left\langle\prod_{i,j}{}' \chi_{\gamma_j^{(i)}}\right\rangle, \tag{11}$$

where the sum is over all sequences $\{\gamma_j^{(i)}\}$, $i = 1,\ldots,n$, of collections of contours such that each of them consists of mutually distinct contours and that all contours from the collections $\{\gamma_j^{(i)}\}$, $j = 1,\ldots,k_i$ encircle the site t_i.
It is easy to verify that for each term in (11) one has the bound

$$r(t_1,\ldots,t_n) \leqslant \sum_{\Gamma\in\hat{\gamma}} |\Gamma| + d\{\gamma_1^{(1)},\ldots,\gamma_{k_1}^{(1)},\ldots,\gamma_{k_n}^{(n)}\},$$

where $\hat{\gamma} \subseteq \gamma$ is the set of mutually distinct contours Γ in the collection $\gamma = \bigcup_{i,j}\gamma_j^{(i)}$. For a fixed collection $\{\Gamma_1,\ldots,\Gamma_m\}$ of distinct contours, each of which encircles some site from the set $t_1,\ldots,t_n$, the number $N_{\{\Gamma_1,\ldots,\Gamma_m\}}(k_1,\ldots,k_n)$ of sequences $\{\gamma_j^{(i)},(i,j)\}$ over which the sum (11) is taken and for which $\hat{\gamma} = \{\Gamma_1,\ldots,\Gamma_m\}$, does not exceed $2^{(k_1+\cdots+k_n)m}$. Indeed, this follows from the representation

$$N_{\{\Gamma_1,\ldots,\Gamma_m\}}(k_1,\ldots,k_n) = \sum_{n_1\geqslant 1,\ldots,n_m\geqslant 1} B_{n_1,\ldots,n_m}^{k_1,\ldots,k_n}, \tag{12}$$

where the numbers $B_{n_1,\ldots,n_m}^{k_1,\ldots,k_n}$ are defined from the expansion

$$\prod_{q=1}^{m}\prod_{i}{}'(1+\lambda_{\Gamma_q})^{k_i} = \sum_{n_1\geqslant 0,\ldots,n_m\geqslant 0} B_{n_1,\ldots,n_m}^{k_1,\ldots,k_n}\lambda_{\Gamma_1}^{n_1}\ldots\lambda_{\Gamma_m}^{n_m}; \tag{13}$$

here, the product $\prod_i'$ is taken over those i, for which the site t_i is encircled by the contour Γ_q. We select further in the collection $\Gamma_1,\ldots,\Gamma_m$ all those contours that encircle the site t_1 and denote their number s_1, then select among the remaining ones all those contours that encircle t_2 and denote their number s_2, etc. In this way we get, with the help of (8), the estimate

$$\begin{aligned}\left|\left\langle\sigma_{t_1}^{\prime k_1},\ldots,\sigma_{t_n}^{\prime k_n}\right\rangle\right| &< (C\kappa^{-1}\beta)^{r(t_1,\ldots,t_n)}(k_1+\cdots+k_n)^{k_1+\cdots+k_n}\\ &\times\sum_{(s_1,\ldots,s_n)}\prod_{i=1}^{n}2^{(k_1+\cdots+k_n)s_i}\sum_{\{\Gamma_1,\ldots,\Gamma_{s_i}\}}\prod_{j=1}^{s_i}\kappa^{(\Gamma_j)},\end{aligned} \tag{14}$$

where the sum $\sum_{\{\Gamma_1,\ldots,\Gamma_{s_i}\}}$ is taken over unordered collections $\{\Gamma_1,\ldots,\Gamma_{s_i}\}$ of mutually distinct contours Γ_j encircling the lattice site t_i and κ is some sufficiently small constant. Furthermore,

$$\sum_{\{\Gamma_1,\ldots,\Gamma_{s_i}\}} \prod \kappa^{|\Gamma_j|} \leqslant \frac{1}{s_i!}\left(\sum_{\Gamma}^{(t)} \kappa^{|\Gamma|}\right)^{s-i} = \frac{C^{s-i}}{s_i!}, \tag{15}$$

with the sum $\sum_{\Gamma}^{(t)} \kappa^{|\Gamma|}$ taken over all contours Γ encircling a given site t (for a suitable choice of κ this series converges). On account of (15), (14), and the inequality

$$(k_1 + \cdots + k_n)^{k_1+\cdots+k_n} < n^{k_1+\cdots+k_n} \prod_{i=1}^{n} k_i!,$$

the estimate (1) with the equality (1′) follows. Theorem 1 is proved.

Proof of Corollary 1. Since

$$(Ce^{-\beta})^{r(t_1,\ldots,t_n)} \leqslant (Ce^{-\beta})^{\hat{d}\{t_1,\ldots,t_n\}} + (Ce^{-\beta})^{\min|\Gamma|}$$

and the sum $\sum_{\{t_2,\ldots,t_n\}}(Ce^{-\beta})^{\hat{d}\{t_1,\ldots,t_n\}}$ converges, as follows by Lemma 10.4.II, it is sufficient to prove that

$$\sum_{\{t_2,\ldots,t_N\}} (Ce^{-\beta})^{\min|\Gamma|} < \infty. \tag{16}$$

Since a fixed contour Γ cannot encircle more than $|\Theta(\Gamma)|^{n-1}$ of different collections $\{t_2,\ldots,t_n\}$, where $\Theta(\Gamma)$ is used to denote the interior of the contour Γ and, since $|\Theta(\Gamma)| < |\Gamma|^2$, we conclude that the sum (16) is bounded by

$$\sum_{\Gamma:0(\Gamma)\geqslant n} |\Gamma|^{2(n-1)}(Ce^{-\beta})^{|\Gamma|} < \infty,$$

where the sum is taken over all contours Γ encircling the site t.

Proof of Corollary 2. The estimate (3) follows from

$$\hat{d}_{\{t_1,\ldots,t_n\}} \leqslant \frac{n\min|\Gamma|}{2}.$$

§6 Scaling Limit of a Random Field

In this section we will show a simple application of the estimates concerning the decay of semi-invariants that is associated with a limit theorem for the so-called renormalization group.

Let $x = \{x(t), t \in Z^\nu\}$ be a translation invariant random field taking values in real numbers and with vanishing means: $\langle x(t)\rangle = 0$. We suppose further that all its moments are finite

$$\langle |x(t)|^k\rangle < \infty, \quad t \in Z^\nu, \quad k > 0, \tag{1}$$

(and, consequently, all mixed moments $\langle (x(t_1))^{k_1} \dots (x(t_n))^{k_n}\rangle$ for arbitrary $t_1, \dots, t_n$ and $k_1 \geqslant 0, \dots, k_n \geqslant 0$, are finite).

Let $\Lambda_0 = \Lambda_0^N = \{0 \leqslant t^{(i)} < N\}$ be a cube Z^ν (N is an integer) and $\Lambda_t = \Lambda_t^{(N)} = \Lambda_0 + Nt$ be its shift by the vector Nt with $t \in Z^\nu$. Clearly, $\bigcup_{t\in Z^\nu} \Lambda_t = Z^\nu$ and $\Lambda_t \cap \Lambda_{t'} = \emptyset$ if $t \neq t'$. We define a new random field $\hat{x} = \hat{x}^{(N)} = \{\hat{x}(t), t \in Z^\nu\}$ by putting

$$\hat{x}(t) = \hat{x}^N(t) = s_t^{(N)}/N^{\nu/2}, \quad s_t^{(N)} = \sum_{t'\in\Lambda_t} x(t'). \tag{2}$$

Clearly, the new random field is translation-invariant, has all moments, and $\langle \hat{x}(t)\rangle = 0$. The mapping (2): $x \mapsto \hat{x}^{(N)}$ in the space of random fields is called a *renormalization* transformation and will be denoted by G_N. It is easy to verify that $G_{N_1}G_{N_2} = G_{N_1N_2}$, i.e., that renormalization transformations form a semigroup.

Let the limit

$$\hat{x} = \lim_{N\to\infty} G_N x \tag{3}$$

exist (and, consequently, also the limit $\lim_{l\to\infty}(G_{N_0}^l)x = \hat{x}$ with an arbitrary field N_0). It is called the *scaling limit*[1] of the original field x. The convergence in (3) is in the sense of the weak local convergence (convergence of finite-dimensional distributions, see §1.I) of probability distributions μ_N on R^{Z^ν} of values of the field $\hat{x}^{(N)}$ towards the distribution μ of the field $\hat{x}$.

Theorem 1. *Let a translation-invariant field $x = \{x(t), t \in Z^\nu\}$ with vanishing mean, $\langle x(t)\rangle = 0$, satisfy the condition (1) and the condition of strong decay of correlations: for any n and $k_1, \dots, k_n$ suppose*

$$\sum_{(t_2,\dots,t_n)} |\langle x(t_1)^{\prime k_1}, \dots, x(t_n)^{\prime k_n}\rangle| < \infty. \tag{4}$$

---

[1] Translator's remark: the term "self-similar field" is also used in the literature.

Then the scaling limit (3) of the field x exists and the limit field $\hat{x} = \{\hat{x}(t), t \in Z^\nu\}$ is an independent, translation-invariant, Gaussian field with vanishing means, $\langle \hat{x}(t) \rangle = 0$, and the variance

$$D = D(\hat{x}(t)) = \lim_{N\to\infty} \frac{Ds_t^{(N)}}{N^\nu} = \sum_{t' \in Z^\nu} \langle x(t), x(t') \rangle. \tag{5}$$

Proof. We use the following lemma.

Lemma 2. *Let μ_N be a sequence of probability distributions on the space R^n with finite moments and vanishing means, $\langle x_i \rangle_{\mu_N} = 0, i = 1, \ldots, n$, such that the semi-invariants*

$$\left\langle x_{i_1}^{\prime k_1}, \ldots, x_{i_s}^{\prime k_s} \right\rangle_{\mu_N} \to 0, \quad N \to \infty \tag{6}$$

for any collections of indices, $1 \leqslant i_1 < i_s < \cdots < i_s \leqslant n$, and numbers $k_1 \geqslant 1, \ldots, k_s \geqslant 1$ such that $k_1 + \cdots + k_s > 2$, and also

$$\langle x_i, x_j \rangle_{\mu_N} \to b_{ij} \quad as \quad N \to \infty. \tag{7}$$

(Here, $(x_1, \ldots, x_n)$ are coordinates in R^n).

Then μ_N weakly converges towards the Gaussian distribution on R^n with vanishing means and the covariance matrix $\{b_{ij}, i, j = 1, \ldots, n\}$.

The proof is based on a similar assertion about convergence of measures with converging moments (see, e.g., [58]) and on the observation that only for a Gaussian measure all higher semi-invariants are vanishing (see §1.II).

The condition (6) for fields $\hat{x}^{(N)}$ of the form (2) follows easily from (4): one must represent the semi-invariant $\langle \hat{x}^N(t_1)^{\prime k_1}, \ldots, \hat{x}^N(t_n)^{\prime k_n} \rangle$ in the form of a sum of semi-invariants of the original field x with the factor $\left(N^{\nu/2(k_1+\cdots+k_n)}\right)^{-1}$, and then observe that the whole sum is of the order $\sim N^\nu$.

Further for $t \neq t'$, one has

$$\langle \hat{x}^{(N)}(t), \hat{x}^{(N)}(t') \rangle = \frac{1}{N^\nu} \sum_{\substack{t_1 \in \Lambda_t^N, \\ t_1 \in \Lambda_{t'}^N}} \langle x(t_1), x(t_1') \rangle. \tag{8}$$

In each cube Λ_t^N, we single out a narrow layer $\partial_{\sqrt{N}} \Lambda_t^N$ of the width $N^{1/2}$ around the boundary $\partial \Lambda_t^N$ and denote $\bar{\Lambda}_t^N = \Lambda_t^N \backslash \partial_{\sqrt{N}} \Lambda_t^N$. Then, for $t_1 \in \Lambda_t^N$, we get by (4):

$$\sum_{t_1' \in \bar{\Lambda}_{t'}^N} \langle x(t_1), x(t_1') \rangle \to 0 \quad \text{as} \quad N \to \infty \tag{9}$$

uniformly in t_1, and

$$\sum_{\substack{t_1 \in \Lambda_t^N, \\ t_1' \in \partial_{\sqrt{N}} \Lambda_{t'}^N}} \langle x(t_1), s(t_1') \rangle \sim N^{\nu - 1/2}. \tag{10}$$

By (8), (9), and (10), we conclude that $\langle \hat{x}^N(t), \hat{x}^{(N)}(t') \rangle \to 0$ as $N \to \infty$. One derives the equality (5) in a similar way. The theorem is proved.

Corollary 1. *In the case of positive variance, $D > 0$, (i.e., $Ds_t^{(N)} = 0(N^\nu)$), the field*

$$\tilde{x}^{(N)}(t) = \frac{s_t^{(N)}}{(Ds_t^{(N)})^{1/2}} = \hat{x}^N(t) \left(\frac{(Ds_t^N)}{N^\nu} \right)^{-1/2}$$

converges towards the Gaussian independent field with the variance one.

Corollary 2. *Let a field in the lattice be of the form $\{f(x(t)), t \in Z^\nu\}$, where $\{x(t), t \in Z^\nu\}$ is a Gibbs translation invariant field studied in Theorem 2.4 and f is a bounded real function on the space S such that $\langle f(x(t)) \rangle_\mu = 0$ with μ being the distribution for the field $\{x(t), t \in Z^\nu\}$. Then the scaling limit of the field $\{f(x(t)), t \in Z^\nu\}$ is a Gaussian independent field with positive variance.*

This follows from the results of Section 4 and also from the expansion

$$\sum_{T'} \langle f(x(t)), f(x(t')) \rangle = \langle f^2(x(t)) \rangle_{\mu_0} + O(\beta),$$

that may be also easily derived from the considerations of Section 4 (here $\langle\rangle_{\mu_0}$ is the mean under the distribution μ_0 of the original independent field). According to the remark at the end of Section 4, this result holds also for an unbounded function f whose all moments are finite: $\left\langle |f(x(t))|^k \right\rangle_{\mu_0} < \infty$.

CHAPTER 7

SUPPLEMENTARY TOPICS AND APPLICATIONS

§1 Gibbs Quasistates

The theory of cluster expansions for classical systems developed in the preceding chapters may be, in may aspects, carried over to the case of quantum systems. Here we will dwell on it only briefly.

Let $\mathfrak{A}$ be an associative algebra with a unit element $\mathbf{1}$ over the field of complex numbers $\mathbf{C}$. A linear functional $\langle\cdot\rangle$ on $\mathfrak{A}$, such that $\langle\mathbf{1}\rangle = 1$, is called a *quasistate*. For the case where an involution $*: x \mapsto x^*$, $x \in \mathfrak{A}$, is defined on the algebra $\mathfrak{A}$, a quasistate is called a *state* if it is nonnegative on positive elements (i.e., on elements of the form xx^*, $x \in \mathfrak{A}$). We recall that an algebra $\mathfrak{A}$ is called a *superalgebra* if it can be represented in the form of a direct sum $\mathfrak{A} = \mathfrak{A}_0 \oplus \mathfrak{A}_1$ where $\mathfrak{A}_0 \subseteq \mathfrak{A}$ is a superalgebra of $\mathfrak{A}$ and $\mathfrak{A}_1 \subset \mathfrak{A}$ a subspace of $\mathfrak{A}$ such that $\mathfrak{A}_0\mathfrak{A}_1 \subseteq \mathfrak{A}_1$, $\mathfrak{A}_1\mathfrak{A}_0 \subseteq \mathfrak{A}_1$, and $\mathfrak{A}_1\mathfrak{A}_1 \subseteq \mathfrak{A}_0$. Elements from $\mathfrak{A}_0$ are called *even* and elements from $\mathfrak{A}_1$ *odd.* Even and odd elements are called *homogeneous.* We notice that each algebra $\mathfrak{A}$ may be regarded as a trivial superalgebra by letting $\mathfrak{A}_0 = \mathfrak{A}$, $\mathfrak{A}_1 = \{0\}$. In this way, all constructions given below for superalgebras can be automatically applied to an arbitrary algebra.

A quasistate $\langle\cdot\rangle$ on a superalgebra $\mathfrak{A}$ is called *even* if $\langle x\rangle = 0$ for each odd $x \in \mathfrak{A}_1$.

A local algebra on a countable set T. Let a (super) algebra $\mathfrak{A}_\Lambda$ with unit $\mathbf{1}_\Lambda$ be associated with each finite subset $\Lambda \subset T$ so that for each pair of sets $\Lambda' \subset \Lambda$, a homomorphic imbedding $\varphi'_{\Lambda',\Lambda}: \mathfrak{A}_{\Lambda'} \to \mathfrak{A}_\Lambda$ is defined. Then the algebra $\mathfrak{A}^0 = \cup_{\Lambda\subset T} \mathfrak{A}_\Lambda$ (more precisely, the inductive limit of the algebras $\mathfrak{A}_\Lambda$, see [10]) is called a *local* (super)algebra in T. For the case where the algebras $\mathfrak{A}_\Lambda$ are normed and the imbeddings $\varphi_{\Lambda,\Lambda'}$ are isometries, one may introduce a norm on $\mathfrak{A}^0$ in a natural way and define the closure $\overline{\mathfrak{A}^0}$ of the algebra $\mathfrak{A}^0$ with respect to this norm called a *quasilocal* algebra (superalgebra).

Superalgebra $\mathfrak{A}_\Lambda$ and imbeddings $\varphi_{\Lambda',\Lambda}$ are usually introduced in the following way. Let $\{\mathfrak{A}_t,\ t \in T\}$ be a family of superalgebras assigned to points $t \in T$, and put

$$\mathfrak{A}_\Lambda = \bigotimes_{t\in\Lambda} \mathfrak{A}_t, \tag{1}$$

where $\bigotimes_{t\in\Lambda}$ is the tensor product of the superalgebras $\mathfrak{A}_t$ (see [33]), with the imbeddings taken in the form

$$\varphi_{\Lambda',\Lambda}(x_{\Lambda'}) = x_{\Lambda'} \otimes \mathbf{1}_{\Lambda\setminus\Lambda'}, \qquad x_{\Lambda'} \in \mathfrak{A}_{\Lambda'}. \tag{2}$$

We do not give the definition of the tensor product $A_1 \otimes A_2$ of superalgebras, observing only that the image $\hat{x} = x \otimes \mathbf{1}_{A_2} \in A_1 \otimes A_2$ of an odd element $x \in A_1$ commutes with the image $\hat{y} = \mathbf{1}_{A_1} \otimes y$ of an even element $y \in A_2$ and anticommutes, $(\hat{x}\hat{y}) = -(\hat{y}\hat{x})$, with the image of an odd element $y \in A_2$. The images of even elements commute.

Examples. 1. Let $\mathfrak{A}_t = A^n$ be the C^*-algebra of $n \times n$ complex matrices with the norm induced by the usual scalar product in C^n (see [17]). Then the C^*-algebra $\overline{\mathfrak{A}^0}$ is called the *spin* algebra in T.

2. Let $\mathfrak{A}_t$ be a Grassmann superalgebra (see [5]) with a finite number of generators (in physical applications, one usually chooses their number to be even). Then $\mathfrak{A}^0$ is also a Grassmann algebra.

3. Let $\mathfrak{A}_t$ be a Clifford superalgebra with unity $\mathbf{1}_t$ and generators a_t, a_t^*:

$$a_t^2 = (a_t^*)^2 = 0, \qquad a_t a_t^* + a_t^* a_t = \mathbf{1}_t.$$

From these relations, it follows that one may choose in $\mathfrak{A}_t$ the basis $(\mathbf{1}, a_t, a_t^*, a_t a_t^*)$, and the algebra $\mathfrak{A}_t$ is thus isomorphic to the superalgebras $M \subset A^2$ of the algebra of 2×2 matrices spanned on the generators $\mathbf{1}, e_1 = \begin{pmatrix} 0 & 1 \\ 0 & 0 \end{pmatrix}, e_2 = \begin{pmatrix} 0 & 0 \\ 1 & 0 \end{pmatrix}$. Hence, one may introduce in the algebras $\mathfrak{A}_t$ and $\mathfrak{A}_\Lambda$ a norm (preserved under imbeddings $\varphi_{\Lambda'\Lambda}$) so that all of them turn into C^*-algebras. The quasilocal Clifford superalgebra $\overline{\mathfrak{A}_0}$ is called the *algebra of canonical anticommutation relations* (or, abbreviated, the CAR-algebra).

4. The case of classical systems is included in the present scheme if we take for $\mathfrak{A}_t = \mathfrak{A}$ the commutative C^*-algebra of bounded continuous functions on a metric space S. Then the algebra $\overline{\mathfrak{A}^0}$ coincides with the algebra of all bounded functions on S^T that are continuous with respect to the Tikhonov topology on S^T.

Independent quasistates on $\mathfrak{A}^0$. Let a local superalgebra in $\mathfrak{A}^0$ be obtained from a family of superalgebras $\{\mathfrak{A}_t,\ t \in T\}$ with the help of the construction described by (1) and (2), and let $\langle\cdot\rangle_t$ be even quasistates on the superalgebras $\mathfrak{A}_t$. We define the quasistate $\langle\ \rangle$ on $\mathfrak{A}^0$ by letting

$$\left\langle \bigotimes_{t\in\Lambda} x_t \right\rangle = \prod_{t\in\Lambda} \langle x_t \rangle_t, \qquad x_t \in \mathfrak{A}_t, \tag{3}$$

and then extending it by linearity onto the whole algebra $\mathfrak{A}^0$ (and, if possible, by continuity onto the algebra $\overline{\mathfrak{A}^0}$).

The quasistate on $\mathfrak{A}^0$ obtained in this way is called *independent*. (We notice that for states $\langle\cdot\rangle_t$ on $\mathfrak{A}_t$ that are not even, the definition (3) is incorrect.)

Semi-invariants. Let $\mathfrak{A}$ be a superalgebra, $\langle\ \rangle$ an even quasistate on $\mathfrak{A}$, and $(\xi_1, \dots, \xi_N)$, $\xi_i \in \mathfrak{A}$, a sequence of (possibility repeating) homogeneous elements.

For any subset $T \subseteq \mathcal{N}$, we write its elements in the order of their growth, $T = (i_1, \dots, i_p)$, $i_1 < i_2 < \dots < i_p$, and in the following, we shall identify T with the sequence $(i_1, \dots, i_p)$. Put

$$\xi_T = \xi_{i_1} \cdot \dots \cdot \xi_{i_p}.$$

We now define *semi-invariants* $\langle \xi'_T \rangle = \langle \xi_{i_1}, \dots, \xi_{i_p} \rangle$ for all $T \subseteq \mathcal{N}$ in the following way. We use $\widehat{T} \subseteq T$ to denote the subset of those $i \in T$ for which the element ξ_i is odd. Put

$$\langle \xi'_T \rangle = 0 \qquad \text{if } |\widehat{T}| \text{ is odd.}$$

For the case where the number $|\widehat{T}|$ is even, the semi-invariant $\langle \xi'_T \rangle$ is defined by an inductive formula analogous to definition (6), §1, Chapter 2:

$$\langle \xi_T \rangle = \sum_{\delta = \{T_1, \dots, T_k\}} \langle \xi'_{T_1} \rangle \dots \langle \xi'_{T_k} \rangle (-1)^{\pi}, \tag{4}$$

where the sum is taken over all partitions $\delta = \{T_1, \dots, T_k\}$ of the set $\mathcal{N}$ such that all $|\widehat{T}_i|$ are even, and $\pi = \pi(\delta)$ is the number of order inversions in the sequence $(\widehat{T}_1, \dots, \widehat{T}_k)$ of indices of even elements written in the same order as they are placed in successive elements T_1, T_2, etc., of the partition δ (due to evenness of $|\widehat{T}_i|$, the sign $(-1)^{\pi}$ does not depend on the order of these elements).

The semi-invariant $\langle \xi'_{\mathcal{N}} \rangle = \langle \xi_1, \dots, \xi_N \rangle$ for a sequence of arbitrary (non-homogeneous) elements $\xi_i \in \mathfrak{A}$ is then defined as a multilinear (unsymmetric) function of elements ξ_i, $i = 1, \dots, N$.

We shall establish a series of properties of semi-invariants.

I. Let a state $\langle\cdot\rangle$ on a local superalgebra $\mathfrak{A}^0$ be independent (and, consequently, even), $\xi_1 \in \mathfrak{A}_{\Lambda_1}, \dots, \xi_N \in \mathfrak{A}_{\Lambda_N}$, and the sets $\Lambda_1, \dots, \Lambda_N \subset T$ be mutually disjoint. Then

$$\langle \xi_1, \dots, \xi_N \rangle = 0. \tag{5}$$

The proof repeats that one of a similar property of semi-invariants of a virtual field (see §1, Chapter 2).

II. The *inversion formula* (for the case of homogeneous elements ξ_i):

$$\langle \xi_1, \dots, \xi_N \rangle = \sum_{\delta = \{T_1, \dots, T_k\}} \langle \xi_{T_1} \rangle \dots \langle \xi_{T_k} \rangle (-1)^{\pi(\delta)} (-1)^{k-1} (k-1)!, \quad (6)$$

where the sum is the same as in (4).

The proof reduces itself to the Möbius inversion formula. For any partition $\delta = \{T_1, \dots, T_k\}$ of the set $\mathcal{N}$, we put

$$f(\delta) = (-1)^{\pi(\delta)} \left\langle \xi'_{T_1} \right\rangle \dots \left\langle \xi'_{T_k} \right\rangle,$$
$$g(\delta) = (-1)^{\pi(\delta)} \langle \xi_{T_1} \rangle \dots \langle \xi_{T_k} \rangle .$$

From (4), we get

$$g(\delta) = \sum_{\beta \leqslant \delta} f(\beta).$$

Hence, by means of the Möbius inversion formula (see §5, Chapter 2), we get (6).

III. *Generalized expansion over connected groups.* Let $(\xi_1, \dots, \xi_N)$ be a sequence of homogeneous elements, and $\delta_0 = \{T_1, \dots, T_k\}$ be a partition: $T_1 = (1, \dots, l_1)$,

$$T_2 = (l_1 + 1, \dots, l_2), \dots, \quad T_k = (l_{k-1} + 1, \dots, N).$$

Then

$$\langle \xi_{T_1}, \dots, \xi_{T_k} \rangle = \sum_{\delta' : \delta' \vee \delta_0 = \underline{1}} \left\langle \xi'_{T'_1} \right\rangle \dots \left\langle \xi'_{T'_q} \right\rangle (-1)^{\pi(\delta')}, \quad (7)$$

where the sum is taken over all partitions δ' that are connected with respect to δ_0 (see §1, Chapter 2).

The proof of (7) is analogous to that of Formula (10), §1, Chapter 2, for ordinary semi-invariants.

IV. *Gaussian quasistates.* Let $\mathfrak{G} = \{\eta_\alpha, \ \eta_\alpha \in \mathfrak{A}\}$ be a collection of generators of a superalgebra $\mathfrak{A}$ with all η_α (except $\mathbf{1}$) being odd. An even quasistate $\langle \cdot \rangle$ on the superalgebra $\mathfrak{A}$ is called *Gaussian* or *generalized free* (with respect to the selected system $\mathfrak{G}$) if all semi-invariants

$$\langle \eta_{\alpha_1}, \dots, \eta_{\alpha_n} \rangle = 0 \quad \text{for} \quad n > 2.$$

In other words, only pair (two-particle) semi-invariants are nonvanishing, and any moment $\langle \eta_1 \dots \eta_N \rangle$, $\eta_i = \eta_{\alpha_i}$, is expressed in terms of them with the help of formula (6):

$$\langle \eta_1 \cdot \dots \cdot \eta_N \rangle = \sum \langle \eta_{i_1}, \eta_{j_1} \rangle \dots \langle \eta_{i_k}, \eta_{j_k} \rangle (-1)^{\pi}, \quad (8)$$

where the sum is taken over all partitions of the set $(1, \dots, N)$ into k pairs $(N = 2k)$: (i_l, j_l), $i_l < j_l$, $l = 1, \dots, k$, and π is the parity of the permutation $(i_1, j_1, i_2, j_2, \dots, i_k, j_k)$.

V. *Formal series.* (a) Consider first the case, where all $\xi_1, \dots, \xi_N$ are even elements of $\mathfrak{A}$. Then the semi-invariant $\langle \xi_1, \dots, \xi_N \rangle$ is the coefficient of the product $\lambda_1 \dots \lambda_N$ in the expansion into the formal power series in $\{\lambda_i, i = 1, \dots, N\}$ of the quantity

$$\ln \left\langle e^{\lambda_1 \xi_1} \dots e^{\lambda_N \xi_N} \right\rangle \tag{9}$$

(here $e^{\lambda \xi}$ is also defined with the help of a formal expansion into a series). This statement is obvious.

Furthermore, the quantity $\ln \langle e^{\xi} \rangle$ where $\xi = \lambda_1 \xi_1 + \dots + \lambda_N \xi_N$ serves as the generating function for symmetrized semi-invariants:

$$\ln \left\langle e^{\xi} \right\rangle = \sum_{\{m_1, \dots, m_N\}} \frac{\lambda_1^{m_1} \dots \lambda_N^{m_N}}{m_1! \cdot \dots \cdot m_N!} \langle \xi_1^{, m_1}, \dots, \xi_N^{, m_N} \rangle^{\text{sym}}, \tag{10}$$

where

$$\langle \xi_1, \dots, \xi_M \rangle^{\text{sym}} = \frac{1}{M!} \sum_{\sigma} \left\langle \xi_{\sigma(1)}, \dots, \xi_{\sigma(M)} \right\rangle, \tag{11}$$

and the sum is over all permuations σ of the set $(1, \dots, M)$.

(b) Let now the elements $(\xi_1, \dots, \xi_N)$ be odd. Then one may obtain the formulas similar to (9) and (10) with the help of formal series in "Grassmann variables." To this end, we introduce the Grassmann algebra $\mathfrak{G}$ with unity $\mathbf{1}_g$ and generators $\eta_1, \dots, \eta_N$ and let $\mathfrak{A} \otimes \mathfrak{G}$ be the tensor product of the superalgebras (in which η_i commutes with even and anticommutes with odd elements of the algebra $\mathfrak{A}$). Any element $A \in \mathfrak{A} \otimes \mathfrak{G}$ can be written in the form

$$A = \sum_{(i_1, \dots, i_k)} a_{i_1, \dots, i_k} \eta_{i_1} \dots \eta_{i_k}, \quad a_{i_1, \dots, i_k} \in \mathfrak{A}.$$

With the help of a quasistate $\langle \cdot \rangle$ on $\mathfrak{A}$, we define the mapping $\mathfrak{A} \otimes \mathfrak{G} \to \mathfrak{G}$ by the formula

$$A \mapsto \langle A \rangle_{\mathfrak{A}} = \sum_{(i_1, \dots, i_k)} \langle a_{i_1, \dots, i_k} \rangle \eta_{i_1} \dots \eta_{i_k}.$$

Consider the polynomial in $\eta_1, \dots, \eta_N$:

$$\ln \left\langle e^{\eta_1 \xi_1} \dots e^{\eta_N \xi_N} \right\rangle_{\mathfrak{A}} \in \mathfrak{G}. \tag{12}$$

Since $\eta_i \xi_i$ are even elements of the superalgebra $\mathfrak{A} \otimes \mathfrak{G}$, we may argue in the same way as when proving (9) to conclude that the term containing the product $\eta_1 \dots \eta_N$ in (12), which we shall denote by $\langle \eta_1\xi_1, \dots, \eta_N\xi_N\rangle_{\mathfrak{A}}$, is of the form

$$\langle \eta_1\xi_1, \dots, \eta_n\xi_N\rangle_{\mathfrak{A}} = \sum_{\delta=\{T_1,\dots,T_k\}} (-1)^{k-1}(k-1)!\,\langle(\eta\xi)_{T_1}\rangle_{\mathfrak{A}}\,\langle(\eta\xi)_{T_k}\rangle_{\mathfrak{A}}\,, \quad (13)$$

where the sum is taken over all partitions $\delta = \{T_1, \dots, T_k\}$ of the set $\mathcal{N}$ to the elements of even cardinality $|T_i|$, and for $T = \{i_1, \dots, i_p\}$, $i_1 < i_2 < \dots < i_p$, we use the notation

$$(\eta\xi)_T = \eta_{i_1}\xi_{i_1} \dots \eta_{i_p}\xi_{i_p} = (-1)^p \xi_T \eta_{\overline{T}},$$

where $\overline{T}$ denotes the sequence T in the reversed order: $\eta_{\overline{T}} = \eta_{i_p} \dots \eta_{i_1}$. Hence,

$$\langle(\eta\xi)_{T_1}\rangle_{\mathfrak{A}} \dots \langle(\eta\xi)_{T_k}\rangle_{\mathfrak{A}} = \langle\xi_{T_1}\rangle \dots \langle\xi_{T_k}\rangle\, \eta_1 \dots \eta_N (-1)^{\pi(\delta)}, \quad (14)$$

and by (4) and (13), we find that

$$\langle \eta_1\xi_1, \dots, \eta_N\xi_N\rangle_{\mathfrak{A}} = \langle \xi_1, \dots, \xi_N\rangle\, \eta_1 \dots \eta_N,$$

i.e., the semi-invariant $\langle \xi_1, \dots, \xi_N\rangle$ is the coefficient of $\eta_1 \dots \eta_N$ in the polynomial (12). If we introduce the quasistate ω_0 on the algebra $\mathfrak{G}$ by the formulas

$$\omega_0(\eta_N) = 1, \quad \omega_0(\eta_T) = 0, \quad \mathcal{N} = \{1, \dots, N\}, \quad T \subset \mathcal{N},$$

(the so-called Berezin anticommuting integral, see[4]), then (14) implies

$$\langle \xi_1, \dots, \xi_N\rangle = \omega_0\left(\ln\left\langle e^{\eta_1\xi_1} \dots e^{\eta_N\xi_N}\right\rangle\right). \quad (15)$$

This formula may be used for estimates of semi-invariants.

Gibbs modifications of a quasistate. Let $\langle\ \rangle_0$ be a quasistate defined on a local algebra $\mathfrak{A}^0$ in T, and $\{U_\Lambda, \Lambda \subset T\}$ be a family of "Hamiltonians": $U_\Lambda \in \mathfrak{A}_\Lambda$. We define the quasistate $\langle\cdot\rangle_\Lambda$ on $\mathfrak{A}^0$ by the formula

$$\langle F\rangle_\Lambda = Z_\Lambda^{-1}\,\langle F \exp\{-U_\Lambda\}\rangle_0\,, \qquad F \in \mathfrak{A}^0, \quad (16)$$

under the assumption that the stability condition is satisfied

$$Z_\Lambda = \langle \exp\{-U_\Lambda\}\rangle_0 \neq 0, \infty,$$

and also that the exponent $\exp\{-U_\Lambda\}$ is defined on the algebra $\mathfrak{A}_\Lambda$ (this condition is satisfied, e.g., for a Banach algebra $\mathfrak{A}_\Lambda$). The quasistate $\langle\cdot\rangle_\Lambda$ is called a finite *Gibbs modification* of the quasistate $\langle\cdot\rangle_0$. For the case where the *thermodynamic limit* exists,

$$\lim_{\Lambda \nearrow T} \langle F\rangle_\Lambda = \langle F\rangle\,,$$

for all $F \in \mathfrak{A}^0$, it defines the quasistate $\langle\cdot\rangle$ on $\mathfrak{A}^0$ called a *limit* Gibbs modification of the quasistate $\langle\cdot\rangle_0$. Hamiltonians U_Λ are most often given, as in the classical case, with the help of a potential $\{\Phi_A, A \subset T\}$, $\Phi_A \in \mathfrak{A}_A$:

$$U_\Lambda = \sum_{\Lambda \subseteq \Lambda} \Phi_A. \quad (17)$$

Many notions and methods of the preceding chapters may be carried over to the case of Gibbs modifications of quasistates (cluster representations of partition functions, cluster expansions of mean values $\langle F_A\rangle$, etc.).

Theorem 1. *Let all (super)algebras $\{\mathfrak{A}_t, t \in T\}$ defining a local algebra $\mathfrak{A}^0$ coincide: $\mathfrak{A}_t = \mathfrak{A}$, where $\mathfrak{A}$ is a Banach (super)algebra and let an independent quasistate $\langle\cdot\rangle_0$ on $\mathfrak{A}^0$ be given in the form of the product of a single even quasistate $\langle\cdot\rangle$ on $\mathfrak{A}$. Furthermore, let Hamiltonians U_Λ be of the form (17) where we suppose that for any point $t \in T$ and any $n \geqslant 2$, the condition ($\|\cdot\|$ is the norm in $\mathfrak{A}^0$)*

$$\sum_{A:\, t\in A, |A|=n} \|\Phi_A\| < c\lambda^n \quad \text{and} \quad \Phi_{\{t\}} = 0, \; t \in T, \tag{18}$$

is satisfied with $0 < \lambda < 1$ and a sufficiently small constant c, $c < c_0(\lambda)$. Then the corresponding limit Gibbs modification $\langle\ \rangle$ on $\mathfrak{A}^0$ exists, and whenever $F = F_B$ is an element of $\mathfrak{A}_B$, $B \subset T$, its mean $\langle F_B\rangle$ admits a cluster expansion

$$\langle F_B\rangle = \sum_{R\subset T} b_R(F_B), \tag{19}$$

similar to the expansion in (8), §3, Chapter 3, or in (11), §3, Chapter 3.

The proof of this theorem repeats that of Theorem (1), §1, Chapter 4; one has to use the expansion

$$\exp\{-U_\Lambda\} = \sum_{n=0}^{\infty} \frac{(-U_\Lambda)^n}{n!} = \sum_{n=0}^{\infty} \frac{(-1)^n}{n!} \sum_{(A_1,\dots,A_n)} \Phi_{A_1}\dots\Phi_{A_n} \tag{20}$$

and then to gather together all terms with the same (unordered) collection of supports $\{A_1, \dots, A_n\}$.

Perturbation of a Hamiltonian and series in means and semi-invariants. Let $\langle\cdot\rangle_0$ be a quasistate on an algebra $\mathfrak{A}^0$, and $\langle\ \rangle_1$, $\langle\ \rangle_2$ be its Gibbs modifications (16) by means of Hamiltonians U_1 and $U_2 = U_1 + V$, respectively.

We notice that, in a general case (as opposed to the classical one), the condition of transitivity of Gibbs modifications is not satisfied, i.e., in general

$$\langle F\rangle_2 \neq \frac{\langle Fe^{-V}\rangle_1}{\langle e^{-V}\rangle_1}, \qquad F \in \mathfrak{A}^0.$$

Nevertheless, one may represent $\langle F\rangle_2$ in the form of a series into which the means (or semi-invariants) of the perturbation V enter. We shall introduce the notation

$$\Gamma(s) = e^{-s(U_1+V)}e^{sU_1}. \tag{21}$$

Then

$$\langle F\rangle_2 = \frac{\langle Fe^{-(U_1+V)}e^{U_1}e^{-U_1}\rangle_0}{\langle e^{-(U_1+V)}e^{U_1}e^{-U_1}\rangle_0} = \frac{\langle F\Gamma(1)\rangle_1}{\langle\Gamma(1)\rangle_1}. \tag{22}$$

It is easy to verify that

$$\frac{d\Gamma}{ds} = -\Gamma(s)V(s), \qquad V(s) = e^{-sU_1}Ve^{sU_1}.$$

Hence,

$$\begin{aligned}\Gamma(1) &= \mathbf{1} - \int_0^1 \Gamma(s)V(s)ds \\ &= \mathbf{1} + \sum_{n=1}^{\infty}(-1)^n \iint\limits_{0\leqslant s_1\leqslant\ldots\leqslant s_n\leqslant 1} V(s_1)\ldots V(s_n)ds_1\ldots ds_n \qquad (23)\end{aligned}$$

(the integral in (23) is understood in the sense of Bochner; see [28]).

Whenever $x = \{s_1, \ldots, s_n\}$ is a finite subset of the segment $[0,1]$ with $s_1 < s_2 < \ldots < s_n$, we define the element

$$V(x) = V(s_1)V(s_2)\ldots V(s_n).$$

Then the formula (23) can be rewritten in the form

$$\Gamma(1) = \int\limits_{\Omega_{\text{fin}}} (-1)^{|x|}V(x)d\nu(x), \qquad (24)$$

where ν is the measure on the space Ω_{fin} of finite subsets of the segment $[0,1]$ defined in §1.

With the help of the representation (22) one may, taking into account (23) and (24), in many cases, obtain a cluster representation for the mean $\langle F\rangle_2$. Moreover, one has a direct expansion for the mean $\langle F\rangle_2$ which is analogous to the expansion in semi-invariants in the classical case (see §1, Chapter 2).

Lemma 2. *Let U_1 and V be even elements of a superalgebra $\mathfrak{A}$ and $\langle\cdot\rangle_0$ be an even quasistate. Then*

$$\langle F\rangle_2 = \langle F\rangle_1 + \sum(-1)^n \iint\limits_{s_1<\ldots<s_n} \langle F, V(s_1), \ldots, V(s_n)\rangle_1 \times ds_1, \ldots, ds_n. \qquad (25)$$

Proof. Introducing the function $\Phi_F(x)$ defined on the subsets $x = \{s_1, \ldots, s_n\}$, $0 \leqslant s_1 < \ldots < s_n \leqslant 1$, by the formula

$$\Phi_F(x) = \langle F, V(s_1), \ldots, V(s_n)\rangle_1,$$

we rewrite (25) in the form

$$\langle F\rangle_2 = \int\limits_{\Omega_{\text{fin}}} (-1)^{|x|}\Phi_F(x)d\nu(x). \qquad (26)$$

On the other hand, by (24) we get

$$\langle F\Gamma(1)\rangle_1 = \int\limits_{\Omega_{\text{fin}}} (-1)^{|x|} G_F(x) d\nu(x),$$

$$\langle \Gamma(1)\rangle_1 = \int (-1)^{|x|} g(x) d\nu(x), \tag{27}$$

where

$$C_F(x) = \langle FV(x)\rangle_1, \quad g(x) = \langle V(x)\rangle_1, \quad x \in \Omega_{\text{fin}}.$$

Furthermore, by definition (4) of semi-invariants (for the case of even elements), it follows that

$$\begin{aligned} G_F(x) &= \langle FV(s_1)\dots V(s_n)\rangle_1 \\ &= \sum_{\{i_1<\dots<i_k\}} \langle F, V(s_{i_1}), \dots, V(s_{i_k})\rangle_1 \, \langle V(s_{j_1})\dots V(s_{j_{n-k}})\rangle_1 \\ &= \sum_{y\subseteq x} \Phi_F(y) g(x\setminus y). \end{aligned}$$

Now by using Lemma (1), §6, Chapter 3, we get

$$\int (-1)^{|x|} G_F(x) d\nu(x) = \int\limits_{\Omega_{\text{fin}}} (-1)^{|y|} \Phi_F(y) d\nu(y) \int\limits_{\Omega_{\text{fin}}} (-1)^{|z|} g(z) d\nu(z).$$

Hence, by (22) and (27), the formula (26) follows. The lemma is proved.

§2 Uniqueness of Gibbs Fields

In §8, Chapter 4, we established the existence and proved a cluster expansion of limit Gibbs fields with real values that were obtained as small perturbations of a Gaussian field in finite sets $\Lambda \subset Z^\nu$ with arbitrary uniformly bounded boundary configurations $y^{\partial\Lambda}$. The following question arises: Is it true that arbitrary (no longer bounded) boundary configurations $y^{\partial\Lambda}$ lead to the same limit measure, i.e., that there is a unique Gibbs distribution in Z^ν (in the sense of DLR, see §2, Chapter 1)? A similar question also appears in connection with the theorem about the uniqueness of a Gibbs field for the case of bounded spins that was proved in §2, Chapter 5: Is it possible to extend the theorem to the case of unbounded spins? A difficulty stems here from the fact that for boundary configurations $y^{\partial\Lambda}$ taking sufficiently large values, it may happen that the measures $\mu_{\Lambda,y^{\partial\Lambda}}$ do not admit cluster expansions anymore. One may overcome this difficulty in a similar way as in the proof of Theorem (1), §2, Chapter 5, where we passed from an arbitrary boundary configuration $y^{\partial\Lambda}$ on the boundary of a large cube Λ to a sufficiently small boundary configuration on the boundary of some random (but sufficiently large) volume $\Lambda' \subseteq \Lambda$. Nevertheless, one needs some a priori estimate on the probability of large values of boundary configurations (increasing with growth of Λ). In other words, we shall prove the uniqueness of Gibbs fields for a certain class of random fields in Z^ν; namely, the fields $\{x(t), t \in Z^\nu\}$ such that

$$\Pr(|x(t)| > M) = o\left[(\ln\ln M)^{\nu-1}\right]^{-1}, \qquad M \to \infty,\ \nu > 1, \tag{1}$$

for each $t \in Z^\nu$ with the estimate $o(\ )$ uniform with respect to $t \in Z^\nu$.

Let the Hamiltonian $U_\Lambda(x/y)$ be of the form

$$\begin{aligned} U_\Lambda(x/y) &= l_0\left[\sum_{t,t'\in\Lambda,|t-t'|=1}(x(t)-x(t'))^2 + \sum_{\substack{t\in\Lambda,t'\in\partial\Lambda\\|t-t'|=1}}(x(t)-y(t'))^2\right.\\ &\quad + m\sum_{t\in\Lambda}x^2(t) + \lambda\sum_{t\in\Lambda}P(x(t)), \end{aligned} \tag{2}$$

where $\Lambda = \Lambda_N$ is a cube of side $N, x \in R^\Lambda, y = y^{\partial\Lambda}$ is a boundary configuration, $l_0 > 0, m > 0, \lambda > 0$, and P is a polynomial of a degree $\geqslant 4$ bounded from below.

We shall use $\mu_{\Lambda,y^{\partial\Lambda}}$ to denote the Gibbs modification of the Lebesgue measure $\mu_\Lambda^0 = (dx)^\Lambda$ on R^Λ by means of the Hamiltonian (2) with the boundary configuration $y^{\partial\Lambda}$:

$$\frac{d\mu_{\Lambda,y^{\partial\Lambda}}}{d\mu_\Lambda^0} = Z_{\Lambda,y^{\partial\Lambda}}^{-1}\ e^{-U_\Lambda(x/y^{\partial\Lambda})}, \tag{3}$$

where

$$Z_{\Lambda,y^{\partial\Lambda}} = \int_{R^\Lambda} e^{-U_\Lambda(x/y^{\partial\Lambda})}(dx)^\Lambda. \tag{3'}$$

Theorem 1. *Let a Hamiltonian of the form (2) be given, and suppose that either*

a) $\lambda > 0$ *is sufficiently small:* $0 < \lambda < \lambda_0(l_0, \nu, m, P)$, *or*

b) l_0 *is sufficiently large:* $l_0 > \bar{l}_0(\lambda, \nu, m, P)$,

and the polynomial P *contains only the powers* x^k *with* $k \geqslant 3$. *Then there exists a unique limit Gibbs field in the lattice* Z^ν *that is generated by the conditional distributions (3) (in the sense of the DLR definition; see §2.1) and that belongs to the class of random fields (1). The distribution* μ *of this field admits a cluster expansion.*

Remark. Case a) is a generalization of results from §8, Chapter 4, and case b) a generalization of Theorem 1, §2, Chapter 5.

The proof of the theorem is based on the following lemma that is of interest in itself. Let us consider a Hamiltonian of a more general type than (2). Namely,

$$\overline{U}_\Lambda(x/y) = l_0 \left[\sum_{\substack{t,t' \in \Lambda, \\ |t-t'|=1}} |x(t) - x(t')|^\alpha + \sum_{\substack{t\in\Lambda, t'\in\partial\Lambda, \\ |t-t'|=1}} |x(t) - y(t')|^\alpha \right] + \sum_{t\in\Lambda} \overline{P}(x(t)), \tag{4}$$

where

$$\overline{P}(x) = \sum_{j=1}^{k} l_j |x|^{\gamma j}$$

(or a similar sum in which some $|x|^{\gamma j}, j < k$, may be replaced by $|x|^{\gamma j} \operatorname{sign} x$),

$$l_0 > 0, \qquad l_k > 0,$$
$$\gamma_k \geqslant \alpha > 1, \qquad \gamma_k > \gamma_{k-1} > \ldots > \gamma_1 > 0.$$

We denote $\overline{\Lambda} = \Lambda \cup \partial\Lambda$ and use $\bar{x}$ to denote the configuration in $\overline{\Lambda}$ that coincides with x in Λ and with $y^{\partial\Lambda}$ in $\partial\Lambda$. Whenever B is a positive number and $\bar{x}$ a configuration, we use $D_B(\bar{x}) = D(\bar{x})$ to denote the set of "B-sites" of this configuration:

$$D_B(\bar{x}) = \{t \in \overline{\Lambda}: |\bar{x}(t)| > B\}.$$

Let $D_1(\bar{x}), \ldots, D_k(\bar{x})$ be 1-connected components of $D(\bar{x})$. A component $D_i(\bar{x})$ is called bordering if it intersects $\partial\Lambda$. We use $D^{\text{bor}}(\bar{x})$ to denote the union of all bordering components and $\mathcal{E}^B_{\overline{\Lambda}} \subset R^{\overline{\Lambda}}$ to denote the set of configurations $x \in R^\Lambda$ for which

$$D^{\text{bor}}(\bar{x}) \cap \Lambda_{N/4} \neq \emptyset,$$

where $\Lambda_{N/4}$ is the cube of side $[N/4] + 1$ that is concentric with the cube Λ_N.

Lemma 2. *Let $\Lambda = \Lambda_N$ be a cube of side N and $\mu_{\Lambda, y^{\partial\Lambda}}$ be the Gibbs modification of the measure μ^0_Λ by means of the Hamiltonian (4):*

1) Let $\gamma_k > \alpha$. Then there exist the numbers $B > 0$ and $\bar{\delta} > 1$ such that if

$$\max_{t\in\partial\Lambda} |y_t| < b^{\bar{\delta} N}, \tag{5}$$

then

$$P(\mathcal{E}^B_\Lambda) < C_1 \exp\{-C_2 N\}, \tag{6}$$

where $C_1 > 0$ and $C_2 > 0$ are constants depending on $B, \bar{\delta}, \overline{P}, l_0$, and ν.

2) For $\gamma_k = \alpha$, the estimate (6) is valid under the condition

$$\max |y_t| < BA^N, \tag{6'}$$

where $A > 1$ is some constant (and C_1 and C_2 in (6) depend on A).

Proof of Theorem 1. In case a), the convergence of $\mu_{\Lambda,y^{\partial\Lambda}}$ to the limit measure μ and also its cluster expansion was proved under the condition that the boundary configurations $y^{\partial\Lambda}$ are uniformly bounded for all Λ. This fact can also be established in case b) by repeating, with some changes, the proof of Theorem 1, §2, Chapter 5. Furthermore, relying on the estimate (6), we get, with the help of similar arguments to those in the proof of Theorem 1, §2, Chapter 5, that for any local function F_A, one has

$$\langle F_A \rangle_{\mu_{\Lambda,y^{\partial\Lambda}}} \to \langle F_A \rangle_{\mu}, \qquad \Lambda \nearrow Z^\nu, \tag{7}$$

with an arbitrary choice of boundary configurations $y^{\partial\Lambda}$ obeying condition (5) (or (6′)). The measure μ belongs to class (1), as may easily be concluded from the estimate

$$\Pr(x(t) > M) < Ce^{-\overline{C}M^n}, \tag{7′}$$

where $C, \overline{C} > 0$ are constants and n is the degree of the polynomial $P(x)$ (the estimate (7′)) is obtained with the help of a cluster expansion). Further, any Gibbs measure $\bar{\mu}$ (from class (1)) satisfies the equality

$$\langle F_A \rangle_{\bar{\mu}} = \int' \langle F_A \rangle_{\mu_{\Lambda,y^{\partial\Lambda}}} d\bar{\mu}(y^{\partial\Lambda}) + \int'' \langle F_A \rangle_{\mu_{\Lambda,y^{\partial\Lambda}}} d\bar{\mu}(y^{\partial\Lambda}), \tag{8}$$

where Λ is a cube containing A, the integration $\int'$ is over the set of configurations $y^{\partial\Lambda}$ obeying condition (5), and the integral $\int''$ is over the complementary set of configurations $y^{\partial\Lambda}$ for which condition (5) fails. Due to (1), the integral $\int''$ tends to zero as $\Lambda \nearrow Z^\nu$ and the integral $\int' \to \langle F \rangle_\mu$ according to (7) and (1). The theorem is proved.

Proof of Lemma 2. We represent the Hamiltonian (4) in the form

$$U_\Lambda(x/y) = \sum_{\substack{t,\bar{t}\in\overline{\Lambda}, \\ |t-t'|=1}} \Phi^{(\Lambda)}_{t,t'}(\bar{x}(t), \bar{x}(t')) + H_\Lambda(y), \tag{9}$$

where the configuration $\bar{x}$ is defined as above, and

$$\Phi^{(\Lambda)}_{t,t'}(x_1, x_2) = l_0|x_1 - x_2|^\alpha + \frac{1}{\nu_t^\Lambda}\overline{P}(x_1) + \frac{1}{\nu_{t'}^\Lambda}\overline{P}(x_2),$$

with ν_t^Λ denoting the number of sites from $\overline{\Lambda}$ neighbouring with t, $H_\Lambda(y)$ depends only on y.

Now we choose $B > 0$ such that if $|x| > B/2$, then

$$\overline{P}(x) > \frac{l_k}{2}|x|^{\gamma_k} \quad \text{and} \quad \overline{P}'(x) > \gamma_k \frac{l_k}{2}|x|^{\gamma_k - 1} \tag{10}$$

and define the mapping $g^B: R \to R$ by

$$g^B(x) \equiv g(x) \equiv \begin{cases} x - B/2, & x > B, \\ x, & |x| < B, \\ x + B/2, & x < -B. \end{cases}$$

Lemma 3. *A) Let $|x_1|, |x_2| \geqslant B$. Then*

$$\Phi^{(\Lambda)}_{t,t'}(x_1, x_2) > \Phi^{(\Lambda)}_{t,t'}(g(x_1), g(x_2)) + C_0 B^{\gamma_k}$$

for any $t, t' \in \overline{\Lambda}$, with $C_0 > 0$ being some constant.
B) Let $|x_2| \geqslant B$ and

$$|x_1| < |x_2| + C_1 |x_2|^\delta, \qquad \delta = \frac{\gamma_k - 1}{\alpha - 1} \geqslant 1, \tag{11}$$

with a sufficiently small constant C_1. Then for all $t, t' \in \overline{\Lambda}$, one has

$$\Phi^{(\Lambda)}_{t,t'}(x_1, x_2) > \Phi^{(\Lambda)}_{t,t'}(x_1, g(x_2)). \tag{12}$$

Proof. A) From (10), we get for $|x_1|, |x_2| \geqslant B$:

$$\begin{aligned} &\Phi^{(\Lambda)}_{t,t'}(x_1, x_2) \\ &\quad \geqslant \Phi^{(\Lambda)}_{t,t'}(g(x_1), g(x_2)) + \frac{1}{\nu^\Lambda_t}\left[\overline{P}(x_1) - \overline{P}(g(x_1))\right] + \frac{1}{\nu^\Lambda_{t'}}\left[\overline{P}(x_2) - \overline{P}(g(x_2))\right] \\ &\quad > \Phi^{(\Lambda)}_{t,t'}(g(x_1), g(x_2)) + \frac{l_k \gamma_k B}{4\nu}\left\{\left|x_1 \pm \frac{B}{2}\right|^{\gamma_k - 1} + \left|x_2 \pm \frac{B}{2}\right|^{\gamma_k - 1}\right\} \\ &\quad > \Phi^{(\Lambda)}_{t,t'}(g(x_1), g(x_2)) + C_0 B^{\gamma_k}. \end{aligned}$$

Here the signs $\pm$ are determined by the signs of x_2 and x_1, and $C_0 = \frac{l_k \gamma_k}{4\nu}\left(\frac{1}{2}\right)^{\gamma_k - 1}$.

B) We shall consider two cases.

1) $\gamma_k > \alpha$. Then

$$\begin{aligned} \Phi^{(\Lambda)}_{t,t'}(x_1, x_2) = &\Phi^{(\Lambda)}_{t,t'}(x_1, g(x_2)) + l_0 |x_1 - x_2|^\alpha \\ &- l_0 \left|x_1 - x_2 \pm \frac{B}{2}\right|^\alpha + \frac{1}{\nu_{t'}}\left[P(x_2) - P(g(x_2))\right]. \end{aligned}$$

From (11), we get $|x_1 - x_2| < \overline{C}_1 |x_2|^\delta, \overline{C}_1 = C_1 - \frac{2}{B^{\delta-1}}$. Consequently

$$\begin{aligned} |x_1 - x_2|^\alpha - \left|x_1 - x_2 \pm \frac{B}{2}\right|^\alpha &> -\alpha \left|\overline{C}_1 |x_2|^\delta + \frac{B}{2}\right|^{\alpha - 1} \frac{B}{2} \\ &> -|x_2|^{\gamma_k - 1} \frac{B}{2} \widetilde{C}, \quad \widetilde{C} = \alpha\left(\overline{C}_1 + \frac{1}{2B^{\delta - 1}}\right)^{\alpha - 1}. \end{aligned}$$

On the other hand,

$$\frac{1}{\nu_{t'}}\left[P(x_2) - P(g(x_2))\right] > C_0|x_2|^{\gamma_k - 1}\frac{B}{2}.$$

Hence, (12) follows for sufficiently small C_1 and large B.

2) $\gamma_k = \alpha$. Let $x_2 > B$ and $x_1 < 0$. Then $|x_2 - x_1|^\alpha > |x_2 - x_1 - B/2|^\alpha$ and (12) is obvious. If $x_2 > B$ and $x_1 > 0$, we have $|x_1 - x_2| < C_1|x_2|$ and the preceding arguments apply. The lemma is proved.

Now let $x \in \mathcal{E}_\Lambda^B$ and $D_1^B(x) \subset D^B(x)$ be a bordering component of the set $D^B(x)$ that reaches the cube $\Lambda_{N/4}$. Consider the sequence of numbers

$$B_0 = B, \quad B_1 = B_0 + C_1 B_0^\delta, \quad B_2 = B_1 + C_1 B_1^\delta, \ldots .$$

For $\delta > 1$, we choose B such that $C_1 B(\delta - 1)/2 > 1$ and put $\tilde{\delta} = 1 + (\delta - 1)/2$. Then it is easy to see that

$$B_n > B^{\tilde{\delta} n}. \tag{13}$$

For the case of $\delta = 1$, we have

$$B_n = B(1 + C_1)^n. \tag{14}$$

We consider now the sequence of crowns

$$S_k = \Lambda_{N/2+k} \backslash \Lambda_{N/2+k-1}, \qquad k = 1, 2, \ldots ,$$

and let

$$T_k(x) = \{t \in S_k \cap D_1^B(x);\ |x(t)| < B_k\}.$$

We use the notation

$$R(x) = D_1^B(x) \cap \left\{\Lambda_N \backslash \bigcup_{k=1}^{[N/2]+1} T_k(x)\right\}$$

and let $\widetilde{R}(x)$ be any 1-connected component of the set $R(x)$ that contains the set $D_1 \cap (\Lambda_{N/2} \backslash \Lambda_{N/4})$.

Clearly,

$$\left|\widetilde{R}(x)\right| > N/4. \tag{15}$$

Further, as follows from (5) and (6′), for $\bar{\delta} = \tilde{\delta}^{1/2}$ for case $\gamma_k > \alpha$ and for $A = (1 + C_1)^{1/2}$ for case $\gamma_k = \alpha$, we have

$$|y(t)| < B_{N/2}.$$

Thus, if $|x(t)| > B_m, t \in \widetilde{R}(x)$, then at all the sites $t' \in \overline{\Lambda}$ neighbouring on t and not belonging to $\widetilde{R}(x)$, we have $|\bar{x}(t)| \leqslant B_{m+1}$. Consider now the mapping $G_{\widetilde{R}}: R^{\Lambda} \to R^{\Lambda}$ defined by

$$\left(G_{\widetilde{R}}x\right)(t) = \begin{cases} x(t) & t \overline{\in} \widetilde{R}(x), \\ g(x(t)), & t \in \widetilde{R}(x). \end{cases}$$

It follows from Lemma 3 that

$$U_{\Lambda}(x/y) > U(G_{\widetilde{R}}x/y) + \frac{C_0 B^{\gamma_k}}{2} \left|\widetilde{R}(x)\right|. \tag{16}$$

For any 1-connected set R intersecting $\partial\Lambda_{N/4}$ we use $\mathfrak{X}(R) \subseteq \mathcal{E}_{\Lambda}^{B}$ to denote the set of all configurations $x \in \mathcal{E}_{\Lambda}^{B}$ for which some of the sets $\widetilde{R}(x) = R$. We shall estimate the probability $p_R = \Pr(x \in \mathfrak{X}(R))$. From (16), we get

$$\begin{aligned} p_R &= \frac{\int_{\mathfrak{X}(R)} \exp\{-U(x/y^{\partial\Lambda})\} d\mu_{\Lambda}^{0}}{Z_{\Lambda, y^{\partial\Lambda}}} \\ &< \frac{\int_{\mathfrak{X}(R)} \exp\{-U(G_R x/y^{\partial\Lambda})\} d\mu}{Z_{\Lambda, y^{\partial\Lambda}}} e^{-\frac{C_0 B^{\gamma_k}}{2}|R|}. \end{aligned} \tag{17}$$

Further, the mapping $G_R: x \mapsto G_R x$, $x \in \mathfrak{X}(R)$, preserves the measure μ_{Λ}^{0}, and thus the substitution $\bar{x} = G_R x$, $x \in \mathfrak{X}(R)$, in the integral (17) leads to the estimate

$$p_R < e^{-\frac{C_0 B^{\gamma_k}}{2}|R|}. \tag{18}$$

The number of sets R of the cardinality M does not exceed const $\cdot|\partial\Lambda_{N/4}|\cdot C^M$ with an absolute constant $C > 0$ (see Lemma 4, §4, Chapter 2). Thus, by (18) and (15), we get estimate (6) for sufficiently large B. Lemma 2 is proved.

§3 Compactness of Gibbs Modifications

In this section, the idea of isolating "clusters" is applied in the proof of the local weak compactness of the set of Gibbs modifications μ_{Λ} (see §1.1). Here we consider Hamiltonians of a general form without any "small" or "large" parameters.

Let $\{x(t), t \in Z^1\}$ be a real field in the one-dimensional lattice Z^1 with a Hamiltonian with (pair) interaction of nearest neighbours:

$$U_{\Lambda}(x) = \sum_{t\in\Lambda} f(x(t)) + \sum_{\substack{t,t'\in\Lambda \\ |t-t'|=1}} \Phi(x(t), x(t')), \tag{1}$$

where $\Lambda \subset Z^1$, f and Φ are functions bounded from below. Moreover, Φ is a locally bounded symmetric function and f obeys the condition

$$\int_{-\infty}^{\infty} e^{-f(x)}dx < \infty. \tag{2}$$

We use μ_Λ to denote the Gibbs modification of the measure $\mu_\Lambda^0 = (dx)^\Lambda$ on the space R^Λ that is given by means of the Hamiltonian (1):

$$\frac{d\mu_\Lambda}{d\mu_\Lambda^0} = Z_\Lambda^{-1}\exp\{-U_\Lambda(x)\}. \tag{3}$$

Theorem 1. *Under the condition specified above, the set of Gibbs modifications* $\{\mu_\Lambda\}$ *is locally weakly compact.*

Remark 1. The theorem and its proof may be directly generalized to the case of an arbitrary pair potential bounded from below and of finite range $\{\Phi_{t,t'}, |t-t'| < d\}$.

Proof. As follows from Lemma 3, §1, Chapter 1, it is enough to prove that for any finite set $A_0 \subset Z^1$ and any $\varepsilon > 0$, there exists N such that for all $\Lambda \supset A_0$, one has

$$p^N \equiv \Pr\left\{x \in R^\Lambda : |x(t)| > N,\ t \in A_0\right\} < \varepsilon, \tag{4}$$

where the probability Pr is computed with the help of the distribution μ_Λ. Furthermore, it is enough to establish (4) for the case of an arbitrary one-point set $A_0 = \{t_0\}$. Let, for definiteness, $t_0 = 0$ and $\Lambda = [L_1, L_2]$, $L_1 < 0 < L_2$.

We shall choose a number $B < N$, and for each configuration $x \in R^\Lambda$, we use $D^B(x)$ to denote the following set of sites of Λ:

$$D^B(x) = \{t \in \Lambda, |x| > B\}.$$

Let further $\Gamma^B(x) \subset D^B(x)$ be the "B-cluster": a 1-connected component of $D^B(x)$ containing the site $0 \in Z^1$ (under the condition that such a component exists).

For each $\Gamma = [r_1, r_2] \subset \Lambda$, $r_1 < r_2$, we use $p_\Gamma^{B,N}$ to denote the probability

$$p_\Gamma^{B,N} = \Pr\left\{x : |x(0)| > N,\ \Gamma^B(x) = \Gamma\right\}. \tag{5}$$

This probability equals: for the case $L_1 + 1 \leqslant r_1$, $r_2 \leqslant L_2 - 1$,

$$p_\Gamma^{B,N} = Z_\Lambda^{-1} \iint_{|y_1|<B, |y_2|<B} \exp\left[-f(y_1) - f(y_2)\right] Z_{\Delta_1}(\emptyset, y_1) \times Z_\Gamma^{B,N}(y_1, y_2) Z_{\Delta_2}(y_2, \emptyset)dy_1dy_2, \tag{6}$$

for the case $L_1 = r_1, r_2 < L_2 - 1$,

$$p_\Gamma^{B,N} = Z_\Lambda^{-1} \int\limits_{|y|<B} Z_\Gamma^{B,N}(\emptyset, y) Z_\Delta(y, \emptyset) e^{-f(y)} dy, \tag{6'}$$

and similarly, for the case $L_1 + 1 \leqslant r_1, r_2 = L_2$ and $L_1 = r_1, L_2 = r_2$. Here we use the notation $\Delta_1 = [L_1, r_1 - 2]$ for $r_1 - 2 \geqslant L_1$, and $\Delta_1 = \emptyset$ for $r_1 = L_1 + 1$; similarly, $\Delta_2 = [r_2 + 2, L_2]$ for $r_2 + 2 \leqslant L_2$, $\Delta_2 = \emptyset$ for $r_1 = L_2 - 1$. $Z_{\Delta_1}(\emptyset, y)$ is the partition function on the segment $\Delta_1 \neq \emptyset$ under the empty boundary condition on the left and the boundary condition y on the right and $Z_\emptyset = 1$; similarly, one defines $Z_{\Delta_2}(y, \emptyset)$. Finally,

$$Z_\Gamma^{B,N}(y_1, y_2) = \int\limits_{S_\Gamma^{B,N}} \exp\left[-U_\Gamma(x/y_1, y_2)\right] d\mu_\Gamma^0, \tag{7}$$

where $U_\Gamma(x/y_1, y_2) = U_\Gamma(x) + \Phi(y_1, x(r_1)) + \Phi(x(r_2), y_2)$, and $S_\Gamma^{B,N} \subset R^\Gamma$ is the set of configurations in Γ,

$$S_\Gamma^{B,N} = \{x \colon |x(t)| > B, \ \ t \in \Gamma, \ |x(0)| > N\}.$$

Similarly, one introduces the partition functions

$$Z_\Gamma^{B,N}(\emptyset, y), \ \ Z_\Gamma^{B,N}(y, \emptyset), \ \ Z_\Gamma^{B,N}(\emptyset, \emptyset).$$

We notice now that

$$Z_\Gamma^{B,N}(y_1, y_2) < V \cdot \delta^{|\Gamma|-1} \delta_1, \tag{8}$$

where

$$V = \max_{x_1, x_2} e^{-\Phi(x_1, x_2)},$$

$$\delta = \delta(B) = V \int\limits_{|y|>B} e^{-f(y)} dy, \qquad \delta_1 = \delta_1(N) = V \int\limits_{|y|>N} e^{-f(y)} dy.$$

Furthermore, the estimate $Z_\Gamma^{B,N}(\emptyset, y) < \delta^{|\Gamma|-1}\delta_1$ holds, and similarly for $Z_\Gamma^{B,N}(y, \emptyset)$ and $Z_\Gamma^{B,N}(\emptyset, \emptyset)$.

For each $\Gamma = [r_1, r_2]$, $r_1 \geqslant L_1 + 1$, $r_2 \leqslant L_2 - 1$, we have

$$Z_\Lambda \geqslant \iint Z_\Lambda(\emptyset, y_1) \overline{Z}_\Gamma^{1}(y_1, y_2) Z_\Lambda(y_2, \emptyset) \exp\left[-f(y_1) - f(y_2)\right] dy_1 dy_2. \tag{9}$$

Similar estimates hold for the case $r_1 = L_1$ or $r_2 = L_2$.

Here, we use $\overline{Z}_\Gamma^1(y_1, y_2)$ to denote the partition function that is analogous to (7) but is computed over the set of configurations $\{x \in R^\Gamma : |x(t)| \leqslant 1, t \in \Gamma\}$.

It is easy to see that for $|y_1| < B$, $|y_2| < B$, one has

$$\overline{Z}_\Gamma^1(y_1, y_2) > C^{-1}\alpha^{|\Gamma|}K^2, \tag{10}$$

where

$$C = \min_{|x_1| \leqslant 1, |x_2| \leqslant 1} e^{-\Phi(x_1, x_2)},$$

$$\alpha = C \int\limits_{|x|<1} e^{-f(x)} dx, \quad K = K(B) = \min_{\substack{|x_1| \leqslant 1, \\ |x_2| \leqslant B}} e^{-\Phi(x_1, x_2)}.$$

From (6) (or (6′)) and the estimates (8) and (10), we conclude that

$$P_\Gamma^{B,N} < M(\delta/\alpha)^{|\Gamma|-1} K^{-2}\delta_1,$$

where $M = VC^{-1}\alpha$. We shall suppose now that B is chosen sufficiently large so that $\delta/\alpha < 1/2$. Then

$$P^N = \sum_\Gamma P_\Gamma^{B,N} < \overline{M}K^{-2}\delta_1.$$

Since K depends only on B, and $\delta_1 \to 0$ as $N \to \infty$, we may choose N sufficiently large so that $MK^{-2}\delta_1 < \varepsilon$. The theorem is proved.

We shall indicate a simple generalization of our method to the case of fields in a many-dimensional lattice Z^ν.

Theorem 2. *Let a Hamiltonian be given of the form (1) and let all the preceding assumptions concerning the potentials f and Φ be fulfilled, and moreover, let the condition*

$$\int\limits_{|x|>B} e^{-f(x)} dx = o\left(\min_{\substack{|x_1| \leqslant 1, \\ |x_2| \leqslant B}} e^{-\Phi(x_1, x_2)} \right), \qquad B \to \infty. \tag{11}$$

be satisfied.

Then the set of Gibbs modifications $\{\mu_\Lambda, \Lambda \subset Z^\nu\}$ is weakly locally compact.

The proof of this theorem is analogous to the proof of Theorem 1. However, it seems that the established result is valid also in a more general case.

§4 Gauge Field with Gauge Group Z_2

We recall (see §0, Chapter 1) that by a lattice gauge Z_2-model we mean the Gibbs field $\{\sigma(\tau), \widetilde{Z}^\nu\}$ defined on the set $\widetilde{Z}^\nu$ of bonds of the ν-dimension lattice Z^ν taking values in the group Z_2 and with the Hamiltonian of the form

$$U_\Lambda = \beta \sum_{p:\, \partial p \subset \Lambda} \sigma_p, \tag{1}$$

where $\Lambda \subset \widetilde{Z}^\nu$ is a finite set of bonds and the sum is taken over all plaquettes of the lattice Z^ν (elementary two-dimensional faces), the boundary of which is in $\Lambda, \partial p \subset \Lambda$, and $\sigma_p = \prod_{\tau \in \partial p} \sigma_\tau$. As a free measure μ_0 in the space $\Omega = (Z_2)^{\widetilde{Z}^\nu}$, one takes the infinite product of the normalized Haar measures $\nu_0(1) = \nu_0(-1) = 1/2$ on Z_2. The Gibbs modification μ_Λ is as usually defined by the formula

$$\frac{d\mu_\Lambda}{d\mu_0} = Z_\Lambda^{-1} \exp\{U_\Lambda\}.$$

If $\beta > 0$, the Hamiltonian (1) is ferromagnetic (see §0.1) and one may prove, with the help of the Griffith inequalities, that the thermodynamic limit $\lim_{\Lambda \nearrow \widetilde{Z}^\nu} \langle \sigma_T \rangle_\Lambda, T \subset \widetilde{Z}^\nu$, of correlation functions exists. Thus, we also have the existence of the limit

$$\lim_{\Lambda \nearrow \widetilde{Z}^\nu} \mu_\Lambda = \mu, \tag{2}$$

similarly as in §0, Chapter 1, for the case of the Ising model. Here, we denoted

$$\sigma_T = \prod_{\tau \in T} \sigma_\tau \tag{3}$$

for any finite set T of bonds.

Besides, the existence of the limit (2) for all small $\beta, |\beta| < \beta_0(\nu)$, follows from the results of §1, Chapter 4. The limit measure μ, moreover, obeys the condition of exponential decay of correlations (see §2, Chapter 6).

Let $C \subset \widetilde{Z}^\nu$ be a closed contour (i.e., a connected set of bonds such that each site $t \in Z^\nu$ belongs to an even number of bonds from C).

The Wilson functional for the contour C is the mean

$$W(C) = \langle \sigma_C \rangle_\mu . \tag{4}$$

For the sake of simplicity, we shall consider only planar contours C without self-intersections (i.e., contours contained entirely in some of the coordinate planes). We use $S(C)$ to denote the set of plaquettes on this plane encircled by the contour C.

Theorem 1. *1. For all sufficiently small $\beta, |\beta| < \overline{\beta}_0$, there exist constants $C_1 = C_1(B,\nu)$ and $b_1 = b_1(\nu)$ such that*

$$|W(C)| < b_1 e^{-C_1|S(C)|} \tag{5}$$

for any arbitrary planar contour C without self-intersections.

2. For $\nu = 3$ and all sufficiently large positive $\beta, \beta > \beta_1 > 0$, there exist constants $C_2 = C_2(\beta_1)$ and b_2 such that

$$|W(C)| > b_2 e^{-C_2|C|} \tag{6}$$

for any planar contour C without intersections.

Remark 1. It is clear from the theorem stated above that the means of form (4) have, for large and small β, different asymptotic behaviour with expanding contours C. It seems that there exists a critical value β_C at which the asymptotic behaviour (5) changes into behaviour (6) (or, possibly, into some intermediate behaviour). The phase transition that thus arises at the point β_C is of a completely different nature than the first-order phase transition for the Ising model described in §0, Chapter 1. In the latter case, at the critical value $h = 0$, there is a discontinuous change of the mean value of the spin at each site of the lattice Z^ν, i.e., of a local functional of the field, while the assumed phase transition for the gauge model reveals itself in a change of the asymptotics of $W(C)$, which is already a nonlocal property of the field.

Remark 2. Theorem 1, and also its proof presented below, may be easily generalized to the case of a gauge field with an arbitrary finite abelian group G and with a Hamiltonian of the form

$$U_\Lambda = \beta \sum_{p:\, \partial p \subset \Lambda} \operatorname{Re} \chi(g_p),$$

where χ is some character of the group G such that $\chi(g) \neq 1$ for $g \neq e$ ($e \in G$ is the unity in G).

Proof of Theorem 1.

1. Clearly

$$\sigma_C = \prod_{p \in S(C)} \sigma_p \equiv \sigma_{S(C)},$$

and thus, for small β, one has the expansion (see §1, Chapter 2, and §2, Chapter 6)

$$W(C) = \sum_{n=0}^{\infty} \frac{(-\beta)^n}{n!} \sum_{(p_1,\dots,p_n)} \langle \sigma_{S(C)}, \sigma_{p_1}, \dots, \sigma_{p_n} \rangle_0, \tag{7}$$

where the summation is over all ordered collections of plaquettes $(p_1, \dots, p_n)$ and the sum converges for $|\beta| < \beta_0$. Using formula (9), §1, Chapter 2, representing semi-invariants in terms of moments, and observing that $\langle \prod_{p \in Q} \sigma_p \rangle = 0$ for any set Q of plaquettes with the nonempty boundary ∂Q (∂Q is the set of bonds $\tau \in \bigcup_{p \in Q} \partial p$ contained in the boundary of an odd number of plaquettes from Q), it is easy to show that

$$\langle \sigma_{S(C)}, \sigma_{p_1}, \dots, \sigma_{p_n} \rangle = 0$$

for $n < S(C)$: From this and from (7), we get the estimate

$$|W(C)| < M(C\beta)^{|S(C)|},$$

which is equivalent to (5).

2. For each configuration $\{\sigma_\tau, \tau \in \widetilde{Z}^\nu\}$ of the gauge field, the collection of variables $\{\sigma_p\}$, $\sigma_p = \prod_{\tau \in \partial p} \sigma_\tau$, defines a configuration of a random field on plaquettes p with values in Z_2. Clearly, for any elementary cube q with the corners at sites of the lattice Z^3, the condition

$$\prod_{p \in \partial q} \sigma_p = 1 \tag{8}$$

is satisfied, with the product being taken over all plaquettes of the cube q. It is easy to show that also conversely, for each $\Lambda \subset \widetilde{Z}^3$ and any function (configuration) $\widehat{S} = \{S_p\}$, defined on the set $\Lambda^{(2)}$ of plaquettes p such that $\partial p \subset \Lambda$ and taking values in Z_2, that obeys the condition (8), a configuration $\{\sigma_\tau, \tau \in \Lambda\}$ of the gauge field can be bound so that $\sigma_p = S_p$ for all $p \in \Lambda^{(2)}$.

Hence,

$$Z_\Lambda = \sum_{S=\{S_p\}}{}' e^{\beta \Sigma S_p} N_\Lambda(S) \tag{9}$$

and

$$W(C) = \frac{1}{Z_\Lambda} \sum_{S=\{S_p\}}{}' \left(\prod_{p \in S(C)} S_p \right) c^{\beta \Sigma S_p} N_\Lambda(S), \tag{10}$$

where $\sum'_{S=\{S_p\}}$ denotes the sum over all configurations $S = (S_p)$ defined on the set $\Lambda^{(2)}$ of plaquettes $p, \partial p \subset \Lambda$, such that the condition (8) is satisfied for all cubes q whose boundary $\partial q \subset \Lambda^{(3)}$ (we use $\Lambda^{(3)}$ to denote the set of those cubes). Such configurations will be called admissible. If S is an admissible configuration defined on plaquettes $p \in \Lambda^{(2)}$, we use $N_\Lambda(S)$ to denote the number of configurations $\{\sigma_\tau, \tau \in \Lambda\}$ of the original field for which the values of functionals σ_p are fixed and

$$\sigma_p = S_p, \qquad p \in \Lambda^{(2)}.$$

Lemma 2. *The number $N_\Lambda(S)$ is constant for all admissible configurations $S = \{S_p, p \in \Lambda^{(2)}\}$.*

Proof. The number $N_\Lambda(S)$ is the number of solutions of the system of equations

$$\prod_{\tau \in \partial p} \sigma_\tau = S_p, \qquad p \in \Lambda^{(2)}. \tag{11}$$

Each solution $\sigma = \{\sigma_\tau, \tau \in \Lambda\}$ of this system is of the form

$$\sigma_\tau = \sigma_\tau^0 \widehat{\sigma}_\tau,$$

where $\sigma^0 = \{\sigma_\tau^0, \tau \in \Lambda\}$ is an arbitrary fixed solution and $\widehat{\sigma} = \{\widehat{\sigma}_\tau, \tau \in \Lambda\}$ is a solution of the system (11) with $S_p \equiv 1$. Hence, $N_\Lambda(S) = N_\Lambda(S \equiv 1)$. The lemma is proved.

Thus, it follows from (9) and (10) that

$$W(C) = \widetilde{Z}_\Lambda^{-1} \sum_S{}' \left(\prod_p S_p\right) \exp\left\{\beta \sum_p S_p\right\} \tag{12}$$

and

$$\widetilde{Z}_\Lambda = \sum_S{}' \exp\left\{\beta \sum_p S_p\right\}. \tag{13}$$

For each admissible configuration $S = \{S_p, P \in \Lambda^{(2)}\}$, we use $Q = Q(S)$ to denote the set of plaquettes for which $S_p = -1$. Let $\widehat{Q}(S) \subset \widehat{\Lambda}^{(2)}$ be the set of bonds of the dual lattice $\widehat{Z}^{(3)}$ (the lattice shifted with respect to Z^3 by the vector $(1/2, 1/2, 1/2)$) that intersects the plaquettes from $Q(S)$. We use $\widehat{\Lambda}^{(2)}$ to denote the set of all bonds from $\widehat{Z}^3$ that intersect plaquettes from the set $\Lambda^{(2)}$, and $\widehat{\Lambda}^{(3)}$ to denote the set of centers of cubes $q \in \Lambda^{(3)}$. According to the condition (8), each site $t \in \widehat{\Lambda}^{(3)}$ belongs to an even number of bonds from $\widehat{Q}(S)$. Thus, the connected components $\Gamma_1, \dots, \Gamma_m$ of $\widehat{Q}(S)$ turn out to be contours on the dual lattice $\widehat{Z}^3$.

The correspondence $S \mapsto \{\Gamma_1, \dots, \Gamma_m\}$ between admissible configurations S in $\Lambda^{(2)}$ and the collections $\{\Gamma_1, \dots, \Gamma_m\}$ of mutually disjoint contours $\Gamma_i \subset \widehat{\Lambda}^{(2)}$ is clearly one-to-one. Moreover,

$$\sum_{p \subset \Lambda^{(2)}} S_p = \left|\Lambda^{(2)}\right| - 2 \sum_{i=1}^m |\Gamma_i|, \tag{14}$$

and

$$\prod_{p \in S(C)} S_p = \prod_{i=1}^m (-1)^{g_C(\Gamma_i)}, \tag{15}$$

where $g_C(\Gamma)$ is the number of intersections of the contour $\Gamma \subset \widehat{\Lambda}^{(2)}$ with the set $S(C)$.

We say that a contour Γ is linked with a contour C if $g_C(\Gamma)$ is odd. Thus, from (12), (13), (14), and (15), we get

$$W(C) = Z_2/Z_1, \tag{16}$$

where

$$Z_i = \sum_{\{\Gamma_1,\dots,\Gamma_m\}} k_i(\Gamma_1)\cdot\ldots\cdot k_i(\Gamma_m), \quad i = 1,2. \tag{17}$$

The sum in both cases is taken over all collections $\{\Gamma_1,\dots,\Gamma_m\}$ of mutually disjoint contours $\Gamma_i \subset \widehat{\Lambda}^{(2)}$, and

$$k_1(\Gamma) = e^{-2\beta|\Gamma|},$$

$$k_2(\Gamma) = \begin{cases} -k_1(\Gamma), & \text{if the contour } \Gamma \text{ is linked with } C \\ k_1(\Gamma) & \text{otherwise.} \end{cases}$$

Consider now the one-parameter family of variables

$$k_\theta(\Gamma) = \begin{cases} e^{i\theta}k_1(\Gamma) & \text{for } \Gamma \text{ linked with } C, \\ k_1(\Gamma) & \text{otherwise,} \end{cases}$$

where $0 \leqslant \theta \leqslant \pi$. Clearly, $k_1(\Gamma) = k_{\theta=0}(\Gamma), k_2(\Gamma) = k_{\theta=\pi}(\Gamma)$.

Thus,

$$\ln W(C) = \ln Z_2 - \ln Z_1 = \int_0^\pi \left(\frac{d}{d\theta}\ln Z_\theta\right) d\theta, \tag{18}$$

where

$$Z_\theta = \sum_{\{\Gamma_1,\dots,\Gamma_m\}} k_\theta(\Gamma)\ldots k_\theta(\Gamma_m). \tag{19}$$

It is easy to see that

$$\frac{d}{d\theta}\ln Z_\theta = i\sum_\Gamma \rho_\theta(\Gamma),$$

where the sum $\sum\limits_\Gamma$ is taken over all contours $\Gamma \subset \widehat{\Lambda}^{(2)}$ linked with C, and

$$\rho_\theta(\Gamma) = \sum_{\{\Gamma_i\}} \Pi k_\theta(\Gamma_i)/Z_\theta,$$

where the sum in the numerator is over all collections $\{\Gamma_i\}$ containing the contour Γ (i.e., $\rho_\theta(\Gamma)$ is the "probability" of the appearance of the contour Γ in the contour ensemble in $\widehat{\Lambda}^{(2)}$). With the help of methods from Chapter 3, it is easy to show that for large $\beta > 0$, one has

$$|\rho_\theta(\Gamma)| < (be^{-2\beta})^{|\Gamma|}, \tag{20}$$

with an absolute constant b.

Furthermore, since each contour Γ linked with C intersects both the interior of the contour C and its exterior (in the plane containing C), we conclude from (18), (19), and (20) that

$$|\ln|W|| < \sum_{\Gamma}(be^{-2\beta})^{|\Gamma|}, \tag{21}$$

where the sum is taken over contours intersecting the plane of the contour C, both inside and outside C.

For such a contour Γ, we use $p(\Gamma)$ to denote anyone of the plaquettes $p \in S(C)$ intersected by Γ, and we notice that the length $|\Gamma| > d(p(\Gamma), C)$, where $d(p, C)$ is the distance of the centre of p from the contour C. Thus, the right-hand side of (21) does not exceed

$$\sum_{p\in S(C)} \sum_{\Gamma: p(\Gamma)=p} (be^{-2\beta})^{|\Gamma|} \leqslant \sum_{p} \sum_{n>d(p,c)} B^n(be^{-2\beta})^n < K \sum_{p\in S(C)} (Bbe^{-2\beta})^{d(p,c)}$$
$$< K \sum_{\tau\in C} \sum_{p} (Bbe^{-2\beta})^{d(\tau,p)} < M|C|, \tag{22}$$

with some constants B, K, and M. Here, B^n yields an estimate of the number of contours of the length n that intersect a given plaquette p (see Lemma 1, §4, Chapter 2). The bound (22) thus implies (6). The theorem is proved.

§5 Markov Processes with Local Interaction

Here, we will examine a particular class of Markov processes (chains) with an infinite space of states, the so-called processes with a *local interaction* for which cluster expansions can be easily derived. We shall consider two cases separately—those of discrete and those of continuous time.

1. Discrete time. Let $T = Z^\nu \times Z_+$ and $\Omega = S^T$, where S is some finite set. For a point $(t, \tau) \in T$, $t \in Z^\nu$, $\tau \in Z_+$, we call t and τ its space and time coordinates, respectively. A set of points $Y_\tau = Z^\nu \times \{\tau\} \subset T$ will be called the layer at the time $\tau \in Z_+$. Each configuration $x = \{x(t,\tau), (t,\tau) \in T\}$ may be regarded as a sequence of configurations $\{x_0, x_1, \dots, x_\tau, \dots\}$ in the layers Y_τ, $\tau = 0, 1, \dots$. We shall often identify the layer Y_τ with Z^ν and view the configurations x_τ in Y_τ as configurations in Z^ν.

Let a finite set $A_0 \subset Z^\nu$ be singled out and let a family $\{p(s/x^{A_0}), s \in S, x^{A_0} \in S^{A_0}\}$ of probability distributions on the set S be given such that they depend on the configuration $x^{A_0} \in S^{A_0}$ in the set A_0. Further, let P_0 be a probability distribution on $S^{Z^\nu} = Y_0$. We shall define a homogeneous Markov chain (i.e., a measure μ on the space $\Omega = S^T$; for the notions from the theory of Markov chains used here, see [51] or [58]), with the initial distribution P_0 and with the transition probability $P(\cdot/\bar{x})$, $\bar{x} \in S^{Z^\nu}$, that equals the infinite product of measures on S:

$$P(\cdot/\bar{x}) = \prod_{t \in Z^\nu} p(\cdot/\bar{x}|_{A_t}), \tag{1}$$

where $A_t = A_0 + t \subset Z^\nu$ is the shift of the set A_0 by the vector $t \in Z^\nu$, and $\bar{x}|_{A_t}$ is the restriction of the configuration $\bar{x}$ onto the set A_t (a configuration $\bar{x}|_{A_t}$ in A_t may be naturally identified with the configuration $\hat{x}^{A_0}$ on the set A_0:

$$\hat{x}^{A_0}(t') = \bar{x}(t' + t), \quad t' \in A_0;$$

the measure $p(\cdot/\bar{x}|_{A_t})$ in (1) may then be understood as $p(\cdot/\hat{x}^{A_0})$).

Definition (1) means that, with respect to the conditional distribution that is induced by the measure μ and is defined on the space S^{Y_τ} of configurations x_τ in the layer $Y_\tau, \tau > 0$, under the condition that a configuration $x_{\tau-1} = \bar{x}$ in the preceding layer $Y_{\tau-1}$ is fixed, the values $x_\tau(t), t \in Y_\tau$, of a configuration x_τ in different points t are independent and distributed with probabilities $p(x_\tau(t)/\bar{x}|_{A_{t,\tau-1}})$, where $A_{t,\tau-1}$ is the set obtained from A_0 by the shift by the vector $(t, \tau - 1) \in Z^\nu \times Z_+$. We use $P_\tau = \mu|_{Y_\tau}$ to denote the probability distribution of configurations x_τ in the layer Y_τ induced by the distribution μ (as agreed above, the distribution P_τ will sometimes be regarded as defined on the space S^{Z^ν}).

We suppose that $p(s/x^{A_0}) \neq 0$ for all $s \in S$ and $x^{A_0} \in S^{A_0}$ and denote

$$\max_{\substack{s \in S, \\ x_1^{A_0}, x_2^{A_0} \in S^{A_0}}} \left| \frac{p(s/x_1^{A_0})}{p(s/x_2^{A_0})} - 1 \right| = \lambda. \tag{2}$$

Theorem 1. *Let λ be sufficiently small: $\lambda < \lambda_0$, where λ_0 does not depend on ν, $|s|$, and $|A_0|$. Then the Markov chain μ is ergodic: For any initial distribution P_0, the distribution P_τ weakly converges, as $\tau \to \infty$, toward some distribution π on S^{Z^ν} (that does not depend on P_0).*

Moreover, for any local function F_B, the rate of convergence of the mean $\langle F_B \rangle_{P_\tau}$ to $\langle F_B \rangle_\pi$ is exponential:

$$\left| \langle F_B \rangle_{P_\tau} - \langle F_B \rangle_\pi \right| < K(F_B)(C\lambda)^\tau, \tag{3}$$

where C is an absolute constant, and $K(F_B)$ is a constant that depends on the function F_B. The measure π on S^{Z^ν} obeys the condition of maximal strong exponential decay of correlations (see §1, Chapter 6).

Proof. For each point $(t, \tau), \tau > 0$, we use $D_{t,\tau} \subset T$ to denote the set $\{(t, \tau)\} \cup A_{t,\tau-1}$ and call the point (t, τ) its vertex. The collection of all sets $D_{t,\tau}$, $(t, \tau) \in T$, $\tau > 0$, will be denoted by $\mathfrak{A}$. Let $\Gamma = \{D_{t_1,\tau_1}, \dots, D_{t_n,\tau_n}\}$ be a finite collection of mutually distinct sets from $\mathfrak{A}$; the vertex (t_i, τ_i) of the set $D_{t_i,\tau_i} \in \Gamma$ will be called *free* if it is not contained in any other of the sets D_{t_j,τ_j}, $j \neq i$. Let $B \subset T$ be a finite set. A collection $\Gamma = \{D_{t_i,\tau_i}, i = 1, \ldots, n\}$ will be called *chronologically-connected* with respect to B if all its free vertices are contained in B. We use Γ_B to denote a maximal collection of sets $D_{t,\tau} \in \mathfrak{A}$ that is chronologically-connected with respect to B. Let $F = F_B$ be a bounded local function defined on Ω, and measurable with respect to the σ-algebra Σ_B, where $B \subset Y_{\tau_0}$ is a finite set contained in the layer Y_{τ_0}. As easily follows from (1), the mean

$$\langle F_B \rangle_\mu = \langle F_B \rangle_{P_{\tau_0}} \tag{4}$$

can be represented in the form

$$\langle F_B \rangle_{P_{\tau_0}} = \sum_{(x_0, \ldots, x_{\tau_0})} P_0^{L_0}(x_0) \prod_{\tau=1}^{\tau_0} \prod_{t \in L_\tau} p\left(x_\tau(t)/x_{\tau-1}|_{A_{t,\tau-1}}\right) F_B(x_{\tau_0}), \tag{5}$$

where we denote $L_\tau = \widetilde{\Gamma}_B \cap Y_\tau$, $\tau = 0, \ldots, \tau_0$, and $P_0^{L_0}$ is the restriction of the original distribution P_0 onto the set S^{L_0}. The summation is over collections of configurations $(x_0, \ldots, x_{\tau_0})$, where $x_\tau \in S^{L_\tau}$ is a configuration in the set L_τ.

Further, we fix any configuration $\bar{x}^{A_0} \in S^{A_0}$ and denote by q the probability distribution on S:

$$q(s) = p(s/\bar{x}^{A_0}),$$

and by $\lambda(s/x^{A_0})$ the quantity

$$\lambda\left(s/x^{A_0}\right) = \frac{p(s/x^{A_0})}{q(s)} - 1.$$

We now introduce a probability distribution on Ω defined by

$$\mu_0 = P_0 \times q^{T \setminus Y_0}$$

as the product of the measure P_0 on S^{Y_0} with the infinite product of measures q on S over all points (t, τ), $\tau > 0$.

Further, for each set $D_{t,\tau} \in \mathfrak{A}$, we denote

$$\lambda(D_{t,\tau}) = \lambda\left(x_\tau(t)/x_{\tau-1}|_{A_{t,\tau-1}}\right), \quad \tau > 0$$

and represent

$$p\left(x_\tau(t)/x_{\tau-1}|_{A_{t,\tau-1}}\right) = q\left(x_\tau(t)\right)\left[1 + \lambda(D_{t,\tau})\right]. \tag{6}$$

First we sum over configurations $x_{\tau_0} \in S^B$ in (5); by inserting (6) into (5) and by multiplying, we get

$$\langle F_B \rangle_{P_{\tau_0}} = \sum_{B_1 \subset B} \sum_{(x_0, \dots, x_{\tau_0 - 1})} \prod_{\tau=1}^{\tau_0 - 1} \prod_{t \in L_\tau(B_1)} p\left(x_\tau(t)/x|_{A_{t,\tau-1}}\right)$$
$$\times \left\langle F(B) \prod_{t \in B_1} \lambda(D_{t,\tau_0}) \right\rangle_{\mu_0^{\tau_0}},$$

where $\mu_0^\tau = q^{Y_\tau}$ is the measure on S_τ^Y, and $L_\tau(B) = \widetilde{\Gamma}_{B_1} \cap Y_\tau$. If we continue this way, we finally get

$$\langle F_B \rangle_{P_{\tau_0}} = \sum_{\Gamma = \{D_{t_1,\tau_1}, \dots, D_{t_s,\tau_s}\}} \langle F_B \Pi \lambda(D_{t_i,\tau_i}) \rangle = \sum_{R \subseteq \Gamma_B} b_R^{\tau_0}(F_B), \tag{7}$$

where the sum $\sum_\Gamma$ is taken over all unordered collections $\Gamma = \{D_{t_1,\tau_1}, \dots, D_{t_s,\tau_s}\}$ of mutually distinct sets from $\mathfrak{A}$ that are chronologically-connected with respect to B, and the $\sum_R$ is taken over all $\mathfrak{A}$-composable sets R such that $R = \widetilde{\Gamma}$ for at least one Γ entering the sum $\sum_\Gamma$. Here also,

$$b_R^{\tau_0}(F_B) = \sum_{\Gamma : \widetilde{\Gamma} = R} \left\langle F_B \prod_{D_{t,\tau} \in \Gamma} \lambda(D_{t,\tau}) \right\rangle_{\mu_0}. \tag{8}$$

Since each collection Γ formed by the set D_{t_i,τ_i} is uniquely defined by the set of their vertices (t_i, τ_i), the number of collections Γ such that $\widetilde{\Gamma} = R$ does not exceed $2^{|R|}$. Hence and from (8), we get

$$|b_R^{\tau_0}(F_B)| < \sup |F_B| \cdot (C\lambda)^{d_R(\mathfrak{A})}, \tag{9}$$

where the quantity $d_R(\mathfrak{A})$ equals the minimal cardinality of collections Γ such that $\widetilde{\Gamma} = R$, and $C = 2^{|A_0|+1}$.

Let us remark that whenever R does not intersect the zero layer Y_0, the quantities $b_R^\tau(F_B)$ depend neither on the initial distribution P_0 nor on τ_0 (here F_B is regarded as a fixed function on the space S^{Z^ν}). Thus, the following limits exist:

$$\lim_{\tau_0 \to \infty} b_R^{\tau_0}(F_B) = b_R(F_B)$$

and

$$\lim_{\tau_0 \to \infty} \langle F_B \rangle_{P_{\tau_0}} = \sum_R b_R(F_B) \equiv \langle F_B \rangle.$$

Hence, it follows that the weak limit $\lim_{\tau \to \infty} P_\tau = \pi$ of measures P_τ exists and

$$\langle F_B \rangle = \langle F_B \rangle_\pi$$

(see §1, Chapter 1).

It is easy to see that π is a stationary distribution of the considered Markov chain, i.e., if $P_0 = \pi$ is chosen as an initial distribution, then for any $\tau > 0$, we have again $P_\tau = \pi$. Furthermore, from the representation (7) and the estimate (9), we may show that for any two initial distributions P_0' and P_0'', one has

$$\left|\langle F_B\rangle_{P_\tau'} - \langle F_B\rangle_{P_\tau''}\right| < (\overline{C}\lambda)^\tau \sup|F_B|,$$

where P_τ' and P_τ'' are the distributions on S^{Y_τ} obtained from the initial distributions P_0' and P_0'', respectively. Setting now $P_0' = \pi = P_\tau'$ and $P_0'' = P_0$, we get (3). Further, from the definition of $b_R(F_B)$ and the estimate (9), it is easy to see that the cluster expansion (7) of the means $\langle F_B\rangle_\pi$ is exponentially regular (see §3, Chapter 3) with respect to the family of sets $\mathfrak{A}$, and by the same token, according to Theorem 1, §3, Chapter 6, the measure π satisfies the condition of maximal strong decay of correlations. Theorem 1 is proved.

2. Continuous time. Let $T = Z^\nu \times R_+$ and $\Omega = S^T$ be the space of step functions $\{x_\tau(t),\, t \in Z^\nu,\, \tau \in R_+\}$ on T with values in S. We define a Markov process with values in S^{Z^ν}, with a transition probability $P_\tau(\cdot/x_0)$ of the form (1) for small τ (up to $o(\tau)$). Let $A_0 \subset Z^\nu$ be a finite set, $0 \in A_0$, and $h(s, x^{A_0})$ be a function defined on the set $S \times S^{A_0}$ so that

$$\begin{aligned} h\left(s, x^{A_0}\right) &\geqslant 0 && \text{if } x^{A_0}(0) \neq s, \\ h\left(s, x^{A_0}\right) &= -\sum_{s' \in S:\, s' \neq s} h\left(s', x^{A_0}\right) && \text{for } s = x^{A_0}(0). \end{aligned} \tag{10}$$

For each site $t \in Z^\nu$, we define an infinite matrix $H_t = \{H_t(x, x')\}$ whose matrix elements $H_t(x, x')$ are "labelled" by the configurations $x, x' \in S^{Z^\nu}$ and are defined by

$$H_t(x, x') = \begin{cases} 0 & \text{if } x(\bar{t}) \neq x'(\bar{t}) \text{ for at least one } \bar{t} \neq t \\ h(x(t), x'|_{A_t}) & \text{otherwise,} \end{cases} \tag{11}$$

where A_t is the shift of A_0 by the vector t and $x'|_{A_t}$ is the restriction of x' onto the set A_t.

For each finite subset $\Lambda \subset Z^\nu$, we define the matrix $H_\Lambda = \sum_{t\in\Lambda} H_t$. We shall now construct a one-parametric (weakly continuous) semigroup of transformations $\{G_\tau^\Lambda; \tau > 0\}$ acting on the space of probability measures defined on $\left(S^{Z^\nu}, \mathfrak{B}_0\right)$ ($\mathfrak{B}_0$ is the Borel σ-algebra in the space S^{Z^ν} equipped with, as usually, the Tikhonov product topology) so that

$$G_\tau^\Lambda = e^{\tau H_\Lambda} \tag{11'}$$

(the sense of this equality will be clarified below).

Let H^*_Λ be the operator acting on the space $C(S^{Z^\nu})$ of continuous functions on S^{Z^ν} according to the formula

$$H^*_\Lambda F = \sum_{t\in\Lambda} H^*_t F,$$

where

$$(H^*_t F)(x) = \sum_{y\in S^{Z^\nu}}{}' H_t(y,x)F(y) = \sum_{s\in S} h(s, x|_{A_t})F(x^s), \tag{12}$$

where $\sum_y{}'$ denotes the sum over all configurations y that coincide outside the site $t \in Z^\nu$ with the configuration x, and x^s is the one among such configurations that takes the value $x^s(t) = s$.

It follows from (11) that

$$\|H^*_\Lambda\| \leqslant \sum_{t\in\Lambda} \|H^*_t\| \leqslant |\Lambda|\cdot|S|\cdot \max_{(s,x^{A_0})} \left|h(s, x^{A_0})\right|,$$

and thus a bounded semigroup

$$\widetilde{G}^\Lambda_\tau = e^{\tau H^*_\Lambda} \tag{13}$$

acting on the space $C(S^{Z^\nu})$ is well defined.

Moreover, as follows from (10), $\widetilde{G}^\Lambda_\tau \mathbf{1} = \mathbf{1}$, and $\widetilde{G}^\Lambda_\tau \cdot F \geqslant 0$ if $F \geqslant 0$. Thus, the adjoint semigroup $(\widetilde{G}^\Lambda_\tau)^*$, acting on the space $C^*\left(S^{Z^\nu}\right)$ of all bounded countably-additive functions on the sets $\{\nu(A), A \in \mathfrak{B}_0\}$ (see, e.g., [16]), leaves invariant the set of probability measures on $\mathfrak{B}_0$. The restriction of the semigroup $(\widetilde{G}^\Lambda_\tau)^*$ to this set will be denoted by G^Λ_τ. The equality (11′) is formally obtained from (13) when passing to adjoint operators.

Let P_0 be an initial probability distribution on S^{Z^ν}. Then the value of the distribution $P^\Lambda_\tau = G^\Lambda_\tau P_0$ for a cylinder set $\mathcal{K} = \mathcal{K}_A \in \Sigma_A$, where $A \subset Z^\nu$ is a finite set, equals

$$P^\Lambda_\tau(\mathcal{K}) = \int_{S^{Z^\nu}} \left(e^{\tau H^*_\Lambda}\chi_{\mathcal{K}}\right)(x)dP_0(x), \tag{14}$$

where $\chi_{\mathcal{K}}$ is the characteristic function of the cylinder set $\mathcal{K}$ (clearly, $\chi_{\mathcal{K}} \in C(S^{Z^\nu})$).

Theorem 2. *There exists $\tau_0 = \tau_0(h,S)$ such that, for each probability measure P_0 on S^{Z^ν} and each $\tau \leqslant \tau_0$, the limit*

$$\lim_{\Lambda\nearrow Z^\nu} G^\Lambda_\tau P_0 = P_\tau$$

exists (in the sense of the weak convergence). Here, P_τ is a probability measure on the σ-algebra $\mathfrak{B}_0$.

The mapping $G_\tau: P_0 \mapsto P_\tau$ is weakly continuous and the family $\{G_\tau, 0 \leqslant \tau \leqslant \tau_0\}$ of these mappings is multiplicative:

$$G_{\tau_1} \cdot G_{\tau_2} = G_{\tau_1+\tau_2}, \qquad \tau_1, \tau_2, \tau_1 + \tau_2 \leqslant \tau_0. \tag{14'}$$

Proof. We shall show that for a sufficiently small $\tau > 0$ and for any local function $F_B \in C\left(S^{Z^\nu}\right)$, there exists the limit $\lim\limits_{\Lambda \nearrow Z^\nu} e^{\tau H^*_\Lambda} F_B$. We write

$$e^{\tau H^*_\Lambda} F_B = \left(E + \tau H^*_\Lambda + \cdots + \frac{\tau^n}{n!}\left(H^*_\Lambda\right)^n + \ldots\right) F_B. \tag{15}$$

Let us observe that $H^*_t F_B = 0$ for $t \overline{\in} B$. Thus,

$$\frac{\tau^n}{n!}\left(H^*_\Lambda\right)^n F_B = \frac{\tau^n}{n!} \sum_{(t_1,\ldots,t_n)} H^*_{t_n} \ldots H^*_{t_1} F_B,$$

where the sum is taken over all sequences $(t_1, \ldots, t_n)$, $t_i \in \Lambda$, such that

$$t_1 \in B \cap \Lambda,\ t_2 \in (B \cup A_{t_1}) \cap \Lambda, \ldots,\ t_n \in (B \cup A_{t_1} \cup \ldots \cup A_{t_{n-1}}) \cap \Lambda.$$

It is easy to see that the number of such terms for all Λ does not exceed

$$\begin{aligned} &|B| \cdot (|B| + |A_0|)(|B| + 2|A_0|) \ldots (|B| + (n-1)|A_0|) \\ &\qquad \leqslant |A_0|^n \kappa(\kappa+1) \ldots (\kappa+n-1) < |A_0|^n \frac{(\kappa+n-1)!}{(\kappa-1)!}, \end{aligned}$$

where $\kappa = \left[\frac{|B|}{|A_0|}\right] + 1$. Thus,

$$\left\| \frac{\tau^n}{n!}\left(H^*_\Lambda\right)^n F_B \right\| \leqslant 2^{\kappa-1} (C_\tau)^n \|F_B\|, \tag{16}$$

where $C = 2|A_0| \cdot |S| \max |h(S, x^{A_0})|$. It follows from the estimate (16) and the expansion (15) that for $\tau < C^{-1}$ the limit

$$\lim_{\Lambda \nearrow Z^\nu} e^{\tau H^*_\Lambda} F_B = F^\tau_B \stackrel{\text{def}}{=} \widetilde{G}_\tau F_B$$

exists for each local function F_B. Moreover, the operator $\widetilde{G}_\tau$ is bounded on any subspace $C_B = C\left(S^{Z^\nu}, \Sigma_B\right)$ of continuous functions F_B measurable with respect to the σ-algebra $\Sigma_B \subset \widetilde{\mathfrak{B}}_0$:

$$\left\| \widetilde{G}_\tau |_{C_B} \right\| < \infty. \tag{17}$$

With the help of formula (14), we conclude that the limit

$$\lim_{\Lambda \nearrow Z^\nu} P_\tau^\Lambda(\mathcal{K}_B) = \int \left(\widetilde{G}_\tau \chi_{\mathcal{K}_B}\right)(x) dP_0 = P_\tau(\mathcal{K}_B)$$

exists for each cylinder set $\mathcal{K}_B$ and that, in view of (17), the values $\{P_\tau(\mathcal{K}_B), \mathcal{K}_B \in \mathfrak{A}\}$ form a probability cylinder measure on the algebra of cylinder sets, $\mathfrak{A} = \cup_B \Sigma_B$ (see §1, Chapter 1). This cylinder measure can be extended to a probability measure P_τ on the σ-algebra $\mathfrak{B}_0$ (the Kolmogorov theorem). The remaining assertions of the theorem follow easily.

For each $\tau > 0$, we define the mapping G_τ in the space of measures by the formula $G_\tau = G_{\tau_0}^k G_{\tau^*}$, where $k = [\tau/\tau_0]$, $\tau^* = \tau - k\tau_0$. It is easy to infer from (14′) that the family of transformations $\{G_\tau, \tau > 0\}$ forms a semigroup.

We notice that if $A_0 = \{0\}$ and the function $h\left(s, x^{A_0}\right) \equiv h(s, s') \neq 0$, $s' = x^{A_0}(0)$, the Markov process $x_\tau = \{x_\tau(t), t \in Z^\nu\}$, $\tau \geqslant 0$ on T constructed above can be represented in the form of a collection $x_\tau^+ = \{x_\tau(t), \tau > 0\}$ of conditionally independent (with fixed initial points $x_0(t)$, $t \in Z^\nu$) ergodic Markov processes with values in S. There exists a stationary distribution $\pi_0 = q^{s^{Z^\nu}}$ (for this process, q is a stationary distribution for each component); we use μ_0 to denote the corresponding distribution of the whole process determined by the initial distribution $P_0 = \pi_0$.

We shall now assume that

$$h\left(s, x^{A_0}\right) = h_0(s, s') + \lambda\left(s, x^{A_0}\right), \qquad s' = x^{A_0}(0), \tag{18}$$

where $h_0(s, s') \neq 0$ obeys the conditions (10) and thus generates a conditionally independent ergodic Markov chain. The following analogy of Theorem 1 holds:

Theorem 3. *In the decomposition (18), let* $\max_{s, x^{A_0}} \left|\lambda\left(s, x^{A_0}\right)\right| < \lambda_0$, *where* $\lambda_0 = \lambda_0(h_0)$ *is sufficiently small. Then the Markov process constructed above in ergodic, its distribution* μ *on the space* Ω *(with an arbitrary initial distribution* P_0*) admits a cluster expansion, and the stationary distribution* π *satisfies the condition of maximal strong decay of correlations.*

§6 Ensemble of External Contours

In §1, Chapter 5, an ensemble of external contours induced by an ensemble of all contours was mentioned. Here we shall consider an ensemble of external contours of a general form and show that, under certain conditions, it is induced by some ensemble of all contours.

Let $\widetilde{Z}^2$ be the set of bonds of the lattice Z^2 and $\Lambda \subset \widetilde{Z}^2$ be a finite subset of bonds. We shall use $\mathfrak{A}_\Lambda$ to denote the set of collections (configurations) $\alpha = \{\Gamma_i\}_{i=1}^s$ of mutually disjoint contours $\Gamma_i \subset \widetilde{Z}^2$ such that $\Gamma_i \cup \operatorname{Int}\Gamma_i \subset \Lambda$, $i = 1, \ldots, s$, where $\operatorname{Int}\Gamma \subset \widetilde{Z}^2$ is the set of bonds encircled by the contour Γ.

In each configuration $\alpha = \mathfrak{A}_\Lambda$, we single out the family $\alpha^{\text{ext}} \subseteq \alpha$ of external contours $\Gamma \in \alpha$, i.e., of those not encircled by any other contour $\Gamma' \in \alpha$. A configuration $\alpha \in \mathfrak{A}_\Lambda$ is called a configuration of external contours, if $\alpha^{\text{ext}} = \alpha$, and we use $\mathfrak{A}_\Lambda^{\text{ext}} \subset \mathfrak{A}_\Lambda$ to denote the set of such configurations.

Let now $\{\Psi(\Gamma)\}$ be a real function defined on all contours $\Gamma \subset \widetilde{Z}^2$ so that

a) $\sup(|\Psi(\Gamma)|/|\Gamma|) < \infty$,

b) $\Psi(\Gamma) > \tau_0|\Gamma|$ for each Γ, where $\tau_0 > 0$ is an absolute constant;

c) $\Psi(\Gamma + t) = \Psi(\Gamma)$, where $t \in Z^2$, and $\Gamma + t$ is the shift of the contour Γ by the vector t.

We shall define the probability distribution on the set $\mathfrak{A}_\Lambda^{\text{ext}}$ by the formula

$$P_\Lambda^{\text{ext}}(\alpha) = \frac{1}{Z_\Lambda^{\text{ext}}} \prod_{\Gamma \in \alpha} \exp(-\Psi(\Gamma) + hV(\Gamma)), \qquad \alpha \in \mathfrak{A}_\Lambda^{\text{ext}}, \tag{1}$$

where

$$Z_\Lambda^{\text{ext}} = \sum_{\alpha \in \mathfrak{A}_\Lambda^{\text{ext}}} \prod_{\Gamma \in \alpha} \exp(-\Psi(\Gamma) + hV(\Gamma)), \tag{1'}$$

$V(\Gamma) = |\operatorname{Int}\Gamma|$, and h is a real parameter. The distribution (1) is called an *ensemble of external contours* in Λ.

We notice that, for $h \leqslant 0$ and a sufficiently large $\tau_0 \gg 1$, the expression in (1) for the partition function Z_Λ^{ext} is a cluster representation (in the sense of the definition from §1, Chapter 3) with the quantities

$$k_\Gamma = k_\Gamma(\Psi, h) = \exp(-\Psi(\Gamma) + hV(\Gamma)), \tag{2}$$

obeying the cluster condition (2), §1, Chapter 3. Hence, in particular, one derives in a standard manner the existence of a *limit ensemble of external contours in* $\widetilde{Z}^2$ (i.e., a probability distribution on the set of finite or countable configurations of external contours $\widetilde{Z}^{(2)}$) such that the correlation functions of finite families (1) converge, as $\Lambda \nearrow \widetilde{Z}^{(2)}$, to the correlation functions of the limit ensemble (see §1, Chapter 5).

Let us now consider the case $h > 0$. Then the quantities (2) do not satisfy the cluster condition any longer and direct application of the cluster technique to the ensemble (1) is not possible. Nevertheless, we shall show that, for sufficiently small h, the ensemble (1) may be reduced to an ensemble of all contours in Λ which admits a cluster expansion.

Let $\{\Phi(\Gamma)\}$ be a function defined on the contours $\Gamma \subset \mathfrak{G}^{(2)}$ obeying the conditions a), b), and c); $\kappa > 0$ is an arbitrary parameter. Let us define a probability distribution on the set $\mathfrak{A}_\Lambda$ of configurations of all contours in Λ by the formula

$$P_\Lambda(\alpha) = \frac{1}{Z_\Lambda} \prod \exp(-\Phi(\Gamma) - \kappa V(\Gamma)) = \frac{1}{Z_\Lambda} \prod k_\Gamma(\Phi, -\kappa), \tag{3}$$

where k_Γ are defined by the formula (2), and

$$Z_\Lambda = \sum_\alpha \prod_{\Gamma \in \alpha} k_\Gamma. \tag{3'}$$

The distribution (3) is called an ensemble of contours in Λ. The distribution (3) induces the probability distribution on the set $\mathfrak{A}_\Lambda^{\text{ext}}$:

$$P_{\Lambda,\text{ind}}^{\text{ext}}(\alpha) = \Pr\{\beta \in \mathfrak{A}_\Lambda : \beta^{\text{ext}} = \alpha\}, \qquad \alpha \in \mathfrak{A}_\Lambda^{\text{ext}}.$$

With the help of (3), it is easy to show that this distribution is of the form

$$P_{\Lambda,\text{ind}}^{\text{ext}}(\alpha) = \frac{1}{Z_\Lambda} \prod_{\Gamma\in\alpha} Z_\Gamma k_\Gamma(\Phi, -\kappa) = Z_\Lambda^{-1} \prod_{\Gamma\in\alpha} \tilde{k}_\Gamma(\Phi, -\kappa), \tag{4}$$

where $Z_\Gamma = Z_{\widetilde{\text{int}\,\Gamma}}$ is the partition function of the ensemble of contours in $\widetilde{\text{Int}\,\Gamma} \subset \widetilde{Z}^2$, which is the set of bonds laying inside the contour Γ and not intersecting Γ, and

$$\tilde{k}_\Gamma(\Phi, -\kappa) = Z_\Gamma k_\Gamma(\Phi, -\kappa). \tag{4'}$$

Theorem 1. *For each function $\{\Psi(\Gamma)\}$ obeying the conditions a), b), and c), there is a number $h_0(\Psi) > 0$ such that, whenever $h \leqslant h_0(\Psi)$, there exists a function $\{\Phi(\Gamma)\}$, $\Phi(\Gamma) = \Phi(\Gamma;\, \Psi, h)$ satisfying the conditions a), b), c) (with the constant $\tau_0' = \tau_0 - c_0 e^{-\tau_0}$ in condition b), where c_0 is an absolute constant) and a parameter $\kappa = \kappa(\Psi, h)$ so that, for each $\Lambda \subset \widetilde{Z}^2$, the distribution $P_{\Lambda,\text{ind}}^{\text{ext}}$ on $\mathfrak{A}_\Lambda^{\text{ext}}$ specified by (4) and induced by the ensemble (3) of contours in Λ coincides with the ensemble P_Λ^{ext} of external contours given by (1).*

Corollary. *For all $h \leqslant h_0(\Psi)$, there exists an ensemble of external contours in $\widetilde{Z}^2$ that is the limit of ensembles of external contours (1) as $\Lambda \nearrow \widetilde{Z}^2$.*

Indeed, for sufficiently large τ_0' and $\kappa \geqslant 0$, the ensemble (3) of contours in Λ admits a cluster expansion. Hence, we again conclude that a limit ensemble of contours exists in $\widetilde{Z}^2$ (see [49] for more details).

Proof of the theorem. With the help of the methods developed in Chapter 3, we infer that for each $\Gamma \subset \widetilde{Z}^2$, the partition function Z_Γ in (4) is of the form

$$Z_\Gamma \exp\left(SV(\Gamma) + \Delta(\Gamma)\right). \tag{5}$$

Here $S = S(\Phi, \kappa)$ is a parameter that is an analytic functional with respect to the variables $\{k_\Gamma(\Phi, -\kappa)\}$:

$$S(\Phi, \kappa) = \sum_\gamma M_\gamma \prod_{\Gamma\in\gamma} k_\Gamma, \tag{6}$$

where the sum is over all connected collections $\gamma = \{\Gamma_i\}$ of contours (possibly with repetitions); for more details, see §4, Chapter 3. The boundary term $\Delta(\Gamma) = \Delta(\Gamma; \Phi, \kappa)$, $\Gamma \subset \widetilde{Z}^2$, obeys the estimate

$$|\Delta(\Gamma)| < C_0 e^{-4\tau} |\Gamma| \tag{7}$$

and admits an expansion similar to (6):

$$\Delta(\Gamma; \Phi, \kappa) = \sum \overline{M}_\gamma(\Gamma) \prod_{\Gamma' \in \gamma} k_{\Gamma'}.$$

Thus, the quantities $\tilde{k}_\Gamma(\Phi, -\kappa)$ equal

$$\tilde{k}(\Phi, -\kappa) = k_\Gamma(\Psi, h),$$

where

$$\Psi(\Gamma) = \Phi(\Gamma) + \Delta(\Gamma; \Phi, \kappa), \tag{8}$$

and

$$h = S(\Phi, \kappa) - \kappa. \tag{8'}$$

For fixed Ψ and h, one may consider these relations as equations with respect to Φ and κ. These equations are called the *Pirogov-Sinaĭ equations* (see [64]).

For any fixed $\kappa \geqslant 0$ (and a sufficiently large τ_0), the equation (8) is uniquely solvable as far as the function $\Phi = \{\Phi(\Gamma)\}$ is concerned, and the solution $\Phi = \Phi(\kappa, \Psi)$ depends continuously on κ. This may be proved by relying on the expansion (7) with the help of the well-known infinite-dimensional version of the theorem on implicit functions due to L. V. Kantorovich (for more details, see [49] or [64]). Substituting the found solution $\Phi(\kappa, \Psi)$ into equation (8′), we get the equation

$$h = S\left(\Phi(\kappa, \Psi), \kappa\right) - \kappa \equiv v_\Psi(\kappa). \tag{9}$$

The function $v_\Psi(\kappa)$ is continuous and, varying κ from 0 to ∞, it ranges from $v_\Psi(0) = S(\Phi(0, \Psi), 0) \equiv h_0(\Psi) > 0$ to $-\infty$ (moreover, it is easy to show that it is monotone). Thus, for $h \leqslant h_0(\Psi)$, equation (9) has the unique solution $\kappa \geqslant 0$. The theorem is proved.

CONCLUDING REMARKS

At the end we briefly, almost in the form of allusions, mention some topics, problems, and examples from the theory of cluster expansions not included in this book, and we also indicate some prospects of the development of this theory.

1. *Lattice systems with an unbounded infinite-range potential.* As an example may serve the problem of development of cluster expansions for the following Gibbs field in the lattice Z^ν with real values $x = \{x_t,\ t \in Z^\nu\}$. In a finite set $\Lambda \subset Z^\nu$, the distribution μ_Λ of this field is given by

$$d_\mu(x) = Z^{-1} \exp\{-\lambda U(x) - \sum_{t \in \Lambda} |x_t|^m\} dx, \tag{1}$$

with

$$U(x) = \sum_{t,t' \in \Lambda} x_1^{2k} x_{t'}^{2k} \frac{1}{|t-t'|^{-(\nu+\varepsilon)}}, \tag{2}$$

$\lambda > 0,\ m > 0,\ \varepsilon > 0,\ k$ an integer.

In [86] it is

$$m > 4k, \qquad \varepsilon > 4\nu, \tag{3}$$

and the parameter λ is sufficiently small.

In [2*], a completely different idea for obtaining a cluster expansion for the interaction (1) was presented. At the same time, the results of [86] were improved.

2. *Gibbs modifications of (generalized) Gaussian fields in R^ν.* These modifications were studied in connection with the problem of construction of Euclidean quantum fields in R^ν (see [15]). A lot of results were obtained and a powerful cluster expansion technique was developed. A lattice analogy of this technique is presented in our book (see Chapters 4 and 5). As regards the ideas, they do not differ from the case of continuous fields.

3. *Duality transformations.* The idea of passing to the so-called dual system often turns out to be fruitful. It enables one to carry over low temperature expansions to high temperature ones, to pass from long-range potentials to short-range ones, and to replace the study of spin systems by the study of continuous gases, and conversely. A remarkable example of application of the

duality, in connection with cluster expansions, is the proof of the so-called screening, i.e., appearance of rapid decay of correlations in neutral systems of charged particles with slowly decaying (Coulomb) interaction [81].

For the duality and screening see the review [142] containing an extensive bibliography.

4. *Equilibrium quantum ensembles.* When studying Gibbs states for quantum gases or quantum spin systems, most often one carries them over, using the Feynman—Kac formulae, to classical ensembles of Markov trajectories (we examine an example of such an ensemble in §4.5). Furthermore, this ensemble is investigated by the standard cluster technique described in our book (for quantum gases see, e.g., the book [6*]; for the transcription of quantum systems in terms of ensembles of random trajectories see an old survey [7*]).

5. *Dynamics of quantum systems in equilibrium states.* The dynamics of infinite particle systems in states close to the equilibrium (both the ground state and that at non-zero temperature) will be the subject of another book by the authors. It turns out that the most important application of cluster expansions in this connection is the study of low lying branches of the spectrum of Hamiltonians of lattice models of quantum field theory (in the ground state). Mathematically, this is reduced to the study of spectral properties of so-called cluster operators (an important class of linear operators, acting in a Fock space, introduced recently).

Moreover, the cluster expansion techniques begin to be used at present also when studying the dynamics of the Gibbs (with temperature) state, and also in the problems regarding the movement of a single particle in a random field of force.

6. *Dynamics of open systems.* The cluster expansions for open quantum systems with an infinite number of particles have not been actually studied. In particular, this concerns general Markov processes with a local interaction and non-compact space of spins, and also their continuous analogies—partial differential equations. The results of §5.7 can be generalized to a number of similar systems.

7. *Communication and queueing networks.* Here, the applications of cluster expansions are now at the very beginning stage.

8. *Combinatorics.* For cluster expansions in the problem of graph enumeration, see [73].

9. *Other applications in probability theory.* The applications of cluster expansions in the classical probability theory are connected with the ability of obtaining asymptotic expansions of the logarithm of the mean, or conditional mean, values of the form $\langle \Pi_{i=1}^{N} \exp \Lambda_i \xi_i \rangle$, with a random vector $(\xi_1, \dots \xi_N)$. The examples of such applications include: central limit theorems, limit theorems on large condensations, ergodic properties of random processes, statistics of random processes, investigation of local properties of Brownian trajectories, and properties of self-avoiding random walks [3*], percolation theory for random fields, ε-entropy of random fields.

10. *Analytic extensions of cluster expansions.* Some cluster expansions can sometimes be extended to a larger domain of parameters than that prescribed in the standard scheme (see Chapter 3). The fact is connected with a group of ideas and methods coming back to the well-known Lee–Yang theorem on zeros of the partition function (an introduction to this topic is in [40]). In a recent work [4*], a new technique of establishing an exact boundary of analyticity has been developed.

11. *Completely integrable systems.* One of the methods of the exact solution of the two-dimensional Ising model consists of a cluster expansion of the logarithm of the partition function which can be explicitly summed up [5*].

12. *Multiscale cluster expansions.* This new method is an essential extension of the devices described in this book. It is adjusted to the study of so-called ultraviolet (behaviour of fields on small scales) and infrared (large scale) problems of quantum field theory and statistical physics. It also allows investigation of critical phenomena and phase transitions for systems with a continuous symmetry group. This method is now at the development stage, and it is too early to evaluate its limitations. One may learn about this method and its achievement from [1∗], see also review [8∗] with complete bibliography.

Bibliographic Comments

These bibliographic remarks do not pretend to be exhaustive. In general, only the works used when writing this book are mentioned.

Chapter 1

§0. The exposition is taken from the lectures of V. A. Malyshev [40].

§1. The thermodynamic limit and limit Gibbs modifications were studied in papers of N. N. Bogolyubov and B. I. Khatset [9] (see also the paper [8] by N. N. Bogolyubov, D. Ya. Petrina, and B. I. Khatset, D. Ruelle [139], and R. A. Minlos [43], [44]. See also the books by D. Ruelle [60] and C. Preston [57]. As to the point fields see the book by M. Kerstan, K. Mattes, and I. Mekke [30].

§2. General definitions of Gibbs fields are introduced in the papers [18], [19], and [20] by R. L. Dobrushin and in the paper [128] by O. Lanford and D. Ruelle. Basic notions of the Markov field theory are presented in the book by Yu. A. Rozanov [59].

Chapter 2

§1. The mixed semi-invariants of random processes were studied, in the Soviet literature, in the paper [35] by V. P. Leonov and A. N. Shiryaev; see also the book by V. P. Leonov [34]. Formal semi-invariants were introduced in Malyshev's paper [38].

§2. A general construction of Itô-Wick polynomials in Gaussian random functions can be found in the review [23] by R. L. Dobrushin and R. A. Minlos (cf. also the books [61] by B. Simon and [15] by J. Glimm and A. Jaffe). For the diagram technique see the book [65] by K. Hepp or also the above mentioned book [15] by J. Glimm and A. Jaffe. The latter book contains the formulae of integration by parts.

§3. The exposition follows V. A. Malyshev [38].

§4. The estimate of the number of connected collections of sets (Lemma 1) is well known. The estimates of the sum over trees can be found in works of many authors; see, e.g., the papers [92] by M. Duneau, D. Iagolnitzer, and B. Souillard, [93] by M. Duneau and B. Souillard, [36] by V. A. Malyshev, [1] by F. G. Abdulla Zade, R. A. Minlos, and S. K. Pogosyan, and [69], [71] by G. A. Battle and P. G. Federbush.

§5. The content of this section is taken from V. A. Malyshev's paper [38].

§6. For the basic notions of the lattice theory, see the book [7] by G. Birkhoff. Theorem 6 is due to G. C. Rota [138].

§7. The exposition follows V. A. Malyshev [38].

§8. The formalism used here has been set forth in D. Ruelle [60].

Chapter 3

§1. A lot of ideas of cluster expansions were used in the physical literature, with the level of exactness accepted in physics a long time ago. Mathematical reformulations of these ideas can be found in a number of papers, see, e.g., N. N. Bogolyubov and B. I. Khatset [9], D. Ruelle [60], R. A. Minlos and Ya. G. Sinaĭ [48], [49], J. Glimm, A. Jaffe, and T. Spencer [48], [109] (cf. the book [15] by J. Glimm and A. Jaffe). The ensembles of subsets and cluster estimates were introduced in [115] by C. Gruber and H. Kunz.

§2. The Kirkwood-Salsburg equation for ensembles of subsets was studied in the papers [115] by C. Gruber and H. Kunz and [14] by J. Glimm, A. Jaffe, and T. Spencer.

§3. An explicit expansion of correlation functions in a series of quantities k_Γ can be found in Malyshev [38]. The notion of an exponentially regular cluster expansion was introduced in the same paper.

§4. Expansions of the logarithm of the partition function, close to that constructed here, were applied in many papers (A. Minlos, Ya. G. Sinaĭ [48], [49], R. L. Dobrushin [22], G. Gallavotti and A. Martin-Löf [103], R. A. Minlos and S. K. Pogosyan [46] and others).

§5. The content of this section is well known. The exposition follows Malyshev's lectures [40].

§6. The results of this section are a slight generalization of well-known facts. The used method appears for the first time here.

Chapter 4

§§1–2. The results of these sections are refinements of the results due to Y. M. Park [137] and V. A. Malyshev [39].

§4. This is a modified exposition of known results; cf. D. Ruelle [60].

§5. The results of this section are new.

§6. Theorem 1 comes from the paper [92] by M. Duneau, D. Iagolnitzer, and B. Souillard. The cluster representation (6c) was used, for the first time, by G. S. Sylvester [142] int the case of Gibbs modifications of an independent field. Theorem 2 is proved in the paper by V. A. Malyshev [38].

§7. The exposition follows the paper [2] by R. R. Akhmitzyanov, V. A. Malyshev, and E. N. Petrova. Close results were obtained in the papers [69], [71] by G. Battle and P. G. Federbush. The method of interpolation used here is due to J. Glimm, A. Jaffe, and T. Spencer [108].

§8. The results of this section are new. The main sense of the method used here consists of the following: it enables an investigation of non-smooth perturbations. The method of interpolation applied in this section is a modification of the method due also to J. Glimm, A. Jaffe, and T. Spencer [14].

Chapter 5

§1. Low temperature expansions (in the contour ensemble) were first obtained in the work of R. A. Minlos and Ya. G. Sinaĭ [47], [48], [49] for the case of the Ising model. The result presented here is due to E. N. Petrova [54], see also the paper [42] by V. A. Malyshev and E. N. Petrova. A similar model was discussed in [11] by F. F. Galeb.

§2. The presented result is a case of a result of G. Zolądek [25]. The method used here was employed in a number of previous papers; e.g. in [50] by Nasr Ali, and also in the review [41] by V. A. Malyshev, R. A. Minlos, E. N. Petrova, and Yu. A. Terletskiĭ [41].

§3. The result of this section is a particular case of results due to J. Glimm, A. Jaffe, and T. Spencer [109] ($P(\varphi)_2$-model), D. C. Brydges [81], and J. Z. Imbrie [119], [120], and [122] (continuous field with a countable number of states). Our constructions are essentially based upon the results of §8.IV. The same case is analyzed in the paper [91] by R. L. Dobrushin and M. Zahradník. In the paper [41] by R. A. Minlos, E. N. Petrova, and Yu. A. Terletskiĭ, an analogous case of a perturbation with a "sharp" minimum is examined.

Chapter 6

§1. The notion of rapid decay of correlations was introduced in [93] by M. Duneau and B. Souillard and in [92], [94], and [95] by M. Duneau, B. Souillard, and D. Iagolnitzer. A number of results concerning the estimates on semi-invariants for point fields is also due to them. As regards the connections of various concepts of decay of correlations introduced here, see J. Slawny [141].

§2. We follow here the paper [39] by Malyshev. Some of the methods, related to those we use, can be found in the paper [95] by M. Duneau, B. Souillard, and D. Iagolnitzer. R. L. Dobrushin suggested a method of a proof of analyticity of partition functions which does not use cluster expansions (unpublished).

§3. We follow here V. A. Malyshev [38].

§4. The results presented here are due to V. A. Malyshev, E. N. Petrova, and L. A. Umirbekova (unpublished). Compare also the paper by M. Duneau, B. Souillard, and D. Iagolnitzer [95].

§5. See the paper [38] by A. Malyshev.

§6. See V. A. Malyshev and B. Tirozzi [134].

Chapter 7

§1. Semi-invariants for algebras were introduced in the book [10] by O. Bratteli and D. W. Robinson. For general super-algebras these notions are introduced for the first time here. Applications of these constructions can be found in the papers [29] and [132] by I. A. Koshapov.

§2. The result of this section is due to V. A. Malyshev and I. Nikolaev [133].

§3. The result is published for the first time here.

§4. One may learn about lattice gauge fields and estimates of Wilson's functional in the reviews [136] by K. Osterwalder and E. Seiler, [74] by Ch. Borgs and E. Seiler, and [140] by E. Seiler.

§5. For Markov processes with local interaction, see the survey [96] by R. Durrett (see also D. Griffeath [113]). The results of this section were obtained by S. A. Pirogov (unpublished).

§6. The result of this section is new. The method applied here is due to S. A. Pirogov and Ya. G. Sinaĭ [55], [56] (cf. also the book [64] by Ya. G. Sinaĭ).

REFERENCES

1. Abdulla-Zade, F. G., Minlos, R. A., Pogosyan S. K. "Cluster estimates for Gibbs random fields and some its applications," in *Many-component random systems.* Moscow: Nauka, 1978, 5–30.
2. Akhmitzyanov, R. R., Malyshev, V. A., Petrova, E. N. "Cluster expansion of Gibbs perturbation of massless Gaussian field," *Teor. mat. fiz.*, 1984, 292–298.
3. Basuev, A. G. "Representation for Ursell functions and cluster estimates," *Teor. mat. fiz.* **39**, 1979, 94–106.
4. Berezin, F. A. *The Method of Second Quantization.* New York: Academic Press, 1966.
5. Berezin, F. A. *Introduction to Superanalysis.* Dordrecht: Reidel, 1987 (transl. from the Russian).
6. Billingsley, P. *Convergence of Probability Measures.* New York: Wiley, 1968.
7. Birkhoff, G. *Lattice Theory.* Providence: American Mathematical Society, 1967.
8. Bogolyubov, N. N., Petrina, D. Ya., Khatset, B. I. "Mathematical description of equilibrium state of classical systems based on the formalism of canonical ensemble," *Teor. mat. fiz.* **1**, 1969, 251–274.
9. Bobolyubov, N. N., Khatset, B. I. "On some mathematical problems of statistical equilibrium," *Dokl. Acad. Sci.* **66**, 1949, 321–324.
10. Bratteli, O., Robinson, D. *Operator Algebras and Quantum Statistical Mechanics I, II.* New York: Springer Verlag, 1979–1981.
11. Galeb, F. F. "Existence of infinitely many phases in certain models of statistical physics," *Trans. Moscow Math. Soc.* **44**, 1982, 111–127.
12. Gertsik, V. M. "Analyticity of correlation functions for lattice models with infinite-range potential for the case of multiple phases," in *Many-Component Random Systems.* Moscow: Nauka, 1978, p. 112–132.
13. *Gibbs States in Statistical Physics. A Collection of Translations.* Moscow: Mir, 1978.
14. Glimm, J., Jaffe, A., Spencer, T. "The particle structure of weakly coupled $P(\phi)_2$ model and other applications of high temperature expansions," in

Constructive Quantum Field Theory (G. Velo, A. S. Wightman, eds.). New York: Springer Verlag, 1973.
15. Glimm, J., Jaffe, A. *Quantum Physics—A Functional Integral Point of View.* New York: Springer Verlag, 1981.
16. Dunford, N., Schwartz, J. T. *Linear Operators, Part I: General Theory.* New York: Wiley, 1958.
17. Dixmier, J. *C^*-algebras.* Amsterdam: North Holland, 1977 (transl. from the French).
18. Dobrushin, R. L. "The description of random field by means of conditional probabilities and conditions of its regularity," *Theor. Probab. Appl.* **13**, 1968, 197–224 (transl. from the Russian).
19. Dobrushin, R. L. (1) "Gibbsian random fields for lattice systems with pairwise interactions," *Funct. Anal. Appl.* **2**, 1968, 292–301 (transl. from the Russian). (2) "The problem of uniqueness of Gibbsian random field and the problem of phase transitions," *Funct. Anal. Appl.* **2**, 1968, 302–312.
20. Dobrushin, R. L. "Gibbsian fields: the general case," *Funct. Anal. Appl.* **3**, 1969, 22–28 (transl. from the Russian).
21. Dobrushin, R. L. "Prescribing a system of random variables by conditional distributions," *Teor. Veroyatn. Primen.* **15**, 1970, 458–486 (transl. from the Russian).
22. Dobrushin, R. L. "Asymptotic behaviour of Gibbs distributions for lattice models and its dependence on the form of the volume," *Teor. Mat. Fiz.* **12**, 1972, 115–134.
23. Dobrushin, R. L., Minlos, R. A. "Polynomials in linear random functions," *Russian Math. Surveys.* **32**(2), 1971, 71–127 (transl. from the Russian).
24. *Euclidean Quantum Field Theory (Markovian Approach). A Collection of Translations.* Moscow: Mir, 1978 (in Russian).
25. Zolądek, G. *Gibbs Fields In the Nondegenerated Case and the Spectrum of Their Stochastic Operators.* PhD thesis, Moscow Univ. 1983.
26. Zhubrenko, I. G. "On strong estimates of mixed semi-invariants of random processes," *Sib. Math. 7.* **13**, 1972, 202–213 (transl. from the Russian).
27. Zagrebnov, V. A. "Bogolyubov-Ruelle theorem: a new proof and generalizations," *Teor. Mat. Fiz.* **51**, 1982, 389–402.
28. Yosida, K. *Functional Analysis.* Berlin: Springer, 1965.
29. Kashapov, I. A. "The spectrum of the transfer-matrix of Fermion fields," *Russian Math. Surveys.* **37**(2), 1982, 213–214 (transl. from the Russian).
30. Kerstsan, M., Matthes, K., Mekke, J. *Infinitely Divisible Point Processes.* New York: Wiley, 1978 (transl. from the German).
31. *Constructive Field Theory. A Collection of Translations.* Moscow: Mir, 1977 (in Russian).
32. Cornfeld, I. P., Sinaĭ, Ya. G., Fomin, S. V. *Ergodic Theory.* Springer, 1982.
33. Leites, D. A. "Introduction to the theory of supermanifolds," *Russian Math. Surveys.* **35**(1), 1980, 1–64 (transl. from the Russian).

34. Leonov, V. P. *Some Applications of Highest Semi-Invariants in the Theory of Stationary Random Processes.* Moscow: Nauka, 1964.
35. Leonov, V. P., Shiryaev, A. N. "On a method of calculating of semi-invariants," *Theor. Probab. Appl.* **4**, 1959, 319–329 (transl. from the Russian).
36. Malyshev, V. A. "Perturbations of Gibbs random fields," in *Many-Component Random Systems.* Moscow: Nauka, 1978, p. 258–276.
37. Malyshev, V. A. "Estimates of coefficients of Mayer expansions on the boundary of infra-red region," *Teor. Mat. Fiz.* **45**, 1980, 235–243.
38. Malyshev, V. A. "Cluster expansions for lattice models in statistical physics and quantum field theory," *Russian Math. Surveys* **35**, 1980, 1–62 (transl. from the Russian).
39. Malyshev, V. A. "Semi-invariants of non-local functionals of Gibbs random fields," *Math. Notes* **34**, 1983, 707–712 (transl. from the Russian).
40. *Elementary Introduction to Mathematical Physics of Infinite-Particle Systems,* Lecture Notes of the Joint Institute for Nuclear Physics. Dubna, 1983.
41. Malyshev, V. A., Minlos, R. A., Petrova, E. N., Terletskii, Yu. A. "Generalized contour models," *J. Soviet Math.* **23**, 1983, 2501–2533 (transl. from the Russian).
42. Malyshev, V. A., Petrova, E. N. "Duality transformation of Gibbs random fields," *J. Soviet Math.* **21**, 1987, 877–910 (transl. from the Russian).
43. Minlos, R. A. "Limiting Gibbs' distribution," *Funct. Anal.Appl.* **1**, 1967, 141–150 (transl. from the Russian).
44. Minlos, R. A. "Regularity of the Gibbs limit distribution," *Funct. Anal. Appl.* **1**, 1967, 206–217 (trans. from the Russian).
45. Minlos, R. A. "Lectures on statistical physics," *Russian Math. Surveys* **23**(1), 1968, 137–201 (transl. from the Russian).
46. Minlos, R. A., Pogosyan, S. K. "Estimates of Ursell functions, group functions, and their derivatives," *Teor. Mat. Fiz.* **31**, 1977, 199–213.
47. Minlos, R. A., Sinaĭ, Ya. G. "The phenomenon of 'phase separation' at low temperatures in some lattice models of a gas I," *Math. USSR-Sb.* **2**, 1967, 335–395 (transl. from the Russian).
48. Minlos, R. A., Sinaĭ, Ya. G. "Some new results on first order phase transitions in lattice systems," *Trans. Moscow Math. Soc.* **17**, 1967, 237–268 (transl. from the Russian).
49. Minlos, R. A., Sinaĭ, Ya. G. "The phenomenon of 'phase separation' at low tempertures in some lattice models of a gas II," *Trans. Moscow Math. Soc.* **19**, 1968, 121–196 (transl. from the Russian).
50. Ali, N. "Gibbsian random fields for the Ising model," *Trans. Moscow Math. Soc.* **32**, 1975, 181–202 (transl. from the Russian).
51. Neveu, J. *Mathematical Foundations of the Calculus of Probability.* San Francisco: Holden-Day, 1965 (transl. from the French).
52. Ore, O. *Theory of Graphs.* Providence, R. I.: Amer. Math. Soc., 1962.

53. Pastur, L. A. "Spectral theory of Kirkwood-Salsburg equations in a finite volume," *Teor. Mat. Fiz.* **18**, 1974, 233–242.
54. Petrova, E. N. "Low-temperature expansions for a $\mathbf{Z}^1$-model," in *Mathematical Models in Statistical Physics.* Tyumeń: Tyumeń Univ., 1982, 79–85.
55. Pirogov, S. A., Sinaĭ, Ya. G. "Phase diagrams of classical lattice systems," *Teor. Mat. Fiz.* **25**, 1975, 358–369.
56. Pirogov, S. A., Sinaĭ, Ya. G. "Phase diagrams of classical lattice systems (cont.)," *Teor. Mat. Fiz.* **26**, 1976, 61–76.
57. Preston, C. *Gibbs States on Countable Sets.* London: Cambridge Univ. Press, 1974.
58. Prokhorov, Yu. V., Rozanov, Yu. A. *Probability Theory.* New York: Springer, 1969 (transl. from the Russian).
59. Rozanov, Yu. A. *Markov Random Fields.* New York: Springer, 1982 (transl. from the Russian).
60. Ruelle, D. *Statistical Mechanics. Rigorous Results.* New York: Benjamin, 1969.
61. Simon, B. *The* $P(\phi)_2$ *Euclidean (Quantum) Field Theory.* Princeton: Princeton University Press, 1974.
62. Szegö, G. *Orthogonal Polynomials.* Providence, R. I.: Amer. Math. Soc., 1975.
63. Serre, J.-P. *Lie Algebras and Lie Groups.* New York: Benjamin, 1965.
64. Sinaĭ, Ya. G. *Theory of Phase Transitions: Rigorous Results.* London: Pergamon Press, 1982.
65. Hepp, K. *Théorie de la Renormalisation.* New York: Springer Verlag, 1969.
66. Shabat, B. V. *Introduction to Complex Analysis. Part II.* Moscow: Nauka, 1976 (in Russian).
67. Bałaban, T. "$(Higgs)_{2,3}$ Quantum Fields in a Finite Volume I. A Lower Bound," *Comm. Math. Phys.* **85**(4), 1982, 603–626.
68. Bałaban, T. "$(Higgs)_{2,3}$ Quantum fields in a finite Volume III. Renormalization," *Comm. Math. Phys.* **88**(3), 1983, 411–445.
69. Battle, G. A., Federbush, P. G. "A phase cell cluster expansion for Euclidean field theories," *Annals of Phys.* **142**, 1982, 95–139.
70. Battle, G. A., Federbush, P. "A phase cell cluster expansion for a hierarchical φ_3^4 model," *Comm. Math. Phys.* **88**(2), 1983, 263–293.
71. Battle, G. A., Federbush, P. "A note on cluster expansions, tree graph identities, extra $1/N!$ factors," *Lett. Math. Phys.* **8**(1), 1984, 55–57.
72. Bellissard, J., Høegh-Krohn, R. "Compactness and the maximal Gibbs, state for random Gibbs fields on a lattice," *Comm. Math. Phys.* **84**(3), 1982, 297–327.
73. Biggs, N. "On cluster expansion in graph theory and physics," *Quart. J. Math.* **29**(114), 1978, 159–173.

74. Borgs, Ch., Seiler, E. "Lattice Yang-Mills theory at non-zero temperature and the confinement problem," *Comm. Math. Phys.* **91**, 1983, 329–380.
75. Bricmont, J., Fontaine, J. R., Lebowitz, J. L., Spencer, T. "Lattice systems with a continuous symmetry. I. Perturbation theory for unbounded spins," *Comm. Math. Phys.* **78**(2), 1980, 281–302.
76. Bricmont, J., Fontaine, J. R., Lebowitz, J. L., Spencer, T. "Lattice systems with a continuous symmetry. II. Decay of correlations," *Comm. Math. Phys.* **78**(3), 1981, 363–371.
77. Bircmont, J., Fontaine, J. R., Lebowitz, J. L., Lieb, E., Spencer, T. "Lattice systems with a continuous symmetry. III. Low temperature asymptotic expansion for the plane rotator model," *Comm. Math. Phys.* **78**(4), 1984, 545–566.
78. Bricmont, J., Fontaine, J. R., "Perturbation about the mean field critical point," *Comm. Math. Phys.* **86**(3), 1982, 337–362.
79. Bricmont, J., Lebowitz, J. L., Pfister, Ch. E. "Low temperature expansion for continuous-spin Ising models," *Comm. Math. Phys.* **78**(1), 1980, 117–135.
80. Bricmont, J., Lebowitz, J. L., Pfister, Ch. E. "Nontranslation invariant Gibbs states with coexisting phases III: analyticity properties," Preprint, Rutgers Univ, 1982.
81. Brydges, D. C. "A rigorous approach to Debye screening in dilute classical Coulomb system," *Comm. Math. Phys.* **58**(3), 1978, 313–350.
82. Brydges, D., Federbush, P. "A new form of the Mayer expansion in classical statistical mechanics," *J. Math Phys.* **19**, 1978, 2064–2067.
83. Brydges, D., Fröhlich, J., Spencer, T. "The random walk representation of classical spin systems and correlation inequalities," *Comm. Math. Phys.* **83**(1), 1982, 123–150.
84. Caginalp, G. "Thermodynamic properties of the φ^4 lattice field theory near the Ising limit," *Ann. Phys.* **126**, 1980, 500–511.
85. Caginalp, G. "The φ^4 lattice field theory as an asymptotic expansion about the Ising limit," *Ann. Phys.* **124**, 1980, 189–207.
86. Cammarota, C. "Decay of correlations for infinite range interactions in unbounded spin systems," *Comm. Math. Phys.* **85**, 1982, 517–528.
87. Cassandro, M., Olivieri, E. "Renormalization group and analyticity in one dimension: a proof of Dobrushin's theorem," *Comm. Math. Phys.* **80**(2), 1981, 255–269.
88. Constantinescu, F. "Strong-coupling expansion of the continuous spin Ising model," *Phys. Rev., Lett.* **43**(22), 1979, 1632–1635.
89. Constantinescu, F., Ströter, B. "The Ising limit of the doublewell model," *J. Math. Phys.* **21**, 1980, 881–890.
90. Dobrushin, R. L., Shlosman, S. B. "The problem of transition-invariance of low-temperature Gibbs fields," *Sov. Math. Rev. Ser. C.* **5**, 1984.
91. Dobrushin, R. L., Zahradník, M. "Pirogov-Sinaĭ method for the continuous spin models," (preprint).

92. Duneau, M., Iagolnitzer, D., Souillard, B. "Decrease properties of truncated correlation functions and analyticity properties for classical lattice and continuous systems," *Comm. Math. Phys.* **31**, 1973, 191–208.

93. Duneau, M., Souillard, B. "Cluster properties of lattice and continuous systems," *Comm. Math. Phys.* **47**, 1976, 155–166.

94. Duneau, M., Souillard, B., Iagolnitzer, D. "Analyticity and strong cluster properties for classical cases with finite range interaction," *Comm. Math. Phys.* **35**, 1974, 307–320.

95. Duneau, M., Souillard, B., Iagolnitzer, D. "Decay correlations for infinite-range interactions," *Journ. of Math. Phys.* **16**, 1975, 1662–1666.

96. Durrett, R. "An introduction to infinite particle systems," *Stochastic Processes and Their Applications.* **11**, 1981, 109–150.

97. Federbush, P. G. "A mass zero cluster expansion. Part 1. The expansion," *Comm. Math. Phys.* **81**(3), 1981, 327–340.

98. Federbush, P. G. "A mass zero cluster expansion. Part II. —Convergence," *Comm. Math. Phys.* **81**(3), 1981, 341–360.

99. Fontaine, J. R. "Bounds on the decay of correlations for $\lambda(\varphi)^4$ models," *Comm. Math. Phys.* **87**(3), 1982, 385–394.

100. Fröhlich, J., Pfister, C. "On the absence of spontaneous symmetry breaking and of crystalline ordering in two-dimensional systems," *Comm. Math. Phys.* **81**(2), 1981, 277–298.

101. Fröhlich, J., Spencer, T. "The Kosterlitz-Thouless transition in two-dimensional Abelian spin systems and the Coulomb gas," *Comm. Math. Phys.* **81**(4), 1981, 527–602.

102. Fröhlich, J., Spencer, T. "Massless phases and symmetry restoration in Abelian gauge theories and spin systems," *Comm. Math.* **83**(3), 1982, 411–454.

103. Gallavotti, G., Martin-Löf, A. "Surface tension in the Ising model," *Comm. Math. Phys.* **25**, 1972, 87–126.

104. Gawędzki, K., Kupiainen, A. "A rigorous block spin approach to massless lattice theories," *Comm. Math. Phys.* **77**(1), 1980, 31–64.

105. Gawędzki, K., Kupiainen, A. "Renormalization group study of a critical lattice model. I. Convergence to the line of fixed points," *Comm. Math. Phys.* **82**(3), 1981, 407–433.

106. Gawędzki, K., Kupiainen, A. "Renormalization group study of a critical lattice model," *Comm. Math. Phys.* **83**(4), 1982, 469–492.

107. Gawędzki, K., Kupiainen, A. "Renormalization group for a critical lattice model. Effective interactions beyond the perturbation expansion or bounded spins approximation," *Comm. Math. Phys.* **88**(1), 1983, 77–94.

108. Glimm, J., Jaffe, A., Spencer, T. "The Wightman axioms and particle structure in $P(\varphi)_2$ quantum field model," *Ann. Math.* **100**, 1974, 585–632.

109. Glimm, J., Jaffe, A., Spencer, T. "A convergent expansion about mean field theory I. The expansion," *Ann. of Phys.* **101**, 1976, 610–630; II. "Convergence of the expansion," *Ann. of Phys.* **101**, 1976, 631–669.
110. Georgii, H. O. "Percolation for low energy clusters and discrete symmetry breaking in classical spin systems," *Comm. Math. Phys.* **81**(4), 1981, 455–473.
111. Göpfert, M., Mack, G. "Iterated Mayer expansion for classical gases at low temperatures," *Comm. Math. Phys.* **81**(1), 1981, 97–126.
112. Göpfert, M., Mack, G. "Proof of confinement of static quarks in 3-dimensional U(1) lattice gauge theory for all values of the coupling constants," *Comm. Math. Phys.* **82**(4), 1982, 545–606.
113. Griffeath, D. "Additive and concellative interacting particle systems," *Lecture Notes, Math.* **724**, New York: Springer, 1979.
114. Gross, L. "Decay of correlations in classical lattice models at high temperature," *Comm. Math. Phys.* **68**, 1979, 9–27.
115. Gruber, C., Kunz, H. "General properties of polymer systems," *Comm. Math. Phys.* **22**, 1971, 133–161.
116. Gruber, C., Hintermann, A. Merlini, D. *Group Analysis of Classical Lattice Systems. Lecture Notes in Physics.* Berlin, Heidelberg, New York: Springer-Verlag, 1977.
117. Holley, R. A., Stroock, D. W. "Applications of the stochastic Ising model to the Gibbs states," *Comm. Math. Phys.* **48**, 1976, 249–265.
118. Imbrie, J. Z. "Mass spectrum of the two-dimensional $\lambda\varphi^4 - \frac{1}{4}\varphi^2 - \mu\varphi$ quantum field model," *Comm. Math. Phys.* **78**(2), 1980, 169–200.
119. Imbrie, J. Z. "Phase diagrams and cluster expansions for low temperature $P(\varphi)_2$ models: I. The phase diagram," *Comm. Math. Phys.* **82**(2), 1981, 261–304.
120. Imbrie, J. Z. "Phase diagrams and cluster expansions for low temperature $P(\varphi)_2$ models. II. The Schwinger functions," *Comm. Math. Phys.* **82**(3), 1981, 305–343.
121. Imbrie, J. Z. "Decay of correlations in the one dimensional Ising model with $J_{ij} = |i-j|^{-2}$," *Comm. Math. Phys.* **85**(4), 1982, 491–516.
122. Imbrie, J. Z. "Debye screening for Jellium and other Coulomb systems," *Comm. Math. Phys.* **87**(4), 1983, 515–565.
123. Ito, K. R. "Construction of Euclidean $(QED)_2$ via lattice gauge theory," *Comm. Math. Phys.* **83**(4), 1982, 537–561.
124. Iagolnitzer, D., Souillard, B. "On the analyticity in the potential in classical statistical mechanics," *Comm. Math. Phys.* **60**, 1978, 131–152.
125. Kirkwood, J. R., Thomas Lawrence, E. "Expansions and phase transitions for the ground state of quantum Ising lattice systems," *Comm. Math. Phys.* **88**(4), 1983, 569–580.
126. Klein, D. "Dobrushin uniqueness techniques and the decay of correlations in continuum statistical mechanics," *Comm. Math. Phys.* **86**(2), 1982, 227–246.

127. Klöckner, K., Ströter, B. "Analyticity properties of the weakly coupled double-well model," *Comm. Math. Phys.* **86**, 1982, 495–508.
128. Lanford, O., Ruelle, D. "Observables at infinity and states with short range correlations in statistical mechanics," *Comm. Math. Phys.* **13**, 1969, 174–215.
129. Lebowitz, J. L., Presutti, E. "Statistical mechanics of systems of unbounded spins," *Comm. Math. Phys.* **78**(1), 1980, 151.
130. Malyshev, V. A., "Uniform cluster estimates for lattice models," *Comm. Math. Phys.* **64**, 1979, 131–157.
131. Malyshev, V. A. "Complete cluster expansion for weakly coupled Gibbs random fields," in *Many Component Systems.* Berlin: Springer, 1979.
132. Malyshev, V. A., Kashapov, I. A. "Complete cluster expansion and the spectrum for fermion lattice models," *Selecta Mathematica Sovietica.* **3**(2), 1984, 151–181.
133. Malyshev, V. A., Nikolaev, I. "Uniqueness of Gibbs fields via cluster expansions," *J. Statist. Phys.* **35**(3), 1984, 375–379.
134. Malyshev, V. A., Tirozzi, B. "Renormalization group convergence for small perturbations of Gaussian random fields with slowly decaying correlations," *J. Math. Phys.* **22**(9), 1981, 2020–2025.
135. Maren Tizian, R. "Probability distribution of random paths in the Ising model at low temperature," *Comm. Math. Phys.* **88**(1), 1983, 105–112.
136. Osterwalder, K., Seiler, E. "Gauge field theories on the lattice," *Ann. Phys.* **110**, 1978, 440–471.
137. Park, Y. M. "The cluster expansion for classical and quantum lattice systems," *J. Stat. Phys.* **27**(3), 1982, 553–576.
138. Rota, G. C. "On the foundations of the combinatorial theory. I. Theory of Möbius functions," *Z. Wahrscheinlichkeitstheorie verw. Geb.* **2**, 1964, 340–368.
139. Rota, G. C. "Correlation functions of classical gases," *Ann. Phys.* **25**(1), 1963, 109–120.
140. Seiler, E. "Gauge theories as a problem of constructive quantum field theory and statistical mechanics," in *Lecture Notes in Physics.* **159**, Berlin, Heidelberg, New York: Springer, 1982.
141. Slawny, J. "Ergodic properties of equilibrium states," *Comm. Math. Phys.* **80**(4), 1981, 477–483.
142. Sylvester, G. S. "Weakly coupled Gibbs measures," *Z. Wahrscheinlichkeitstheorie verw. Geb.* **50**, 1979, 97–118.
143. Wagner, W. "Analyticity and Borel-summability of the perturbation expansion for correlation functions of continuous spin-systems," *Helvetica Physica Acta.* **54**, 1981, 341–363.
144. Wagner, W. "Borel-summability of the high temperature expansion for classical continuous systems," *Comm. Math. Phys.* **82**(2), 1981, 183–189.
145. Zagrebnov, V. A. "Spectral properties of Kirkwood-Salsburg and Kirkwood-Ruelle operator," *J. Stat. Phys.* **27**(3), 1982, 577–591.

1*. *Scaling and Self-Similarity in Physics. Renormalization in Statistical Mechanics and Dynamics*, ed. J. Fröhlich. Birkhäuser, 1983.
2*. Akhmitzyanov, R. R., Malyshev, V. A., Petrova, E. N. "Cluster expansion for unbounded nonfinite potential," in *Workshop in Random Fields.* Köszeg, Hungary: Birkhäuser, 1984.
3*. Westwater, J. "On Edward's model for polymer chains. I.," *Comm. Math. Phys.* **72**, 1980, 131–174.
4*. Isakov, S. N. "Nonanalytic features of the first order phase transition in the Ising model," *Comm. Math. Phys.* **95**, 1984, 427–443.
5*. Sherman, S. S. "Combinatorial aspects of the Ising model for ferromagnetism. I. A Conjecture of Feynman on paths and graphs," *J. Math. Phys.* **1**, 1960, 202–217.
6*. Kelberg, M. Ya., Sukhov, Yu. M. "Properties of weak dependence of a complete random field describing the state of commutation network," *Problems of Transmission of Information.* **3**, 1985.
7*. Ginibre, J. "Some applications of functional integration in statistical mechanics," in *Statistical Mechanics and Quantum Field Theory*, eds. C. DeWitt, R. Stora. New York: Gordon and Breach, 1971.
8*. Malyshev, V. A. "Ultraviolet problems in field theory and multiscale expansions," in *Itogy Nauki itechniki Probability Theory* **24**. *Moscow: VINITI, 1986, 111–186.*

SUBJECT INDEX